Behavior Computing

Longbing Cao · Philip S. Yu
Editors

Behavior Computing

Modeling, Analysis, Mining and Decision

 Springer

Editors
Longbing Cao
Advanced Analytics Institute
University of Technology Sydney
Sydney, NSW, Australia

Prof. Philip S. Yu
Department of Computer Science
University of Illinois at Chicago
Chicago, IL, USA

ISBN 978-1-4471-6206-3
DOI 10.1007/978-1-4471-2969-1
Springer London Heidelberg New York Dordrecht

ISBN 978-1-4471-2969-1 (eBook)

Printed on acid-free paper

Springer is part of Springer Science+Business Media (www.springer.com)

Preface

'Behavior' is an increasingly important concept in the scientific, societal, economic, cultural, political, military, living and virtual world. In dictionaries, 'Behavior' refers to manner of behaving or acting, and the action or reaction of any material under given circumstances. In Wikipedia, 'behavior' refers to the actions and mannerisms made by organisms, systems or artificial entities in conjunction with its environment, which includes the other systems or organisms around as well as the physical environment. It is the response of the system or organism to various stimuli or inputs, whether internal or external, conscious or subconscious, overt or covert, and voluntary or involuntary.

Behavior is ubiquitous. Besides the common terms such as consumer behaviors, human behaviors, animal behaviors, and organizational behaviors, behaviors appear everywhere at any time. Behaviors in the physical world are explicit, and have been studied from many different aspects. With the fast development and deep engagement of social and digitalized life, family, city and planet with advanced computing technology, in particular, virtual reality, multimedia information processing, visualization, machine learning, pattern recognition, behaviors in the virtual and social world are emerging increasingly. In addition, behaviors in the traditional spheres are becoming more and more complex with the involvement and marriage with the virtual and social world.

In different applications and scenarios, behaviors present respective characteristics and features. For instance, in stock markets, trader's behaviors are embodied through trading actions and action properties, such as placing a buy quote at a certain time, price and volume on a target security. The action, response or presentation associated with the corresponding properties forms a concrete and rich object—behavior.

The representation, modeling, analysis, data mining and decision-making of behaviors are becoming increasingly *useful*, *essential*, and *challenging* in ubiquitous behavioral applications and problem-solving. They form into a new computing opportunity, necessity and technology innovation, we refer to it as *behavior computing* or *behavior informatics*.

Behavior computing, or behavior informatics, consists of methodologies, techniques and practical tools for representing, modeling, analyzing, understanding and utilizing human, organismal, organizational, societal, artificial and virtual behaviors, behavioral interactions and relationships, behavioral networks, behavioral patterns, behavioral impacts, the formation and decomposition of behavior-oriented groups and collective intelligence, and the emergence of behavioral intelligence. Behavior computing contributes to the in-depth understanding, discovery, applications and management of behavior intelligence.

The above observations and discussions motivate the editing of this book *Behavior Computing: Modeling, Analysis, Mining and Decision*. The edited book reports state-of-the-art advances in methodologies, techniques, systems and applications of behavior computing. Although there are some newly established conferences and workshops, as well as special issues on behavior modeling and analysis of social networks, this edited book creates an important opportunity to broaden current research to areas that consist of behaviors. It aims to serve as the first dedicated source of references for the theory and applications of behavior informatics and behavior computing, establishing state of the art research, disseminating the latest research discoveries, and providing a ground-breaking textbook to senior undergraduate and postgraduate students.

The book is composed of 23 chapters, which are selected from the 2010 and 2011 International Workshop on Behavior Informatics, submissions to this edited book, partial submissions to the Special Issue on Behavior Computing, and invited chapters. The book consists of four parts, covering behavior modeling, behavior analysis, behavior mining and behavior applications.

In Part I, the book reports attempts and efforts in developing representation and modeling methods and tools for capturing behavior characteristics and dynamics in areas such as social media, soccer game, and software packaging. This involves new techniques such as modeling influential behaviors in social media, a behavior ontology system called SAPMAS representing social activity process, using narrative knowledge representation language to represent behaviors, and applying semi-Markov models to represent user behaviors.

Part II selects a number of the corresponding techniques for behavior analysis. This involves great efforts to develop effective techniques and tools for emergent areas and domains in analyzing behaviors, including a group buying behavior recommendation system, simultaneously modeling reply networks and contents to generate user's profiles on web forum, analyzing information searching behaviors by reinforcement learning, estimating conceptual similarities by distributed representation and extended backpropagation, scoring and predicting risk preferences, and creating simulated feedback.

Part III features behavior mining. The selected chapters address issues including clustering trajectory routes, linking behavioral patterns to personal attributes, mining causality from non-categorical numerical data, mining high utility itemsets, modeling and detection of suspicious activities, a behavioral modeling approach to discover unauthorized copying of large-scale documents, and analyzing twitter user behaviors and topic trends.

Six case studies are reported in Part IV on behavior applications. They cover domains and areas including telecom user behaviors, event detection in calling records, predicting the next call for smart phones, 3D handwriting recognition on handheld devices, medical student search behaviors, and evaluation of software testing strategies.

The intended audience of this book will mainly consist of researchers, research students and practitioners in behavior studies, including in the communities of computer science, behavioral science, and social science. In particular, this book fits interests from behavior informatics, behavioral science, data mining, knowledge representation, machine learning, and knowledge discovery. The book is also of interest to researchers and industrial practitioners in areas such as marketing analytics, consumer behavior analysis, social analytics, online behavior analysis, business analytics, human-computer interaction, artificial intelligence, intelligent information processing, decision support systems, and knowledge management.

Readers who are interested in behavior computing and behavior informatics are encouraged to refer to the special interest group: *Behavior Informatics*. The SIG on Behavior Informatics is a dedicated online research portal and repository, presenting research outcomes and opportunities on theoretical, technical and practical issues in behavior computing and behavior informatics.

We would like to convey our appreciation to all contributors including the accepted chapters' authors, and many other participants who submitted their chapters that cannot be included in the book due to space limits. Our special thanks to Ms. Helen Desmond and Mr. Ben Bishop from Springer UK and Ms. Melissa Fearon from Springer US for their kind support and great efforts in bringing the book to fruition. In addition, we also appreciate all reviewers, as well as Mr. Zhong She's assistance in formatting the book.

Sydney, Australia   Longbing Cao
Chicago, USA   Philip S. Yu

Contents

Contributors

Nitin Agarwal Information Science Department, University of Arkansas, Little Rock, USA

Pervaiz K. Ahmed School of Business, Monash University, Bandar Sunway, Malaysia

K. Ajay Babu Informations & Communications Technology, Melbourne Institute of Technology, Melbourne, Australia

Saadat M. Alhashmi School of Information Technology, Monash University, Bandar Sunway, Malaysia

Chen Bai School of Economics and Management, Nanjing University of Science and Technology, Jiangsu, China

Robert P. Biuk-Aghai Data Analytics and Collaborative Computing Group, Department of Computer and Information Science, Faculty of Science and Technology, University of Macau, Macau, China

Luca Cagliero Dipartimento di Automatica e Informatica, Politecnico di Torino, Torino, Italy

Fatih Cakmak Faculty of Arts and Social Sciences, Sabancı University, Istanbul, Turkey

Longbing Cao Advanced Analytics Institute, University of Technology, Sydney, NSW, Australia; QCIS Centre, Faculty of Engineering and Information Technology, University of Technology, Sydney, Australia

Yonghua Cen School of Economics and Management, Nanjing University of Science and Technology, Jiangsu, China

Chia-Ching Chen Department of Computer Science and Information Engineering, Ming Chuan University, Taoyuan County, Taiwan

Ming-Syan Chen National Taiwan University, Taipei, Taiwan

Yi-Yuan Chiang Department of CSIE, Vanung University, Chungli, Taoyuan, Taiwan

Ram Dantu Computer Science and Engineering, University of North Texas, Denton, TX, USA

Ayhan Demiriz Department of Industrial Engineering, Sakarya University, Sakarya, Turkey

Peter Dreisiger Maritime Operations Division, Defence Science and Technology Organisation, Perth, Australia; School of Computer Science & Software Engineering, The University of Western Australia, Perth, Australia

Gürdal Ertek Faculty of Engineering and Natural Sciences, Sabancı University, Istanbul, Turkey

António Falcão Uninova, Caparica, Portugal

Alessandro Fiori Dipartimento di Automatica e Informatica, Politecnico di Torino, Torino, Italy

Simon Fong Data Analytics and Collaborative Computing Group, Department of Computer and Information Science, Faculty of Science and Technology, University of Macau, Macau, China

Liren Gan School of Economics and Management, Nanjing University of Science and Technology, Jiangsu, China

Huiji Gao Computer Science, SCIDSE, Arizona State University, Phoenix, USA

Wang-Hsin Hsu Department of CSIE, Vanung University, Chungli, Taoyuan, Taiwan; Department of EE, National Central University, Chungli, Taoyuan, Taiwan

Jen-Wei Huang Yuan Ze University, Chung-Li, Taiwan

Chih-Chieh Hung Department of Computer Science, National Chiao Tung University, Hsinchu, Taiwan

Yun-Hui Hung National Taiwan University, Taipei, Taiwan

Anushia Inthiran School of Information Technology, Monash University, Bandar Sunway, Malaysia

Evgeny E. Ivashko Institute of Applied Mathematical Research, Karelian Research Centre of the RAS, Petrozavodsk, Karelia, Russia

Murat Kaya Faculty of Engineering and Natural Sciences, Sabancı University, Istanbul, Turkey

Cemre Kefeli Faculty of Engineering and Natural Sciences, Sabancı University, Istanbul, Turkey

Shamanth Kumar Computer Science, SCIDSE, Arizona State University, Phoenix, USA

Nithin Kumar M Search and Information Extraction lab, International Institute of Information Technology, Hyderabad, India

Yue-Shi Lee Department of Computer Science and Information Engineering, Ming Chuan University, Taoyuan County, Taiwan

Huan Liu Computer Science, SCIDSE, Arizona State University, Phoenix, USA

Wei Liu School of Computer Science & Software Engineering, The University of Western Australia, Perth, Australia

Gabriel Lopes FCT/Universidade Nova de Lisboa, Caparica, Portugal

Cara MacNish School of Computer Science & Software Engineering, The University of Western Australia, Perth, Australia

K. Madhuri Department of Electronics and Computer Engineering, K.L.E.F University Vaddeswaram, Guntur (dist), A.P., India

Howard Michel University of Massachusetts, Dartmouth, MA, USA

Saravanan Mohan Ericsson R&D, Chennai, India

Prateeti Mohapatra The Flash Center for Computational Science, University of Chicago, Chicago, IL, USA

Natalia N. Nikitina Institute of Applied Mathematical Research, Karelian Research Centre of the RAS, Petrozavodsk, Karelia, Russia

Özge Onur Faculty of Engineering and Natural Sciences, Sabancı University, Istanbul, Turkey

Wen-Chih Peng Department of Computer Science, National Chiao Tung University, Hsinchu, Taiwan

Avinash Polepally Ericsson R&D, Chennai, India

Weining Qian Institute of Massive Computing, East China Normal University, Shanghai, P.R. China

Yain-Whar Si Data Analytics and Collaborative Computing Group, Department of Computer and Information Science, Faculty of Science and Technology, University of Macau, Macau, China

Joaquim Silva FCT/Universidade Nova de Lisboa, Caparica, Portugal

M. Suman Department of Electronics and Computer Engineering, K.L.E.F University Vaddeswaram, Guntur (dist), A.P., India

Kerem Uzer School of Management, Sabancı University, Istanbul, Turkey

Vasudeva Varma Search and Information Extraction lab, International Institute of Information Technology, Hyderabad, India

Can Wang QCIS Centre, Faculty of Engineering and Information Technology, University of Technology, Sydney, Australia

Ling-Yin Wei Department of Computer Science, National Chiao Tung University, Hsinchu, Taiwan

Jung-Shyr Wu Department of EE, National Central University, Chungli, Taoyuan, Taiwan

Peng-Fan Yan Data Analytics and Collaborative Computing Group, Department of Computer and Information Science, Faculty of Science and Technology, University of Macau, Macau, China

Show-Jane Yen Department of Computer Science and Information Engineering, Ming Chuan University, Taoyuan County, Taiwan

Philip S. Yu Department of Computer Science, University of Illinois, Chicago, IL, USA

Reza Zafarani Computer Science, SCIDSE, Arizona State University, Phoenix, USA

Gian Piero Zarri LiSSi Laboratory, University Paris-Est/UPEC, Vitry sur Seine, France

Huiqi Zhang Computer Science and Engineering, University of North Texas, Denton, TX, USA

Zhao Zhang Institute of Massive Computing, East China Normal University, Shanghai, P.R. China

Aoying Zhou Institute of Massive Computing, East China Normal University, Shanghai, P.R. China

Part I
Behavior Modeling

Chapter 1
Analyzing Behavior of the Influentials Across Social Media

Nitin Agarwal, Shamanth Kumar, Huiji Gao, Reza Zafarani, and Huan Liu

Abstract The popularity of social media as an information source, in the recent years has spawned several interesting applications, and consequently challenges to using it effectively. Identifying and targeting influential individuals on sites is a crucial way to maximize the returns of advertising and marketing efforts. Recently, this problem has been well studied in the context of blogs, microblogs, and other forms of social media sites. Understanding how these users behave on a social media site and even across social media sites will lead to more effective strategies. In this book chapter, we present existing techniques to identify influential individuals in a social media site. We present a user identification strategy, which can help us to identify influential individuals across sites. Using a combination of these approaches we present a study of the characteristics and behavior of influential individuals across sites. We evaluate our approaches on several of the popular social media sites. Among other interesting findings, we discover that influential individuals on one site are more likely to be influential on other sites as well. We also find that influential users are more likely to connect to other influential individuals.

N. Agarwal (✉)
Information Science Department, University of Arkansas, Little Rock, USA
e-mail: nxagarwal@ualr.edu

S. Kumar · H. Gao · R. Zafarani · H. Liu
Computer Science, SCIDSE, Arizona State University, Phoenix, USA

S. Kumar
e-mail: shamanth.kumar@asu.edu

H. Gao
e-mail: huiji.gao@asu.edu

R. Zafarani
e-mail: reza@asu.edu

H. Liu
e-mail: huan.liu@asu.edu

L. Cao, P.S. Yu (eds.), *Behavior Computing*,
DOI 10.1007/978-1-4471-2969-1_1, © Springer-Verlag London 2012

1.1 Introduction

Social media, also known as the Social Web, consists of myriad applications including blogs, social networking websites, wikis, social bookmarking or folksonomies,
online media sharing, social news, social games, etc. Through reactive interfaces,
low barrier to publication, and zero operational costs, made possible by the new
Web 2.0 paradigm, social media has observed a phenomenal growth in user participation leading to participatory web or citizen journalism. The blogosphere, for
instance, has been growing at a phenomenal rate of 100% every 5 months,[1] Blog-
Pulse, another blog indexing service, says it has tracked over 165 million blogs by
July 2011.[2] Facebook recorded more than 750 million active users as of July 2011;[3]
Twitter amassed nearly 200 million users in March 2011;[4] and other social media
sites like Digg, Delicious, StumbleUpon, Flickr, YouTube, etc. are also growing at
terrific pace. This clearly shows the awareness of social media sites among individuals.

Individuals use different social media sites for different reasons. For instance,
they use Facebook to keep in touch with their friends, make new ones, share their
updates and get updates of their friends; Flickr to upload and share photos with
friends and others; Twitter to update their status; Delicious to bookmark, tag, and
share their favorites with friends or others; Digg to rate and promote the content
that they feel is relevant to the society; etc. Some individuals are active on a few
social media sites and some are active on many of them. Essentially they try to be
sociable or gregarious by making friends on these social media sites. Another type of
behavior that can be observed on these social media sites is the influential behavior.
Individuals try to lead the community or the conversation in the social media sites.
In this chapter, we focus on the latter behavior exhibited by individuals.

There is a significant body of work on identifying such influential and/or active
members [2, 15] on one social media site. It would be more interesting to study
the behavior of these individuals across different social media sites. Social media
sites could be alike or different in terms of functionality, which is discussed in more
detail in Sect. 1.2. It would be interesting to study the different types of behaviors
users exhibit on sites which are alike and sites which are different. This entails
studying the relationship between individuals' behavior and the social media sites
in the sense that there could be: same behavior on similar social media site; same
behavior on different social media sites; different behavior on similar social media
site; and different behavior on different social media site.

Studying these behavioral patterns of users across different social media sites
have many applications. If an individual exhibits similar behavior on various social
media sites then this information could be used to predict his behavior on other

---

[1]http://technorati.com/blogging/feature/state-of-the-blogosphere-2008/

[2]http://www.blogpulse.com/

[3]http://www.facebook.com/press/info.php?statistics

[4]http://www.aolnews.com/2011/03/21/twitter-celebrates-5-years-and-200-million-users/

Table 1.1   A taxonomy of social media applications based on their functionalities

Category	Social media sites
Social Signaling	Blogs (Wordpress, Blogger, Blogcatalog, MyBlogLog), Friendship Networks (MySpace, Facebook, Friendfeed, Bebo, Orkut, LinkedIn), Microblogging (Twitter, SixApart)
Media Sharing	Flickr, Photobucket, YouTube, Multiply, Justin.tv, Ustream
Social Health	PatientsLikeMe, DailyStrength, CureTogether
Social Bookmarking	Del.icio.us, StumbleUpon
Social News	Digg, Reddit
Social Collaboration	Wikipedia, Wikiversity, Scholarpedia, Ganfyd, AskDrWiki
Social Games	Farmville, MafiaWars, SecondLife, EverQuest

social media sites by studying his behavior on few sites. These social media sites can also be clustered based on the behavioral patterns of individuals. These clusters can help discover helpful and valuable trends. The activity of individuals can also help in explaining which social media sites are likely to get more activity for various groups of people. These patterns could then be used to explore marketing opportunities and study the movement of individuals on social media sites to focus on niche sites for unique opportunities.

The rest of the chapter is organized as follows: Sect. 1.2 elaborates on the social media taxonomy, Sect. 1.3 discusses the influential behavior of individuals along with models to quantify influence, Sect. 1.4 highlights the challenges and opportunities of data collection across multiple social media, Sect. 1.5 presents experiments and interesting findings on cross media behavior, and Sect. 1.6 presents conclusions with future directions for research.

1.2 Social Media Taxonomy

Individuals participate in different social media applications with different intentions and expectations. Based on a multitude of such functionalities, social media applications can be organized into a taxonomy as illustrated in Table 1.1. In this section, we briefly describe each category and its functionalities:

Social Signaling refers to a collection of applications that allows individuals to express interactions, opinions, ideas, thoughts, and connect with fellow individuals through blogs, microblogs, and social friendship networks. Blogs, or web logs, are collections of articles written by people arranged in reverse chronological order. These individual articles are known as blog posts. The blogs are collectively referred to as the blogosphere. A blog can be maintained by an individual, known as an individual blog or by a group of people, known as a community blog. The authors of blogs are known as bloggers. Some blog cataloging services such as BlogCatalog[5]

---

[5]www.blogcatalog.com

also allow users to create friendship networks. Microblogging sites, as the name suggests, are similar to blogs except for the fact that the length of the articles is limited. In the case of Twitter,[6] the posts (or tweets in this case) can be 140 characters or less. These sites are typically used to share what you are doing and diffuse information simulating a word-of-mouth scenario. Social Friendship Networks allow people to stay in touch with their friends and also create new friends. Individuals create their profile on these sites based on their interests, location, education, work, etc. Usually the ties are non-directional, which means that there is a need to reciprocate the friendship relation between two nodes.

Media Sharing sites allow people to upload and share their multimedia content on the web, including images, videos, audio clips, etc., with other people. People can watch the content shared by others, enrich them with tags, and share their thoughts through comments. Some media sharing sites also allow users to create friendship networks.

Social Health applications strategically use various social media tools in revolutionizing healthcare process. They cut costs for both patients and providers by fostering patient communities for psychological support through social networking opportunities, building knowledge portals with vertical search capabilities, and promoting telehealth and telemedicine opportunities. Internet and software giants such as Google and Microsoft have both launched health services Google Health and Microsoft Health Vault providing interfaces accessible from mobile devices to the cloud.

Social Bookmarking sites allow people to tag their favorite webpages or websites and share it with other users. This generates a good amount of metadata for the webpages. People can search through this metadata to find relevant or most favorite webpages/websites. People can also see the most popular tags or the most recently used tags and emerging website/webpage in terms of user popularity. Some social bookmarking sites like StumbleUpon[7] allow people to create friendship networks.

Social News sites help people share and promote news stories with others. News articles that receive positive votes emerge as the popular news stories. People can tag these news stories as well. They can search for most popular stories, fastest upcoming stories for different time periods, and share their comments.

Social Collaboration applications are publicly edited encyclopedias. Anyone can contribute articles to wikis or edit existing ones. However, most of the wikis are moderated to protect them from vandalism. Wikis are a wonderful medium for content management, where people with very basic knowledge of formatting can contribute and produce rich information sources. Wikis also maintain history of all the changes to a page and are capable of rollbacks. Popular wiki sites like Wikipedia[8] also allow people to classify articles under one of the following categories: Featured, Good, Cleanup, and Stub.

---

[6]www.twitter.com

[7]www.stumbleupon.com

[8]www.wikipedia.org

Social Games offer a medium for individuals to express their interactions and provide an opportunity for researchers to gain detailed insights into their behavior in a simulated environment. These could be casual games where individuals play to achieve an objective governed by incentives or could be as complicated as massively multiplayer online role-playing games (MMORPGs) where users can self-portray as avatars, create objects, and interact with other individuals and objects. Avatars can be displayed in various forms including text, 2D, and 3D images with rich graphics and intricate detail. The almost real like nature of these virtual interactions offers ways which were previously impossible to simulate and explore the unexplained landscape of human behavioral psychology. Social Games have been used in many diverse areas such as military training, movie theaters, and scientific visualization.

Next, we discuss the influential behavior of individuals along with models to quantify influence.

1.3 Influence in Social Media

Influence is reflected by the degree to which an individual is able to affect other individuals in the form of shaping or changing their attitudes or overt behavior in a desired fashion with relative frequency [11, 18, 21]. Accordingly, identifying influential blog sites in the blogosphere involves studying how few blog sites affect other blogs [7] and the external world in terms of behavior or opinion towards the information. The blogosphere, however, follows a power law distribution [6] with very few influential blog sites forming the short head of the distribution and a large number of non-influential sites forming the long tail, where abundant new business, marketing, and development opportunities can be explored [4]. Regardless of the blog being influential, there could exist influential bloggers.

Different social media websites provide various types of information including link and content information. In this chapter, we use a generic model of computing influence scores of individuals based on both link and content information with tunable weights [2]. The choice of model presented in [2] over others like [1, 9, 10, 13, 20] is exercised due to its flexibility to adapt to various social media sites depending on the availability of content or network information. Next we give a brief explanation of the model which uses both content-driven statistics and graph information to identify influential individuals.[9] Some of the desirable properties of an influential individual are summarized as follows:

Recognition: An influential individual is recognized by many. His writings, p are referenced by other individuals. The more influential the referring individuals are, the more influential the referred individual becomes. Recognition is measured through the inlinks (ι). Here ι denotes the set of inlinks to an individual's writings p.

---

[9]Interested readers can find more details in [2].

Activity Generation: An individual's capability of generating activity can be indirectly measured by how many comments he receives, the amount of discussion he initiates. A large number of comments (γ) indicates that the individual *affects* many, such that they care enough to write comments, and therefore, the individual can be influential. Some of these comments could be spam which could be eliminated using the existing work in [14, 17].

Novelty: Novel ideas exert more influence as suggested in [12]. Hence, the outlinks (θ) is an indicator of novelty. If an individual's writing refers to many other articles it indicates that it is less likely to be novel. An individual's writing p is less novel if it refers to more influential articles than if it referred to less influential articles.

Eloquence: An influential individual is often eloquent [12]. Given the informal nature of the social media, there is no incentive for an individual to write a lengthy piece. Hence, a long writing often suggests a necessity of doing so. Therefore, we use the length of the blog post (λ) as a heuristic measure for computing eloquence.[10]

Influence of an individual can be visualized in terms of an influence graph or *i-graph*. Each node of an i-graph represents an individual's writing characterized by the four properties (or parameters): ι, θ, γ and λ. i-graph is a directed graph with ι and θ representing the incoming and outgoing influence flows of a node, respectively. Hence, if I denotes the influence of a node p, then *InfluenceFlow* across that node is given by,

$$InfluenceFlow(p) = w_{in} \sum_{m=1}^{|\iota|} I(p_m) - w_{out} \sum_{n=1}^{|\theta|} I(p_n), \qquad (1.1)$$

where w_{in} and w_{out} are the weights that can be used to adjust the contribution of incoming and outgoing influence, respectively. p_m denotes all the nodes that link to p, where $1 \leq m \leq |\iota|$; and p_n denotes all the nodes that are referred by p, where $1 \leq n \leq |\theta|$. $|\iota|$ and $|\theta|$ are the total numbers of inlinks and outlinks of p. *InfluenceFlow* measures the difference between the total incoming influence of all inlinks and the total outgoing influence by all outlinks of p. From Eq. (1.1), it is clear that the more inlinks a node acquires the more recognized it is, hence the more influential it gets; and an excessive number of outlinks jeopardizes the novelty of the node which affects its influence. We illustrate the concept of *InfluenceFlow* in the i-graph displayed in Fig. 1.1. This shows an instance of the i-graph with a single node p. Here we are measuring the *InfluenceFlow* across p. Towards the right of p are the outlinks and inlinks are towards the left of p.

The influence (I) of an individual is also proportional to the number of comments (γ_p) posted on his(her) writing. The influence of p can be defined as,

$$I(p) \propto w_{com} \gamma_p + InfluenceFlow(p), \qquad (1.2)$$

---

[10]This property is most difficult to approximate using some statistics. Eloquence of an article could be gauged using more sophisticated linguistic based measures.

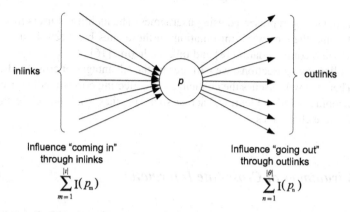

inlinks

outlinks

p

Influence "coming in"
through inlinks

$$\sum_{m=1}^{|r|} I(P_m)$$

Influence "going out"
through outlinks

$$\sum_{n=1}^{|\theta|} I(P_n)$$

Fig. 1.1 *i-graph* showing the *InfluenceFlow* across blog post p

where w_{com} denotes the weight that can be used to regulate the contribution of the number of comments (γ_p) towards the influence of p.

Although there are many measures that quantify the quality of a writing such as fluency, rhetoric skills, vocabulary usage, and content analysis, for the sake of simplicity, here we use the length of the writing p as a heuristic measure of its quality. We define a weight function, w, which rewards or penalizes the influence score of a p depending on its length (λ). The weight function could be replaced with appropriate content and literary analysis tools. Combining Eq. (1.1) and Eq. (1.2), the influence of p can thus be defined as,

$$I(p) = w(\lambda) \times (w_{com}\gamma_p + \textit{InfluenceFlow}(p)). \qquad (1.3)$$

The above equation gives an influence score to each writing of an individual. Now we consider how to use I to determine whether an individual is influential or not. An individual can be considered influential if he has at least one influential piece of writing, p. We use the p with maximum influence score as the representative and assign its influence score as the *influence index* or *iIndex* of the individual. For an individual B, we can calculate the influence score for each of B's N writings and use the maximum influence score as the individuals *iIndex*, or

$$iIndex(B) = \max(I(p_i)), \qquad (1.4)$$

where $1 \le i \le N$. With *iIndex*, individuals on a social media site can be ranked according to their influence. The top k among all the individuals are the most influential ones. Next, we expand our study of identifying influential behavior patterns across multiple social media websites.

1.4 Data Collection

Collecting data from one social media site is considered to be a straightforward task [19]. In most cases, the social network graphs are collected. In these datasets identities are usually represented using usernames. The identity of users from social media sites and their network information on these sites has been used for tasks, such as movie recommendation [8] and link prediction [16].

In the following subsections, we will present the advantages of cross media information. Then, we will discuss the challenges of collecting corresponding user identities (a mapping between identities) across social media sites, followed by methods to address these challenges.

1.4.1 Advantages of Cross-Site Information

Individuals use different social media services for varying purposes and exhibit diverse behaviors on every one of them. We use Flickr to share our pictures with friends, Twitter to update our status, Facebook to keep in touch with friends, and Blogs to express our interests, opinions, and thoughts. It is hence evident that by consolidating this complementary information, a more comprehensive profile of an individual can be built. Few existing studies have considered the prospects of utilizing such information on problems in social media, such as recommending new friends and enhancing user experience. Studying behavioral patterns of users across different social media sites has many applications. As an example, if an individual exhibits the same behavior on various social media sites, then it can help predict his or her behavior on other social media websites. The activity patterns of individuals can also help explain why some social media sites are likely to get more activity.

1.4.2 Challenges of Collecting Cross-Site Information

As identified above, there are several advantages of using user information from multiple sites. However, the task of identifying a user across sites is not so straightforward. Websites do not talk to each other and therefore a user has to create separate user credentials on each social media site. Although many sites now support the usage of credentials from other sites for login, for example, users on the media sharing website DailyMotion[11] can use their Facebook credentials to log onto the website, this information cannot be collected through APIs or other means. Therefore, the task of identifying a user across sites is a challenge.

A simple method for gathering data across social networks is to conduct surveys and ask users to provide their usernames across social networks. Using these usernames, data can be collected across social networks. However, in addition to being expensive, small data size and population sampling are big challenges with this method. Another method for identifying user identities across sites is finding

---

[11]http://www.dailymotion.com

users on these websites manually. Users more often than not provide personal information such as their real names, E-mail addresses, location, sex, profile photos, and age on these websites. This information can be employed to map users on different sites to the same individual. However, even manually, finding users on these sites can be quite challenging. Many users intentionally hide their identities by limiting the amount of personal information they share or by providing fake information, as reported in our prior research [22]. In [3], the authors address a similar problem in the context of cross-social media sites by looking at the content generation behavior of the same individuals on different social media sites. They define this category of users as "serial sharers" and claim that the compression signature of an author of multiple pages on web is unique across all his authored pages. By computing this signature using the measure Normalized Compression Distance (NCD) as described in [5], the authors show that it is indeed possible to identify pages from the same author on the web.

Fortunately, there exist websites where users have the opportunity of listing their identities (user IDs or screen names) on different social networks. Below, we describe two of these websites, the type of information they provide, and our data collection procedures:

1. *BlogCatalog*:[12] BlogCatalog is a comprehensive directory of blogs that not only provides useful information about various weblogs, but also comprises of different facilities for users to interact within its community. Users in BlogCatalog are provided with a feature called "My Communities". This feature enables users to list their usernames in other social networks.
2. *MyBlogLog*:[13] MyBlogLog is a social network for the blogger community. It provides a popular web widget that many members have installed on their blogs and is essentially a site based on the interactions that are facilitated by this widget. Users have the "My Sites and Services" feature in their profile for listing their usernames on different social networks.

Users on these websites voluntarily disclose their identities from other websites. This provides blog authors with an opportunity to interact effectively on appropriate channels with their readers. Thus, users on these websites have a valid motivation to publish their identities and these identities can be considered to be reliable. For the experiments described in the next section, we collected user indentities of 96,000 users from BlogCatalog. We identified three popular social media sites, viz., Delicious, StumbleUpon, and Twitter. Using APIs when available and screen scraping in other cases, we collected the activity and profile information of the users on these sites. Note that, not necessarily all the users collected from BlogCatalog had usernames in all of these sites. In Table 1.2, we present a brief overview of the information collected from these sites. Further, it should be recognized that only publicly shared information was collected from these sites. No private or protected

---

[12]http://www.blogcatalog.com/

[13]http://www.mybloglog.com/

Table 1.2 Information
gathered from the selected
social media sites

Social media site	No of users	Profile attributes
Delicious	8,483	10
StumbleUpon	8,935	13
Twitter	13,819	15

information was collected. The dataset was anonymized after collection for privacy reasons.

In the next section, we present preliminary evaluations of the data and envisage future directions to cross social media studies and influence in social media.

1.5 Studying Influentials Across Sites

In this section, we present a study of the behavior of the influentials across the three sites described above. Influential behavior here refers to their actions towards their network, and the difference in the amount and the type of their activity. This section will categorize the behavior of influential individuals across various social media sites and attempt to address issues such as, sustenance of influence, differences in the sphere of influence, and differences in the influence homophily across different sites.

1.5.1 Sustenance of Influence

In this section, we study the tendency of an influential on one network to remain influential on another social network. This is defined as the sustenance of the influence of an individual on one site across other sites where he is also a member. The motivation behind doing this study is to identify if there exists a pattern in the characteristics of those people who are influential in a network. After reading this section, reader will have a better understanding of starting points to search for individuals who might be influential within a network given some of their identifiable information from other networks. This can also be very useful in tasks such as, finding individuals who can help promote a product in a network where accessing the network information of a large number of individuals is expensive.

We investigate a user's influence sustenance across a pair of social media sites. Three pairs of social media datasets are used in this experiment: Delicious VS StumbleUpon, Delicious VS Twitter, StumbleUpon VS Twitter. We capture the sustenance of influence through a influence intersection ratio, which is defined as the proportion of users who have the same influence position (i.e. top 10% of the influence list) across a pair of datasets. As an example, to capture the influence sustenance on site A and site B, we first calculate the influence score of each user

Fig. 1.2 Delicious VS
StumbleUpon

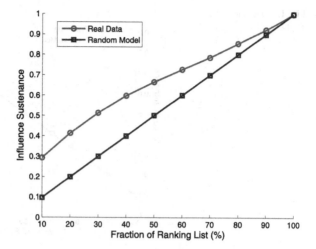

on both sites. Then, we obtain two ranking lists based on these influence scores. After that, we compute the influence intersection ratio among different proportion of the ranking lists, which is the ratio of users whose influence falls into the top $x\%$ ($x \in [10, 20, 30, 40, 50, 60, 70, 80, 90]$) of both ranking lists. Here, a user $\mathcal{U}$'s influence score is defined as:

$$I(\mathcal{U}) = S_d(\mathcal{U}) + S_m(\mathcal{U}). \tag{1.5}$$

$S_d(\mathcal{U})$ is the degree score of user $\mathcal{U}$, which is the ratio of user $\mathcal{U}$'s indegree over the maximum user indegree. $S_m(\mathcal{U})$ is the message score of user $\mathcal{U}$, which is the ratio of message amount published by user $\mathcal{U}$ over the maximum message amount published by other users.

To evaluate the influence behavior, we compare the influence intersection ratio with the null model. Here the null model consists of two shuffled lists of users not ranked by any measure and then compute the intersection ratio for users in these lists. The results are presented in Figs. 1.2, 1.3, and 1.4. We observe that the observed influence intersection ratio is always higher than the null model (referred as the Random Model in the figures) on the three social media sites studied. Our results indicate that a user who is influential on one site has a tendency to be influential on other sites where he is also a member.

1.5.2 Sphere of Influence

The sphere of influence of an influential individual is defined by his closest connections within a network. The influence of a user can be assumed to be strongest on those users who are directly connected to the user. Previous studies [10] have modeled the diffusion of a user's influence beyond his neighbors. However, in this study, we will concentrate our efforts on the immediate neighbors of the user. Intuitively,

Fig. 1.3 Delicious VS Twitter

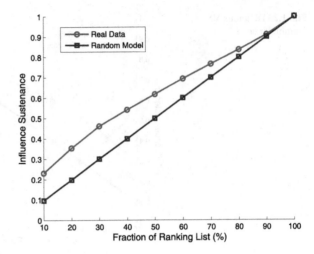

Fig. 1.4 StumbleUpon VS Twitter

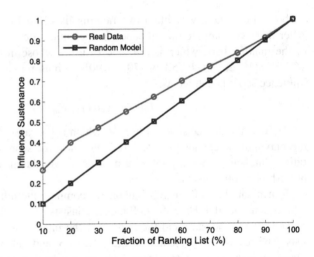

we expect to observe a significant overlap in the spheres of influence of a user across social networks. This can be explained by the tendency of a user to connect with his established friends on other networks.

The sphere of influence of a user can be defined differently depending on the nature of the network. Some networks permit the formation of directed links, such as Twitter, where a person whom you consider your friend may not reciprocate the same feeling towards you. Other networks only consist of undirected links where the feeling of friendship towards another individual is mutual, as long as both users agree to connect. Given this context, the sphere of influence of an individual would consist of his friends (outlinks) in a directed network and his contacts in the case of an undirected network. The goal of this segment of the study is to analyze the retention of a user's network when we observe him on different sites.

Table 1.3 Sphere of influence on Delicious, StumbleUpon and Twitter

Dataset	Delicious	StumbleUpon	Twitter
Delicious	–	0.2031	0.4241
StumbleUpon	0.0166	–	0.3101
Twitter	0.0058	0.0422	–

In this section, we analyze the overlap of a user's influence sphere on social media sites. As before, three pairs of datasets are tested. We are interested in a user's common friends on both sites, i.e., how many friends of user $\mathcal{U}$ on site A are also his friends on site B. To estimate this overlap, we first extract a user $\mathcal{U}$'s friends on the two sites A and B, namely, $F_A(\mathcal{U})$ and $F_B(\mathcal{U})$. We then calculate the intersection of $F_A(\mathcal{U})$ and $F_B(\mathcal{U})$, represented by $C_{AB}(\mathcal{U})$, which consists of the common friends of user $\mathcal{U}$ on the two sites. The overlapping ratio is defined as:

$$O_{A \to B}(\mathcal{U}) = \frac{C_{AB}(\mathcal{U})}{size(F_B)}. \tag{1.6}$$

Similarly, the overlapping ratio of user $\mathcal{U}$'s friendship from site B to site A is:

$$O_{B \to A}(\mathcal{U}) = \frac{C_{AB}(\mathcal{U})}{size(F_A)}. \tag{1.7}$$

Note that the ratio is not symmetric because of the difference in the size of a user's network on sites A and B, respectively. The total overlapping ratio from site A to site B is:

$$O_{A \to B} = \frac{1}{n} \sum_{i=1}^{n} O_{A \to B}(\mathcal{U}_i), \tag{1.8}$$

where, n is the total number of users on site A. The overlapping results from site A to site B are presented in Table 1.3, where rows represent sites A and columns represent sites B. For example, the overlapping ratio from Twitter to Delicious is the lowest, i.e., 0.0058, which on the contrary (i.e., overlapping ratio from Delicious to Twitter) is the highest in our study. The results indicate that people tend to have different friends on different sites depending on the user's interests and the site's purpose. For example, Delicious may contain more information about life and entertainment, while Twitter may serve as a real-time news channel. These results could be further analyzed to study the likelihood of an individual joining a social media site given the presence of his friends on the site. These results could also offer great insights on the dependencies between different online social networks.

1.5.3 Influence Homophily Across Different Sites

In this section, we investigate the influence of an influential user's friends. We want to analyze whether an influential user tends to connect to influential friends on a

Table 1.4 Influence position across two sites

Dataset	Delicious	StumbleUpon	Twitter
Delicious	0.7289	0.8974	0.7999
StumbleUpon	0.7193	0.9016	0.7928
Twitter	0.7516	0.8843	0.8249

social web site. Furthermore, we want to investigate whether an influential user on one web site tends to connect to influential friends on other sites as well. We define this phenomenon as influence homophily. Individual influence homophily is a directed property of a user $\mathcal{U}$ from site A to site B. That is, for every influential user $\mathcal{U}$ on site A, we compute the proportion of his influential friends on site B. Then we get the average influence homophily from site A to site B by averaging for all the influential users on site A. Here, user $\mathcal{U}$ is considered to be influential on a site as long as his influence score exceeds the average influence score of users on that site. An individual's influence homophily from site A to site B is defined as,

$$IH_{A \to B}(\mathcal{U}) = \frac{IF_B(\mathcal{U})}{size(F_B(\mathcal{U}))}, \tag{1.9}$$

where, $IF_B(\mathcal{U})$ is the number of user $\mathcal{U}$'s influential friends on site B, and $F_B(\mathcal{U})$ is user $\mathcal{U}$'s friends on site B.

The results are presented in Table 1.4. All the influence homophily scores are higher than 70% for Delicious, StumbleUpon and Twitter datasets, which indicates that an influential user on site A tends to have influential friends on site B. In the situation where $A = B$, i.e. A and B are the same site (i.e., the diagonal entries of Table 1.4), the results show that influential users tend to make friendship with other influential individuals on the same site as well, which is consistent with homophily that explains the tendency of individuals to associate and bond with similar others.

1.5.4 Observed Influential Behavioral Patterns

To Summarize, we've observed that an influential user $\mathcal{U}$ on site A is likely to be influential on another site B in Sect. 1.5.1. In Sect. 1.5.2, we observed that an individual generally doesn't have many common friends across two social media sites. From the diagonal entries of Table 1.4, we find that an influential user is likely to be connected to other influential individuals on a site. Therefore, the results of the non-diagonal entries of Table 1.4 indicate that influence homophily does exist, whereby, influential users are more likely to be connected to other influential users of the network and even across network. This observation can be used to design advertising strategies in virtual marketing, in whom only a selective set of influential users need to be targeted to propagate the news of the product across sites through their network.

1.6 Conclusions

In this chapter, we studied the influential behavioral patterns of individuals spanning across multiple social media websites. Although the problem of identifying influential individuals has been extensively studied in sociology literature, the problem is relatively new in the context of online/virtual spaces, especially social media. The peculiarities specific to online environments are introduced in the chapter along with models to quantify influence. We introduce a formal definition of influence and also propose a model that uses user content and network information to measure the influence of an individual.

This chapter primarily focuses on a new avenue of research, namely, cross site study of behavior in social media sites. It is known that individuals use different social media sites for different purposes. We provide a novel approach to address the challenges of cross site data collection through the use of blog directory sites, where users voluntarily provide their online identities. We introduce the idea of cross site study to the problem of analyzing behavior of influential individuals through a case study on three popular social media sites, viz., Delicious, StumbleUpon, and Twitter. From our study, we conclude that:

1. Influential individuals have a higher probability to remain influential at most of the sites they are a member of.
2. An individual generally doesn't have many common friends across two social media sites.
3. The principle of 'influence homophily' exists in the formation of ties on social media sites. In other words, influential individuals are more likely to befriend other influential individuals.

Influential individuals are also commercially important nodes in a network because of their information diffusion capabilities. Analyzing the influence of these influential nodes across social media sites gives us a good starting point in the analysis of an unknown social media site. The study could have far-reaching implications on targeted advertising and social customer relationship management (Social CRM) at large. We envisage the challenges, opportunities, analysis, and findings presented in this chapter will open doors for innovation in this burgeoning area of social media analytics, especially across social media studies with significant contributions to various disciplines such as, computational sociology, cultural anthropology, and behavioral psychology, among others.

Acknowledgements This research was funded in part by the National Science Foundation's Social-Computational Systems (SoCS) program and Human Centered Computing (HCC) program within the Directorate for Computer and Information Science and Engineering's Division of Information and Intelligent Systems (award numbers: IIS - 1110868 and IIS - 1110649) and by the U.S. Office of Naval Research (award number: N000141010091). We gratefully acknowledge this support.

References

1. Adar, E., Adamic, L.A.: Tracking information epidemics in blogspace. In: WI'05: Proceedings of the 2005 IEEE/WIC/ACM International Conference on Web Intelligence (WI'05), Washington, DC, USA, pp. 207–214. IEEE Comput. Soc., Los Alamitos (2005). doi:10.1109/WI.2005.151
2. Agarwal, N., Liu, H., Tang, L., Yu, P.S.: Identifying the influential bloggers in a community. In: Proceedings of the International Conference on Web Search and Web Data Mining, pp. 207–218. ACM, New York (2008)
3. Amitay, E., Yogev, S., Yom-Tov, E.: Serial sharers: Detecting split identities of web authors. In: Workshop on Plagiarism Analysis, Authorship Identification, and Near-Duplicate Detection, Amsterdam, Netherlands, July (2007). http://einat.webir.org/SIGIR_PAN_workshop_2007.pdf
4. Anderson, C.: The Long Tail: Why the Future of Business Is Selling Less of More. Hyperion Books, New York (2008)
5. Cilibrasi, R., Vitanyi, P.M.B.: Clustering by compression. IEEE Trans. Inf. Theory 51(4), 1523–1545 (2005)
6. Faloutsos, M., Faloutsos, P., Faloutsos, C.: On power-law relationships of the internet topology. In: Proceedings of the Conference on Applications, Technologies, Architectures, and Protocols for Computer Communication, pp. 251–262. ACM, New York (1999)
7. Gill, K.E.: How can we measure the influence of the blogosphere. In: WWW 2004 Workshop on the Weblogging Ecosystem: Aggregation, Analysis and Dynamics, New York. Citeseer (2004)
8. Golbeck, J., Hendler, J.: Filmtrust: Movie recommendations using trust in web-based social networks. In: Proceedings of the IEEE Consumer Communications and Networking Conference, vol. 96. Citeseer (2006)
9. Goldenberg, J., Libai, B., Muller, E.: Talk of the network: A complex systems look at the underlying process of word-of-mouth. Mark. Lett. 12(3), 211–223 (2001)
10. Gruhl, D., Guha, R., Liben-Nowell, D., Tomkins, A.s.: Information diffusion through blogspace. In: Proceedings of the 13th International Conference on World Wide Web, pp. 491–501. ACM, New York (2004)
11. Katz, E.: The two-step flow of communication: An up-to-date report on an hypothesis. Public Opin. Q. 21(1), 61–78 (1957)
12. Keller, E., Berry, J.: One American in Ten Tells the Other Nine How to Vote, Where to Eat and, What to Buy. They Are the Influentials. The Free Press, New York (2003)
13. Kempe, D., Kleinberg, J., Tardos, É.: Maximizing the spread of influence through a social network. In: Proceedings of the Ninth ACM SIGKDD International Conference on Knowledge Discovery and Data Mining, pp. 137–146. ACM, New York (2003)
14. Kolari, P., Finin, T., Joshi, A.: SVMs for the blogosphere: Blog identification and splog detection. In: AAAI Spring Symposium on Computational Approaches to Analyzing Weblogs (2006)
15. Kumar, R., et al.: On the bursty evolution of blogspace. In: 12th International Conference on World Wide Web (2003)
16. Liben-Nowell, D., Kleinberg, J.: The link-prediction problem for social networks. J. Am. Soc. Inf. Sci. Technol. 58(7), 1019–1031 (2007)
17. Lin, Y.-R., Sundaram, H., Chi, Y., Tatemura, J., Tseng, B.L.: Detecting splogs via temporal dynamics using self-similarity analysis. ACM Trans. Web 2(1), 1–35 (2008)
18. Merton, R.K.: Social Theory and Social Structure. Free Press, New York (1968)
19. Mislove, A., Marcon, M., Gummadi, K.P., Druschel, P., Bhattacharjee, B.: Measurement and analysis of online social networks. In: Proceedings of the 7th ACM SIGCOMM Conference on Internet Measurement, p. 42. ACM, New York (2007)
20. Richardson, M., Domingos, P.: Mining knowledge-sharing sites for viral marketing (2002)

21. Rogers, E.M., Shoemaker, F.F.: Communication of innovations; a cross-cultural approach (1971)
22. Zafarani, R., Liu, H.: Connecting corresponding identities across communities. In: Proceedings of the 3rd International Conference on Weblogs and Social Media (ICWSM09) (2009)

Chapter 2
Modeling and Analysis of Social Activity Process

Can Wang and Longbing Cao

Abstract Behavior modeling has been increasingly recognized as a crucial means for disclosing interior driving forces and impact in social activity processes. Traditional behavior modeling in behavior and social sciences that mainly relies on qualitative methods is not aimed at deep and quantitative analysis of social activities. However, with the booming needs of understanding customer behaviors and social networks etc., there is a shortage of formal, systematic and unified behavior modeling and analysis methodologies and techniques. This paper proposes a novel and unified general framework, called Social Activity Process Modeling and Analysis System (SAPMAS). Our approach is to model social behaviors and analyze social activity processes by using model checking. More specifically, we construct behavior models from sub-models of actor, action, environment and relationship, followed by the translation from concrete properties to formal temporal logic formulae, finally obtain analyzing results with model checker SPIN. Online shopping process is illustrated to explain this whole framework.

2.1 Introduction

Behavior refers to the action or reaction of any material under given circumstances and environment.[1] Human behavior has been increasingly highlighted for social activities in many areas such as social computing [20], intrusion detection, fraud detection [8], event analysis [21], and group decision making, etc. In both natural and social sciences and applications, multiple behaviors from either one or multiple

---

[1] http://dictionary.reference.com

The work of C. Wang is sponsored by Australian Research Discovery Grant (DP0988016).

C. Wang (✉) · L. Cao
QCIS Centre, Faculty of Engineering and Information Technology, University of Technology, Sydney, Australia
e-mail: cawang@it.uts.edu.au
url: http://datamining.it.uts.edu.au/group/

L. Cao
e-mail: lbcao@it.uts.edu.au

L. Cao, P.S. Yu (eds.), *Behavior Computing*,
DOI 10.1007/978-1-4471-2969-1_2, © Springer-Verlag London 2012

actors often interact with one another. Such behavior interactions may form interior driving forces that impact underlying social activities or situations, and may even cause challenging problems. Take the online shopping process as an example, the customer and the merchant communicate with each other to guarantee the success of an online transaction through the inspection of a trusted third party. Similar behavior communications are widespread in many applications, such as interactions in social communities and multi-agent systems.

To the best of our knowledge, along with qualitative research in behavior sciences [16], behavior representation has been a typical topic in the AI community. Major efforts on action reasoning [10] and composition [17], behavior coordination [15] and planning [6], and modeling systems rather than behaviors [18], have been made. For instance, Serrano and Saugar [18] exploited the application-independent software connector to specify multi-agent societies rather than agent behaviors. Gu and Soutchanski [10] discussed reasoning about action based on a modified Situation Calculus. Sardina et al. [17] considered behavior compositions when failure presents. In addition, many works on 'behavior modeling' actually refer to behavior recognition [9] and simulation [19] instead of representation and checking, which is different from our focus here. Limited work can be identified on representation [2] and checking [4] complex behavior structures and interactions.

Modeling complex behaviors and their interactions are challenging. In the existing work on behavior modeling strategies, there are major issues: (1) traditional behavior modeling that mainly relies on qualitative methods from behavior and social sciences [16] often leads to ineffective and limited analysis in understanding social activities deeply and accurately; (2) traditional behavior expressiveness is too weak to reveal the fact that behavior plays the key role of an internal driving force [3] for social activities; (3) the existing behavior modeling approaches for social activities are bottom-up techniques, which have too many styles and forms according to distinct situations; (4) the existing work often overlooks the checking of behavior modeling, which weakens the soundness and robustness of behavior modeling methods. Consequently, a unified quantitative representation and checking approach for behavior modeling and analysis is in great demand for social activity studies. In this paper, we take great advantage of the model checking techniques to build a novel and unified modeling and analysis system for social activity processes.

There are different types of formal verification, from the manual proof of mathematical arguments to interactive computer aided theorem proof, and automated model checking. Manual proofs are time-consuming, error-prone, and often not economically viable. Computer-aided theorem provers still require significant expert knowledge. However, a considerable number of works in recent years have been devoted to studying theoretical aspects and application fields of the famous technique model checking [4], which outperforms the manual proof and test with simulation in terms of nondeterminism and automation. Model checking is a system verification technique that is being applied to the design of Information and Communication Technology Systems, ranging from software controlling storm surge barriers [14], through space craft controllers [11] to integrated circuits. Moreover, model checker

SPIN [12] becomes a general tool for verifying the correctness of distributed software models in a rigorous and mostly automated fashion. Model checking is currently attracting considerable attentions, and was given the ACM Turing Award 2007 in recognition of the paradigm-shifting work on this topic initiated a quarter century ago.

This verification technology provides an algorithmic means of determining whether an abstract model, i.e. a hardware or software design, satisfies a formal specification expressed as a temporal logic formula. Based on this principle, we could develop our novel framework, that is *Social Activity Process Modeling and Analysis System* (SAPMAS). Specifically, when given a social activity process in terms of actor, action, environment and relationship sub-models, together with the negative forms of desired properties to be verified, transformations are conducted to convert them into a graphical model and combination of patterns, respectively. Subsequently, corresponding transition system (TS) and temporal logic formulae (TLF) can be obtained by semantic mappings. Then, model checker SPIN is used to output either "no sequence found" or "targeted activity sequences" with the inputs: TS and TLF. Afterwards, distinguished activity patterns can be extracted from the activity sequences attained. In the final stage, we illustrate the proposed system through one social activity process on analyzing online shopping behavior interactions, which shows the promising potential of our new system SAPMAS for handling similar issues in behavior modeling.

Overall, according to the theories and techniques we have used, the most outstanding advantages of our method lie on the following four aspects to overwhelm traditional approaches:

– All the involved procedures complete automatically with solid theoretical backgrounds for effective analysis.
– Logics can easily express concurrent properties accurately to expose the vital role that behavior plays.
– This uniform system can be constructed for varied processes with little or no relevance, such as social network and multi-agent systems.
– The involvement of checking make our behavior model solid and stable.

The paper is organized as follows. In Sect. 2.2, the concepts of behavior model, including sub-models of actor, action, environment and relationship, are specified. Section 2.3 introduces the behavior property that contains combination of patterns and temporal logic. Section 2.4 proposes the whole framework of SAPMAS. A case study of online shopping process is exemplified in Sect. 2.5. We conclude the paper in Sect. 2.6 with research issues.

2.2 Behavior Model

Within the scope of Behavior Informatics (BI) [3], behaviors refer to those activities that present as actions, operations or events as well as activity sequences conducted by human beings within certain contexts and environments in either a virtual or physical organization. In the sequel, we will establish both graphical and formal ways to model behaviors. The core of behavior modeling contains concepts that enable representation and reasoning about behaviors. Four dimensions are identified to represent a behavior.

- *Actor Sub-model*: The dimension allows for describing the behavior subjects and objects, for example, organizations, departments, systems and people involved in an activity or activity sequence.
- *Action Sub-model*: The dimension allows for tracing the activities, operations and events, etc. happening in an activity or activity sequence.
- *Environment Sub-model*: The dimension allows for depicting the contexts and circumstances surrounding an activity or activity sequence.
- *Relationship Sub-model*: The dimension allows for representing the connections among activities or activity sequences.

2.2.1 Actor Sub-model

The basic concept in actor sub-model is "actor", which designates an entity that conducts and implements activities. More systematically, actors are considered to be structured, that is to say, they may contain other actors. Also, actors are related to each other as so to guarantee the occurrence of actions, if one of them is the subject, then the other one must be the object. It induces three critical attributes of the actor sub-model as follows:

- *Subject(s)*: The entity that issues the activity or activity sequence;
- *Object(o)*: The entity on which a behavior is imposed;
- *Structure(st)*: The hierarchy to reflect the roles of the entities involved.

As we know, for a behavior, a subject must make an action on an object. In this way, a predicate logic form can be induced as *Action(Subject, Object)*, where subject and object are terms, while action is a predicate indicating the relationships between subject and object. Moreover, a hierarchy can be utilized over actors such that an actor can be defined in terms of other actors. Therefore, we could formally represent actor sub-model as *Action(Structure(Subject, Object))*, where structure is a binary function to reveal the specific structures between multiple subjects and objects. In addition, the corresponding subject and object are equipped with communication points, indicating physical or logical relationships among actors involved. The symbols are shown in Fig. 2.1.

2.2.2 Action Sub-model

The basic concept in action sub-model is "action", for one action, it has several associated attributes to describe the features of the so-called action. We could divide

Fig. 2.1 Symbols of the actor sub-model

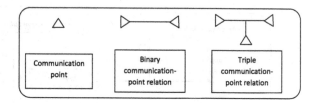

them into two categories as follows:

- Objective Attribute
  - *Time(t)*: When the behavior occurs;
  - *Place(w)*: Where the behavior happens;
  - *Status(u)*: The stage where a behavior is currently located;
  - *Constraint(c)*: What conditions impact on the behavior.
- Subjective Attribute
  - *Belief(b)*: Information and knowledge of the subject about the world;
  - *Goal(g)*: Targets that the behavior subject would like to accomplish;
  - *Plan(l)*: Sequences of actions that the behavior subject can perform to achieve one or more of its intentions;
  - *Operation(o)*: What the subject has chosen to do or operate;
  - *Effect(e)*: The results or influence led by the execution of a behavior on the object or the context.

In action sub-model, from the perspectives of both predicate logic and temporal logic, we could conduct a formal expression of all these features as follows:

$$Belief(X) \rightarrow \Diamond Goal(X) \rightarrow \Diamond Plan(X) \rightarrow \Diamond Operation(X) \rightarrow \Diamond Effect(X),$$

where X denotes a function as $X = f(Actor, Time, Place, Status, Constraint)$ with five items, eventually $\Diamond$ is one of temporal modalities. Here, the above formula means that if item X has belief, then eventually will set up a goal, this goal will lead to a plan in the future, followed by the actual operation afterwards, then effect will generate finally. It is similar to the well-known BDI (Belief-Desire-Intention) model [22] developed in multi-agent systems, the techniques involved can be adapted in this model as well for our further research.

2.2.3 Environment Sub-model

The basic concept in environment sub-model is "environment", two categories are focused on here. One is related to physical condition, while the other one matters the social relationship. Specifically, they are

- *Context(e)*: The environment in which a behavior is operated;
- *Associate(a)*: Other behavior instances that are associated.

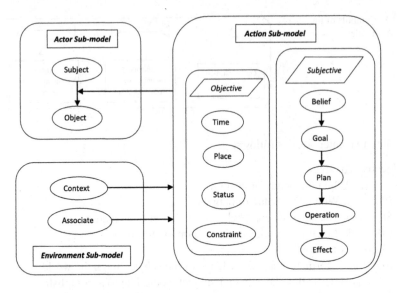

Fig. 2.2 Attributes and links among actor, action, and environment sub-models

In fact, they have been embedded into the above action sub-model. *Context* is essentially the enabling condition, i.e. *constraint* in action sub-model, while *associate* is the relevant *operation* in action sub-model. In logic language, that is (*Context* → *Action*) ∧ (*Associate* → *Action*).

Overall, for one behavior, all of the 13 attributes and their links among actor, action and environment sub-models are depicted in the following Fig. 2.2.

2.2.4 Relationship Sub-model

The basic concept in relationship sub-model is "relationship", all of the above three sub-models focus on only one behavior, but for multiple behaviors, we should concentrate on the couplings among them.

As we know, an action can only happen when its enabling condition is satisfied. To some extent, this kind of enabling condition is a constraint which can be formulated in terms of other actions having occurred yet or not. There are several kinds of relationships [7] shown in Fig. 2.3.

In this figure, initiative actions are actions without causality conditions to start the activity sequences. The enabling relationship requests that behavior A must perform before behavior B, while the disenabling relationship means that if behavior A happens, behavior B cannot take place. The or-split relationship represents the situation that one of the latter actions can be chosen after the former one has been done, while the and-split relationship expresses that all of the latter actions will be conducted once the former action has occurred. Similarly, or-join means after selecting

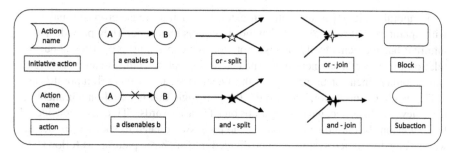

Fig. 2.3 Symbols of the relationship sub-model

Table 2.1 Logic representations of relationship sub-model

Relationship	*enable*	*disenable*	*or-split*	*and-split*	*or-join*	*and-join*
Logic Form	$a \rightarrow b$	$\neg a \rightarrow b$	$a \rightarrow (b \vee c)$	$a \rightarrow (b \wedge c)$	$(a \vee b) \rightarrow c$	$(a \wedge b) \rightarrow c$

one of the former actions to be done, the latter one follows, whereas and-join depicts the scenario that the latter action can only be taken when all of the former ones have completed. These are the basic elements to describe the multiple types of relationships in social activity processes, for some special cases, we may add some symbols or operators if possible. By using them, an instance graphical model is constructed as exemplified in Sect. 2.5. The corresponding logic forms are formalized in the simplest case with merely actions a, b, c as shown in Table 2.1.

However, extensive modeling of all relevant actions and their enabling conditions will lead to a huge, disorderly and unsystematic graphical model. Thus, we need to develop methods to express all of these information in a systematic and hierarchical way. Here, two means are used to tackle the complexity of models. One is to divide the complex graph into blocks according to different stages, while the other is to make decomposition based on distinct actors, which will be illustrated in Sect. 2.5.

2.3 Behavior Property

In order to analyze and mine some special behaviors in an existing system, such as frequent sequence, exceptional sequence and hidden sequence, we must firstly define the properties of those patterns to be mined accurately. State differently, if we aim to discover some frequent patterns, we should give a specific definition of the term "support", then frequent pattern is an activity sequence with support greater than or equal to a given minimum threshold support, while exceptional behavior is concerned on the condition that both "Intentional Interestingness" and "Exceptional Interestingness" are well defined [3], what hidden means must be confirmed before mining hidden sequences. In this section, we take advantage of temporal logic to express properties of the desired behavior sequences.

Temporal logic [4] is basically an extension of traditional propositional logic with operators that refer to the behavior of systems over time. They provide a very intuitive but mathematically precise notation for expressing properties about the relationship between distinct states. The underlying nature of time in temporal logics can be either linear or branching. In the linear view, it is Linear Temporal Logic (LTL), while Computation Tree Logic (CTL) is logic that is based on a branching-time view. At this stage, we mainly focus on LTL, afterwards CTL can be considered for extension. In our consideration, all the processes related to social activities, as well as any distributed system, can be divided into six basic patterns as follows.

- *Tracing*: Different actions with sequential order, i.e. $\{a_1, a_2, \ldots, a_n\}$.
- *Consequence*: Different actions have causalities in occurrence, i.e. $\{a_i \rightarrow a_j\}$.
- *Synchronization*: Actions occur at the same time, i.e. $\{a_1 \leftrightarrow, \ldots, \leftrightarrow a_n\}$.
- *Combination*: Different actions occur in concurrency, i.e. $\{a_1 \| a_2 \|, \ldots, \| a_n\}$.
- *Exclusion*: Different actions occur mutually exclusively, i.e. $\{a_1 \oplus, \ldots, \oplus a_n\}$.
- *Precedence*: Some actions have required precedence, i.e. $\{a_i \Rightarrow a_j\}$.

Note that the *Consequence* means that the occurrence of action a_i will lead to action a_j, while the *Precedence* represents that the occurrence of action a_j must require the happening of action a_i. The above items are the six fundamental patterns in social activities, while in application quantifiers and temporal modalities are necessary to express a variety of properties. Specifically, the involved quantifiers are an existential quantifier $\exists$ and a universal quantifier $\forall$, and the temporal modalities are basic operators such as next $\bigcirc$, until $\bigcup$, eventually $\Diamond$ and always $\square$.

The sequential or parallel combinations of the above basic patterns, quantifiers and temporal modalities have a great power to express diverse properties. The sequential combination, intuitively, describes that the actions occur in a successive order for a property. The parallel combination, however, represents that all the actions involved in a property take place simultaneously. Symbols $\times$ and $\otimes$ are used to denote sequential and parallel combinations, respectively. In addition, nested combination is of great necessity to reveal the inner and outer relationships of actions. For extension, we could consider the fuzzy or probabilistic combinations of those relationships.

2.4 Behavior Analysis

In order to analyze the expressiveness, robustness and stability of our behavior model, we take great advantage of model checking to test the behavior model with the given properties. In this section, we interpret how to use model checking appropriately in behavior analysis.

Model checking [4] is an automated technique that, given a finite-state model of a system and a formal property, systematically checks whether this property holds for that model. Specifically, two essential prerequisites for model checking are a system under consideration and required properties. On having them at hand, the

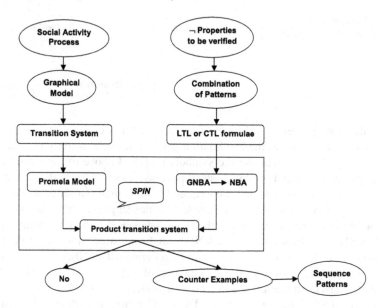

Fig. 2.4 The steps performed in SAPMAS

model checker examines all relevant system states to check whether they satisfy the desired property. If a state is encountered that violates the property, the model checker provides a counterexample that indicates how the model could reach the undesired state. Stated in details, the system will be represented or transformed as a transition system (TS), which are basically direct graphs where nodes denote states and edges model transitions (state changes), to describe the behavior of systems. The requirement will be formalized as a property specification in the form of temporal logic formula, then the negation of temporal-formula property is accepted by a Generalized Nondeterministic Büchi Automaton (GNBA), afterwards transformation will be conducted to generate a Nondeterministic Büchi Automaton (NBA) to replace the GNBA. When a TS and a NBA are ready, a persistence checking by nested depth-first search is led to verify whether the property holds for this system, then relevant results will be obtained: either "Satisfied" or "No" with a counter example.

Here, we mainly take advantage of model checking to analyze behaviors. At this stage, we only give a general framework of the whole process, more details will be discussed for our future work. We have introduced the behavior model and behavior properties, which will be mapped to a transition system (TS) and temporal logic formulae (TLF), respectively. Below, Fig. 2.4 illustrates the steps performed in Social Activity Process Modeling and Analysis System (SAPMAS).

Specifically, at first, a social activity process and the properties that are the features of the desired patterns must be given. However, the negative forms of the properties are considered rather than the original ones, since the counter examples obtained against ¬properties are just the desired activity sequences in accor-

Fig. 2.5 The graphical actor
sub-model of online shopping

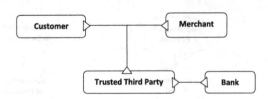

dance with those properties. Then the particular patterns can be extracted based on the attainable counter examples. So after getting the social activity process and ¬properties ready, we could use the symbols proposed above to transform them into a graphical model and combinations of patterns, respectively. In the next stage, TS and TLF are further obtained. Subsequently, TS is then translated to Promela using the approach discussed in [13], note that the language Promela, which is short for "process metalanguage", is the input language for the prominent model checker SPIN [12]. Simultaneously, a GNBA can be constructed to accept the negative forms of LTL or CTL formulae, and then converted to NBA. Afterwards, a product of TS is made up from the combination of a Promela Model and a NBA. These three steps are achieved within a famous model checker SPIN. Thus, two input elements of SPIN are corresponding TS and TLF, then accordingly, the outputs of SPIN are two alternative answers, one is "no" which means there is no activity or action sequence possessing the desired properties, while in contrast, counter examples will be given out in accordance with the obtainment of the targeted sequence patterns.

2.5 Case Study

In this section, we use the increasingly popular e-Business process, i.e., online shopping process [1, 5] to illustrate the whole procedure of the system SAPMAS.

For *actor* sub-model, messages are communicated among a customer, a merchant and a trusted third party (TTP), along with the bank. Figure 2.5 depicts this typical actor sub-model, showing the parties involved in the online shopping process. There are four actors in total, i.e., a customer, a merchant, a TTP and the bank. It includes triple communication-point relation among the customer, the merchant and the TTP. The legend has been given in Fig. 2.1.

With regard to *action* sub-model, for the action "Send product", *subject* is the merchant, *object* is the customer, this action occurs at the *time* when the merchant has accepted the purchase order (PO) and the *place* is online, the *status* is in the main stage of online shopping process, it has the *constraint* of the acceptance of PO. This action has the *belief* that the customer will buy the product finally, and aims that the product will not be corrupted during transit. To achieve this *goal*, the merchant draws up a *plan* to send the copy of the encrypted product together with a signed cryptographic checksum. Intuitively, the *operation* is just to send the product, and this operation leads to the *effect* of requesting the decrypting key from the customer. Moreover, for *environment* sub-model, the *context* is that both the product ID has been validated by the customer and the PO has been accepted by the merchant, while the *associate* is all the operations with direct link to the action "Send product", for instance, Accept PO, Receive product and Forward PO.

In the sequel, we will show how the transaction process takes place in a graphical model for the *relationship* sub-model. The whole procedure can be explained as follows: First, the customer browses the product catalog online located at TTP and chooses a favorable product. After that, the customer downloads the encrypted product together with the product identifier. Then if the identifier of the encrypted product file corresponds to the identifier in the product identifier file exactly, the transaction proceeds, otherwise advice is sent to the TTP and the customer waits for correct encrypted product ID. Subsequently, the customer prepares a cryptographic checksum and PO, which are sent to the merchant. Once receiving the PO, the merchant examines its contents. If the merchant is satisfied with the PO, the merchant digitally signs the cryptographic checksum of the endorsed PO and forwards to the TTP as well as a single use decrypting key for the product. Meanwhile, the merchant then sends a copy of the encrypted product to the customer, together with a signed cryptographic checksum. Next, the customer validates whether the first and second copies of the product are identical so as to confirm the ordered product. Afterwards, the customer forwards to the TTP the PO and a signed payment token, together with its cryptographic checksum to ask for decrypting key from the TTP. To verify the transaction, the TTP first compares the digest included in the PO from the customer with the digest of that from the merchant. If the two do not match, the TTP aborts the transaction. If they do match, the TTP proceeds by validating the payment token with the customer's financial institution. If the token is not validated, the TTP aborts the transaction and advises the merchant simultaneously. Otherwise, the TTP sends the decrypting key and the payment token, both digitally signed with TTP private key, to the customer and the merchant, respectively. Finally, after a somewhat complex process, the customer will receive the decrypted product if everything is all right, or else the transaction will be aborted by the TTP.

In stage-based Fig. 2.6, we divide the whole graphical model into three parts, namely *Preliminary Stage*, *Main Stage* and *Final Stage*. The *Preliminary Stage* is the phase for the customer to choose his or her favorable product, while the critical activities and communications happen in the *Main Stage*, then the TTP makes the last validation to determine the transaction in the *Final Stage*. This figure reveals the distinct phases of the whole procedure.

In the actor-based one in Fig. 2.7, the whole graph model has been divided into four sections according to the four actors involved. As mentioned before, the participating actors are the *Customer*, the *Merchant*, the *TTP* and the *Bank*. Note that all the actions involved in the communications of two actors have been cut into two sub-actions with symmetrical shapes. It shows that this kind of action only can happen with all of its related sub-actions enabled simultaneously. This figure reveals the different roles of distinct actors in a certain process or a series of activities.

Now, we have got the online shopping process visualized to a graphical model as shown in Figs. 2.6 and 2.7. Here, we will analyze the behavior properties. For the

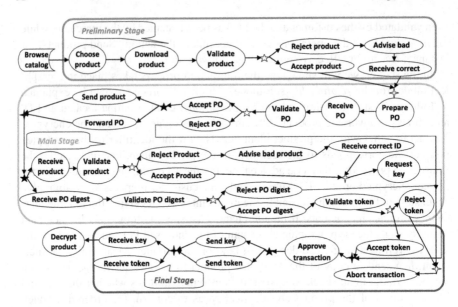

Fig. 2.6 The graphical action sub-model of online shopping based on stages

simple property stated as "*the actions choose product, download encrypted product and validate product ID must happen one after another in order*" can be written as $\{CP, \bigcirc DC, \bigcirc VP\}$, where the elements are the abbreviated forms of the corresponding actions, they are the so-called "atomic proposition" in logic language, and the same denotation applies for the following representations. Further, for the more complicate case that "*after accepting the PO, the merchant will send product to the customer and forward PO to the TTP*", accordingly, can be represented as $\{AP\} \times \bigcirc \{SP \| FPO\}$. Moreover, the property described as "*customer's rejecting product and accepting product cannot happen simultaneously while the former one leads to advise bad ID and the latter one goes directly to the stage of requesting key from the TTP*", similarly, can be depicted as $\{\{RP \to \bigcirc AID\} \oplus \{AP, \bigcirc RK\}\}$. Besides, formula $\{\{\forall RWOC\} \to \{\bigcirc AT\}\}$ means that "*all the rejected actions without proper correction will lead to undesirable transaction abortion*". Note that all of the above expressions of properties are validate for the graphical model. But those following properties presented as $\{ST, AP\}$ and $\{\{\forall AA\} \to \{\lozenge DP\}\}$, which mean "*accepting product after sending token*" and "*all the accepted actions will lead to receiving decrypted product eventually in the end*" respectively, will not hold. Note that expressions *RWOC* and *AA* denote the sets of corresponding atomic propositions. All of the mentioned examples are rather simple instances just to illustrate how to express properties in logic language.

For behavior analysis, two properties chosen are "*customer's rejecting product and accepting product cannot happen simultaneously while the former one leads to advise bad ID and the latter one goes directly to the stage of requesting key from the TTP*" and "*all the accepted actions will lead to receiving decrypted product eventually in the end*", represented as $\{\{RP \to \bigcirc AID\} \oplus \{AP, \bigcirc RK\}\}$ and

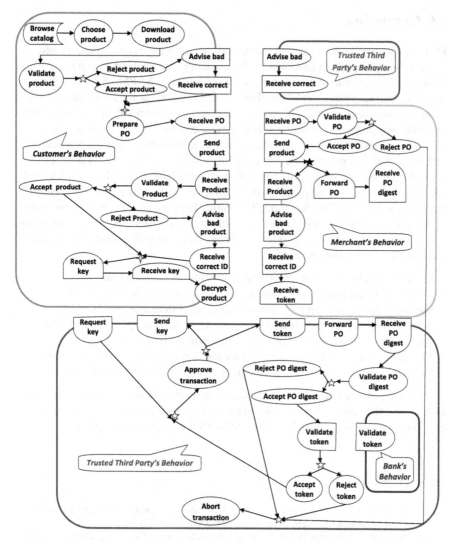

Fig. 2.7 The graphical action sub-model of online shopping based on stages

$\{\{\forall AA\} \rightarrow \{\Diamond DP\}\}$, respectively. After that, accurate transition system semantics of the graphical model and transformation rules are given to form corresponding TS and TLF. Then by using model checker SPIN, we could easily get the result that all the activity sequences following the online shopping process are the desired sequences for the former property, while for the latter one, the answer is no as there exists no activity sequence satisfying this property. Although this is rather a simple and naive example for analysis, it has shown the basic procedures of our modeling and analyzing system SAPMAS.

2.6 Conclusion

In this paper, we build a general framework, i.e., Social Activity Process Modeling and Analysis System (SAPMAS), for modeling and analyzing social activities. Different from existing behavior representation systems, SAPMAS utilizes the current advanced technique model checking, which is based on solid mathematical and computational backgrounds. It provides a graphical model to capture behavioral elements and properties within a social activity process, and summarizes the combinations of patterns to categorize the properties to be verified. The graphical model and combinational patterns are further transformed to a transition system and the logic form respectively to verify and refine social behavior models. SAPMAS outputs the desired activity sequence patterns after verification by SPIN. We exemplify the success use of SAPMAS in modeling online shopping process for our case study.

As a new research topic in behavior informatics, SAPMAS consists of many open issues that are worthy of systematic investigation as well as case studies. These issues include the following aspects:

- Combine quantitative and qualitative properties for advanced mining.
- Construct accurate transition system semantics for the graphical model.
- Establish transformation rules between combined formulae and TLF.
- Adapt the model checker SPIN to improve the functionality and applicability.
- Build extraction rules from counter examples to desired sequence patterns.
- Extend existing transition system to probabilistic or fuzzy transition system.
- Consider state explosion problems existing in model checking techniques.

These are all the unresolved and challenging issues for this new framework and system, and they are also our research issues in the future work for completing and enriching our newly proposed system SAPMAS with real world applications.

References

1. Bonnie, B.A., James, V.H., Paul, B.L., Scott, L.S.: Model checking for design and assurance of e-business processes. Decis. Support Syst. **39**, 333–344 (2004)
2. Brachman, R.J., Levesque, H.J.: Knowledge Representation and Reasoning. Elsevier, Amsterdam (2004)
3. Cao, L.: In-depth behavior understanding and use: the behavior informatics approach. Inf. Sci. **180**, 3067–3085 (2010)
4. Christel, B., Joost, P.K.: Principles of Model Checking. MIT Press, Cambridge (2008)
5. Chung, S.Y., Park, C.: Online shopping behavior model: a literature review and proposed model. In: Proceedings of the 11th International Conference on Advanced Communication Technology, pp. 2276–2282 (2009)
6. Edelkamp, S., Kissmann, P.: Optimal symbolic planning with action costs and preferences. In: Proceedings of IJCAI-09, pp. 1690–1695 (2009)
7. Eertink, H., Janssen, W.P.M., Oude Luttighuis, P.H.W.M., Teeuw, W., Vissers, C.A.: A business process design language. In: Proceedings World Congress on Formal Methods, Toulouse. Springer LNCS (1999)
8. Fast, A., Friedland, L., Maier, M., Taylor, B., Jensen, D., Goldberg, H., Komoroske, J.: Relational data preprocessing techniques for improved securities fraud detection. In: KDD (2007)

 9. Gabaldon, A.: Activity recognition with intended actions. In: Proceedings of IJCAI-09, pp. 1696–1701 (2009)
10. Gu, Y., Soutchanski, M.: Decidable reasoning in a modified situation calculus. In: Proceedings of IJCAI-07, pp. 1891–1897 (2007)
11. Havelund, K., Lowry, M., Penix, J.: Formal analysis of a space craft controller using spin. In: Holzman, G., Najm, E., Serhrouchni, A. (eds.) Proceedings of the 4th International SPIN Workshop, pp. 147–167 (1998)
12. Holzmann, G.J.: The Spin Model Checker: Primer and Reference Manual. Lucent Technologies (2004)
13. Janssen, W., Mateescu, R., Mauw, S., Springintveld, J.: Verifying business processes using SPIN. In: Holzman, G., Najm, E., Serhrouchni, A. (eds.) Proceedings of the 4th International SPIN Workshop, pp. 21–36 (1998)
14. Kars, P.: The application of Promela and Spin in the BOS project. In: Proceedings of Second Spin Workshop (1996)
15. Koenig, S., Keskinocak, P., Tovey, C.: Progress on agent coordination with cooperative auctions. In: Proceedings of AAAI-10, pp. 1713–1717 (2010)
16. Pierce, W.D., Cheney, C.D.: Behavior Analysis and Learning. Lawrence Erlbaum Associates, 3rd edn. (2004)
17. Sardina, S., Patrizi, F., de Giacomo, G.: Behavior composition in the presence of failure. In: Proceedings of KR-08, pp. 640–650 (2008)
18. Serrano, J.M., Saugar, S.: An architectural perspective on multiagent societies. In: Proceedings of AOSE-10 at AAMAS-10, pp. 85–90 (2010)
19. Subramanian, K.: Task space behavior learning for humanoid robots using Gaussian mixture models. In: Proceedings of AAAI-10, pp. 1961 (2010)
20. Wang, F.Y., Carley, K.M., Zeng, D., Mao, W.J.: Social computing: from social informatics to social intelligence. IEEE Intell. Syst. **22**(2), 79–83 (2007)
21. Weiss, G., Hirsh, H.: Learning to predict rare events in event sequences. In: KDD-98, pp. 359–363 (1998)
22. Wooldridge, M.: Reasoning About Rational Agents. MIT Press, Cambridge (2000)

Chapter 3
Behaviour Representation and Management Making Use of the Narrative Knowledge Representation Language

Gian Piero Zarri

Abstract This chapter illustrates some of the different knowledge representation and inference tools used by a high-level, fully implemented conceptual language, NKRL (Narrative Knowledge Representation Language), to deal with the most common types of human "behaviours". All possible kinds of multimedia "narratives", fictional or non-fictional, can be seen in fact as streams of elementary events that concern the behaviours, in the most general meaning of this term, of some specific characters. These try to attain a specific result, experience particular situations, manipulate some (concrete or abstract) materials, send or receive messages, buy, sell, deliver, etc. Being able to deal in a correct (and computer-usable) way with narratives implies then being able to deal correctly with the behaviours of the concerned characters.

3.1 Introduction

"*Behaviour computing*" (or "behavioural informatics" etc.) [17] is a young discipline aiming at investigating *formal methods and workable tools* for human behaviour representation and processing—(at least partially) related disciplines are "sentiments analysis" see, e.g., [1] and "affective computing" see, e.g., [2]. In this chapter, we have focused our attention on a fundamental aspect of behaviour computing, *behaviour modelling and its translation into computer-usable tools*, by supplying an extremely condensed overview of a (wholly implemented) knowledge representation language and computer system, NKRL—Narrative Knowledge Representation Language [3]—which deals with the *behaviour* of (human and non-human) characters *when this is expressed under the form of "narratives"*.

"Narratives" is a generic term used to denote real-life or fictional stories involving concrete or imaginary characters. A narrative can then be synthetically defined

G.P. Zarri (✉)
LiSSi Laboratory, University Paris-Est/UPEC, 120-122 rue Paul Armangot, 94400 Vitry sur Seine, France
e-mail: zarri@noos.fr

G.P. Zarri (✉)
e-mail: gian-piero.zarri@u-pec.fr

L. Cao, P.S. Yu (eds.), *Behavior Computing*,
DOI 10.1007/978-1-4471-2969-1_3, © Springer-Verlag London 2012

as a *sequence of logically structured "elementary events"* (*a non-linear 'stream' of elementary events*). An elementary event describes in turn the *behaviour* (according to the most general meaning of this term) and the *mutual relationships* among specific (human or non-human) entities involved in the narrative. Some essential properties of narratives are summarized below; see also [3, 2–13] for more details:

- A key feature that defines the *connected* nature of the elementary events of the stream concerns the fact that these are *chronologically related*, i.e., narratives *extend over time* (they have a *beginning*, an *end* and some *form of development*).
- *Space* is also very important, given that *the elementary events occur generally in well defined locations*, real or imaginary ones. The connected events that make up a narrative are then both *temporally and spatially bounded*.
- Chronological successions of elementary events cannot be defined as 'narratives' without some sort of *semantic coherence* and *uniqueness of the theme* of these elementary events. If this *logical coherence* is lacking, the events pertain to different narratives: a narrative can also include a single elementary event.
- When the elementary events are verbalized, their *coherence* is expressed through syntactic features like causality, goal, co-ordination, etc. In this context, we use the term "*connectivity phenomena*" to denote what, within the stream, (i) leads to a *global meaning* that goes beyond the simple addition of the *meanings* conveyed by the single events; (ii) explains the influence of the context.
- Characters that have a specific "*role*" in the global narrative are not necessarily human beings: we can have a narrative concerning, e.g., the vicissitudes in the journey of a nuclear submarine (the actor, character etc.).
- Eventually, a well-known, universal characteristic of narratives concern their *highly 'pervasive' nature*—narratives are, in fact, *ubiquitous* see, e.g., [4].

We can add, with respect to NKRL, that this language/environment has been especially used for dealing with the so-called "*non-fictional narratives*", see [5], like those typically embodied into corporate memory records, news stories, normative and legal texts, medical records etc., but also in multimedia supports as surveillance videos, actuality photos for newspapers and magazines, material for eLearning and Cultural Heritage, etc. We can note that this preference is only due to practical constraints—to make use, e.g., of the financial support of the European Commission—and, as it will appear in the following, nothing (apart from considerations about time and interest) could prevent us from dealing with the whole "Gone with the wind" *fictional-narrative* novel according to an NKRL approach.

In the following, Sect. 3.2 will supply first a short overview of the main representational features of NKRL. Section 3.3 will present a series of examples intended to emphasize the *behavioural* properties of NKRL. Section 3.4 will supply some information about NKRL's inference procedures. Section 3.5 will mention some related work; Sect. 3.6 is a short "Conclusion".

3.2 A Short Presentation of the NKRL Language

3.2.1 General Principles

NKRL innovates with respect to the current ontological paradigms—see Protégé
[6], OWL [7], OWL-2 [8] etc.—by adding to the usual (binary) "ontology of concepts" (called HClass, hierarchy of classes, in an NKRL context) an *"ontology of
elementary events"*. This last is *a new sort of hierarchical organization where the
nodes correspond to n-ary structures called "templates", represented schematically
according to the syntax of Eq. (3.1) below*. The ontology of elementary events is then
denoted as HTemp (hierarchy of templates) in NKRL; templates can be conceived
as the canonical, formal representation of generic classes of elementary events like
"move a physical object", "be present in a place", "having a specific attitude towards
someone/something", "produce a service", "send/receive a message", etc.

$$\left(L_i \left(P_j (R_1 a_1)(R_2 a_2) \ldots (R_n a_n)\right)\right). \tag{3.1}$$

In Eq. (3.1), L_i is the *"symbolic label"* identifying the particular *n-ary* structure corresponding to a specific template, P_j is a *"conceptual predicate"*, R_k is a
generic *"functional role"* and a_k the corresponding *"predicate arguments"*. When a
template following the syntax of Eq. (3.1) is *instantiated* to provide the representation of a simple elementary event like "Bill gives a book to Mary", the predicate
P_j (of the GIVE or MOVE type) will introduce its *three* arguments a_k ("individuals", instances of HClass "concepts") JOHN_, MARY_ and BOOK_1 through *three
functional relationships* (R_k roles) as SUBJECT (or AGENT), BENEFICIARY and
OBJECT. The whole *n*-ary construction is *reified through the symbolic label L_i and
necessarily managed as a coherent block at the same time*. The *concrete instantia-
tions* of structures in the style of Eq. (3.1) are called *"predicative occurrences"* and
correspond then to the NKRL representation of *specific elementary events*. To avoid
the ambiguities of natural language and the possible *"combinatorial explosion"*
problems—see [3, 56–61]—both the (unique) conceptual predicate of Eq. (3.1) and
the associated functional roles are *"primitives"*. Predicates P_j pertain to the set
{BEHAVE, EXIST, EXPERIENCE, MOVE, OWN, PRODUCE, RECEIVE}, and the
functional roles R_k to the set {SUBJ(ect), OBJ(ect), SOURCE, BEN(e)F(iciary),
MODAL(ity), TOPIC, CONTEXT}.

Note also that single arguments a_k of a template/predicative occurrence, or tem-
plates/occurrences as a whole, may be characterised by *"determiners"* (attributes)
that (i) *introduce further details/precisions* about the 'meaning' of these arguments
or templates/occurrences, but that (ii) are *never strictly necessary for their basic semantic interpretation in NKRL terms*, see [3, 70–86]. Determiners are represented
mainly by (i) "locations", lists (in general) of concepts or individuals associated
with the *arguments* a_k of templates/occurrences through the "colon" operator, ":",
see Table 3.1 below; (ii) "modulators", which apply to a *full template or occurrence* to *particularize* their meaning according to the modulators used: they pertain
to three categories, *temporal*, *deontic* and *modal modulators*; (iii) the two *temporal*

Fig. 3.1 An (extremely reduced) image of the *external structure* of HClass, the 'standard' ontology of concepts of NKRL

determiners date-1 and date-2, used only in association with predicative occurrences in order to introduce the temporal information associated with an elementary event.

Several predicative occurrences (elementary events)—denoted by their symbolic labels L_i—can be associated within the scope of *second order structures* called *"binding occurrences"* [3, 91–98]. These are *labeled lists* made up of a *"binding operator Bn"* like CAUSE, GOAL, COND(ition) and their arguments, which are used to deal with the *"connectivity phenomena"* mentioned in the previous section, A binding occurrence can then be expressed as:

$$(Bn_k \quad arg_1 \quad arg_2 \quad \ldots \quad arg_n). \tag{3.2}$$

Equation (3.2) supplies, among other things and in agreement with the intuitive definition in Sect. 3.1 above, the *formal representation of a (whole) narrative*. The arguments arg_i of Eq. (3.2) can, in fact, (i) correspond *directly* to L_i labels—i.e., they can denote simply the presence of particular elementary events represented formally as predicative occurrences, see above—or (ii) *correspond recursively to sets of labeled lists in Eq. (3.2) format*, i.e., to complex combinations of CAUSE, GOAL, COND etc. clauses.

3.2.2 Additional Details

As already stated, HClass is the (standard) '*ontology of concepts*' of NKRL: a particular fragment of HClass is reproduced in Fig. 3.1. HClass includes presently (September 2011) more than 7,500 concepts.

Emphasis on "semantic roles" in Fig. 3.1 allows us to introduce the differentiation between *functional* and *semantic roles* proper to NKRL. Functional roles correspond to the seven R_k *primitive symbols* introduced in the context of Eq. (3.1). According then to the linguistic traditions, functional roles like SUBJ(ect) or CONTEXT are seen as *relations* linking a semantic predicate to its arguments and allowing us to specify exactly the function of each of them—e.g., JOHN_ and BOOK_1 in the previous example—with respect to the predicate. Semantic roles on the contrary, in conformity with the ontological practice, are equated with ordinary *concepts* to be inserted into a standard ontology. With respect then to Fig. 3.1, daughter_ is an NKRL example of extended_family_role and traveler_ of transitory/generic_role—in NKRL, concepts like daughter_ are denoted in capital letters, and the instances of concepts (individuals) like JOHN_ (an instance of individual_person) in lower case. A recent paper where the differentiation functional/semantic is discussed in depth is [9].

We can add here that the main architectural principle underpinning the HClass' *upper level* concerns the partition between sortal_concept and non_sortal_concept, see Fig. 3.1. This corresponds to the differentiation between "*(sortal) notions that can be instantiated directly into enumerable specimens (individuals)*", like chair_ (a physical object) and "*(non-sortal) notions which cannot be instantiated directly into specimens*", like gold_ (a substance), white_ (a color) or student_ (a semantic_role property)—see [3, 123–137] for further details.

Figure 3.2 reproduces a fragment of the HTemp hierarchy that displays, in particular, the *conceptual labels*—see Table 3.1 below for the *structural details*—of some off-springs of the Behave: sub-hierarchy. As it appears from this figure, HTemp is structured into *seven branches*, where each one includes *only* the templates organized—following the syntax of Eq. (3.1)—around one of the *seven predicates* (P_j) admitted by the NKRL language. HTemp includes presently (September 2011) more than 150 templates, very easy to specialize and customize according to the application domain.

When a *specific elementary event* pertaining to one of the *general classes of events* denoted by templates must be represented, the corresponding template is *instantiated*, giving then rise to one of the *predicative occurrences* introduced before. To represent then a simple elementary event like: "British Telecom will offer its customers a pay-as-you-go (payg) Internet service in autumn 1998", we must select firstly in the HTemp hierarchy the template corresponding to "supply a service to someone", represented in the upper part of Table 3.1. This template is a specialization of the particular MOVE template corresponding to "transfer of resources to someone".

In a template, the arguments of the predicate (the a_k terms in Eq. (3.1)) are concretely represented *by variables with associated constraints*: these are expressed as HClass concepts or combinations of concepts, i.e., the two ontologies, HTemp and HClass, are strictly intermingled. When creating a predicative occurrence like c1 in Table 3.1, the role fillers in this occurrence must conform to the constraints of the father-template. For example, BRITISH_TELECOM

Fig. 3.2 An image of the
external structure of HTemp,
the NKRL ontology of
elementary events, where the
Behave: branch has been
partially unfolded

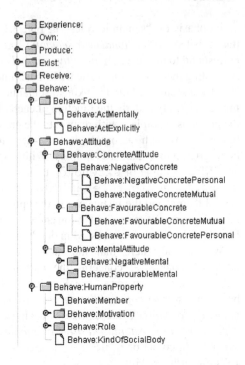

is an individual, instance of the HClass concept company_, specialization in turn of human_being_or_social_body; payg_internet_service is a specialization of service_, a specific term of social_activity, etc. The meaning of the expression "BENF (SPECIF customer_ BRITISH_TELECOM)" in c1 is: the beneficiaries (role BENF) of the service are the customers of—SPECIF(ication)—British Telecom. The "attributive operator", SPECIF(ication) is, like ALTERN(ative) or COORD(ination), one of the operators used for the set up of *structured arguments* of the predicates, see [3, 68–70]. The terms included within square parentheses, "[ ]", are 'possible/optional', which means that they can be found or not in the corresponding occurrences when the original template will be instantiated—see the roles SOURCE, MODAL etc. in Table 3.1.

To supply an at least intuitive idea of how a *complete narrative*—according to the informal definition of Sect. 3.1 and to Eq. (3.2)—is represented in NKRL, and returning to the Table 3.1 example, let us suppose we would now state that: "We can note that, on March 2008, British Telecom *plans to offer* to its customers, in autumn 1998, a pay-as-you-go (payg) Internet service...". In this example, see Table 3.2, the specific event corresponding to the offer is still represented by occurrence c1 in Table 3.1. To encode correctly the new information, we must introduce first an additional predicative occurrence labelled as c2, see Table 3.2, meaning that: "at the specific date associated with c2 (March 1998), it can be noticed, modulator obs(erve), that British Telecom *is planning* to act in some way"—note that, to express the notion of "planning", we make use of one of the Behave: templates of Fig. 3.2 (Behave:ActExplicitly). obs(erve) is a *temporal modulator* used

Table 3.1 Deriving a predicative occurrence from a template

---

name: Move:TransferOfServiceToSomeone

father: Move:TransferToSomeone

position: 4.11

natural language description: "Transfer or Supply a Service to Someone"

```
MOVE    SUBJ            var1: [var2]
        OBJ             var3
        [SOURCE         var4: [var5]]
        BENF            var6: [var7]
        [MODAL          var8]
        [TOPIC          var9]
        [CONTEXT        var10]
        {[modulators]}
```

var1 = human_being_or_social_body

var3 = service_

var4 = human_being_or_social_body

var6 = human_being_or_social_body

var8 = process_, sector_specific_activity

var9 = sortal_concept

var10 = situation_

var2, var5, var7 = geographical_location

```
c1)     MOVE    SUBJ    BRITISH_TELECOM
                OBJ     payg_internet_service
                BENF    (SPECIF customer_ BRITISH_TELECOM)
                date-1: after-1-september-1998
                date-2:
```

---

to identify a *particular timestamp* within the temporal interval of validity of an elementary event. We will then add a *binding occurrence* c3 labelled with a GOAL *Bn* operator, used to link c2 (the planning activity) to c1 (the intended result). The global meaning of c3 can be verbalized as: "The activity described in c2 is focalized towards (GOAL) the realization of c1". More details about this "acting to obtain a given result" construction are supplied in Sect. 3.3.1 below.

3.3 "Behavioural" Information and the NKRL Templates

We are now ready to examine some important 'behavioural' features that characterize the HTemp templates. Given the space limitations, this description will be necessarily quite limited: the interested reader will find many additional details in [3, 149–177].

Table 3.2 Binding and predicative occurrences

```
c2)   BEHAVE  SUBJ    BRITISH_TELECOM
              MODAL   planning_
              {obs}
              date1:  march-1998
              date2:
Behave:ActExplicitly (1.12)
*c1)  MOVE    SUBJ    BRITISH_TELECOM
              OBJ     payg_internet_service
              BENF    (SPECIF customer_ BRITISH_TELECOM)
              date-1: after-1-september-1998
              date-2:
Move:TransferOfServiceToSomeone (4.11)
c3)   (GOAL c2 c1)
```

3.3.1 *"Behave:" Templates*

The Behave: templates (see Fig. 3.2 above) appear, obviously, as particularly appropriate as vectors of *"behavioural"* features—even if the presence in a template of a particular conceptual predicate like BEHAVE (or MOVE, EXIST etc.) does not imply, by itself, the assertion of any particular 'conceptual meaning': a full meaning can only arise when the four main elements of Eq. (3.1) are *all present together*.

The Behave: templates can be grouped in two main classes according to the mandatory/forbidden presence of the OBJ(ect) role. Filling the OBJ(ect) role is *forbidden*, +(OBJ), in the predicative occurrences derived from templates pertaining to the two branches Behave:HumanProperty and Behave:Focus of Fig. 3.2. The Behave:HumanProperty templates are used in general in situations where *one or more characters perform according to a specific, proper 'function', 'task' or 'role'*: hence, the most important among them are those represented by Behave:Role and its specializations. The Behave:Focus templates are employed when a character or group of characters *would like, concretely or as a desire, intention, etc., to make a given situation happen*. In the two cases, the presence in the derived occurrences of a 'direct object' of the SUBJ(ect)'s *behaviour* is logically inconsistent.

A second class of Behave: templates corresponds to the Behave:Attitude branch of Fig. 3.2. They are used to model situations where a SUBJ(ect) manifests *directly* a given *behaviour*, real or purely speculative, *in favour or against a person, a social body, a situation/activity etc.* In the derived occurrences, filling the OBJ role is then *mandatory*; the BEN(e)F(iciary) role is now 'forbidden', +(BENF), given that the 'direct object' of the 'attitude' corresponds here to the OBJ's filler.

The Behave:HumanProperty Templates The general schema of these templates is shown in Table 3.3; the "/" symbol indicates the presence of syntactic alternatives, "+" means "forbidden". The constraints on the variables (and their different

Table 3.3 The HumanProperty sub-domain of Behave

[Behave:HumanProperty]	BEHAVE	SUBJ	*var1*: [(*var2*)]
		+(OBJ)	
		[SOURCE	*var3*: [(*var4*)]]
		[BENF	*var5*: [(*var6*)]]/+(BENF)
		MODAL	*var7*
		TOPIC	*var8* / [TOPIC *var8*]
		[CONTEXT	*var9*]

options) have been suppressed for simplicity's sake; see, however, the template in Table 3.4 below.

Filling the MODAL role is then *mandatory*—as signalled in Table 3.3 by the absence, for this role, of the "possible/optional" code ("[ ]", see the previous section)—in all the predicative occurrences derived from all the Behave: HumanProperty templates. For generality's sake, filling the BEN(e)F(iciary) role has been left as possible/optional, see again Table 3.3, for some of these templates, like the two Behave:Motivation templates (willing/unwilling about the execution of a given task) of Fig. 3.2. Filling the BEN(e)F(iciary) role is, on the contrary, *strictly forbidden*, +(BENF), in the Behave:Role predicative occurrences—the Behave:Role template is reproduced in full in Table 3.4. We can note, moreover, that filling the TOPIC role is *possible/optional* for the occurrences derived from the generic Behave:Role template, but it is *necessarily required* in the occurrences derived from all the specializations of this template, like Behave:User or Behave:Believer. More in general, it is required when—in agreement with the conceptual function of the TOPIC role—it is necessary to give *additional precisions about a specific role/function/task*. The lower part of Table 3.4 reproduces then two examples of use of Behave:Role, where the first does not imply the use of TOPIC, which is needed, on the contrary, in the second.

Acting to Obtain a Given Result The templates of the Behave:Focus sub-hierarchy are used to translate the general idea of *acting to obtain a given result*—see also the example of Table 3.2 above—according to the following modalities:

- A predicative occurrence, which must necessarily be an instance of a Behave:Focus template—see, e.g., the template Behave:ActExplicitly in Table 3.2—is used to express the '*acting*' *component*, i.e., it allows us to identify the SUBJ(ect) of the action, the temporal information, possibly the MODAL(ity) or the instigator (SOURCE) of this component, etc. In this occurrence, the OBJ role is '*empty*', in conformity with what stated at the beginning of Sect. 3.3.1.
- A second occurrence—a single predicative occurrence or a binding occurrence denoting several predicative occurrences—is used to express the '*intended result*' *component*. This second occurrence, which happens 'in the future' with respect to the first, i.e., the Behave:Focus one, must necessarily be marked as *hypothetical*. This implies *directly adding* to the second occurrence, if this is a predicative

Table 3.4 Examples of use of Behave:Role

name: Behave:Role

father: Behave:HumanProperty

position: 1.33

NL description: 'A Human Being or a Social Body Acts in a Particular Role'

```
BEHAVE      SUBJ           var1: [(var2)]
            +(OBJ)
            [SOURCE        var3: [(var4)]]
            +(BENF)
            MODAL          var5
            [TOPIC         var6]
            [CONTEXT       var7]
            {[modulators], ≠(abs)}
var1 = human_being_or_social_body
var3 = human_being_or_social_body
var5 = role_
var6 = entity_
var7 = situation_, symbolic_label
var2, var4 = geographical_location
```

```
mod33.c9)    BEHAVE    SUBJ     ARIEL_BROWN
                       MODAL    journalist_
                       {obs}
                       date-1:  13-june-1999
                       date-2:
```

Behave:Role (1.33)

On June 13, 1999, we can remark (obs) *that Ariel Brown is journalist* (specific term of professional_role).

```
sent4.c17)   BEHAVE    SUBJ     JOHN_KERRY
                       MODAL    chairman_
                       TOPIC    (SPECIF foreign_relations_committee
                                US_SENATE)
                       {obs}
                       date-1:  9-march-2009
                       date-2:
```

Behave:Role (1.11)

On March 9, 2009, we can note that John Kerry is the chairman of the Foreign Relations Committee of the Senate of the USA.

one, an *uncertainty validity code* '*', see Table 3.2 above and [3, 70–71]. If the second occurrence is a binding one, all the *included predicative occurrences* must be characterised by the addition of this code.

Table 3.5 The "attitude" sub-domain of Behave

[Behave:Attitude]	BEHAVE	SUBJ	var1: [(var2)]
		OBJ	var3: [(var4)]
		[SOURCE	var5: [(var6)]]
		+(BENF)	
		MODAL	var7
		[TOPIC	var8]
		[CONTEXT	var9]
		{for / against}	

- A third occurrence, *a "binding" one, which makes necessarily use of a GOAL operator*, is then used to link the previous two, see again Table 3.2.

The general syntax of the NKRL expressions used to code the "acting to obtain a given result" situations is then given by:

c_α) BEHAVE SUBJ <human_being_or_social_body>
*c_β) <predicative occurrence(s), with any syntax>
c_γ) (GOAL c_β c_γ)

The addition of a ment(al) "modal" modulator—see [3, 73–75]—to the BEHAVE occurrence, c_α, that introduces an "acting to obtain a result" construction implies, on the contrary, that *no concrete initiative* has actually been taken by the SUBJ of BEHAVE in order to fulfil the result. To return to the British Telecom example, this would be the case if, e.g., the British Telecom's move represented by occurrence c2 of Table 3.2 was only an (actual) project. With the addition of ment, the 'result', *c_β, reflects then only the *planned intentions of the SUBJ(ect)*; note that, in this last case, the template to be used for c_α should be now Behave:ActMentally instead of Behave:ActExplicitly, see Fig. 3.2 and Table 3.2. For additional information about the representation of the general *"motivational attitudes"* domain (goals, wants, desires, preferences, wishes, choices, intentions, commitments, behaviours, plans), see [3, 153–155]. An example of use of the modal modulator "wish" is given in Table 3.8.

The Behave:Attitude Templates The templates corresponding to the Behave:Attitude branch of Fig. 3.2 follow the general schema of Table 3.5.

As already stated at the beginning of Sect. 3.3.1, filling the OBJ role is now *strictly mandatory* in their derived occurrences; for this class of templates, also the MODAL role is mandatory. Moreover, unlike the Behave:Focus occurrences discussed in the previous sub-Section, these predicative occurrences cannot be included within binding occurrences of the GOAL type. A very simple example of 'positive' attitude is represented in Table 3.6. As shown in Table 3.5, the occurrences derived from templates of the Behave:Attitude type *must* necessarily be associated with one of the two modal modulators "for/against". The *global* meaning of occurrence cob1.c1 is then: the government has a *specific attitude* about the sale

Table 3.6 Example of use of Behave:FavourableConcretePersonal

```
cob1.c1)   BEHAVE   SUBJ      (SPECIF GOVERNMENT_1 (SPECIF
                              CARLO_AZEGLIO_CIAMPI
                              prime_minister)):

ROME_

                    OBJ       (SPECIF sale_ CRDI_)
                    MODAL     COMMITMENT_1
                    CONTEXT   italian_privatisation_programme
                    {for}
                    date-1:   (1-august-1999), 7-september-1999
                    date-2:

Behave:FavourableConcretePersonal (1.2122)
```

(Before September 7, 1999), the government of Carlo Azeglio Ciampi has pledged to sell the Credito Italiano SpA (CRDI) bank as part of Italy's privatization program.

of the bank, which is defined as *favourable* thanks to the association of the modal modulator "for" to the *whole occurrence*. In the date-1 temporal attribute, the first element of the temporal interval is a "reconstructed date", see [3, 85–86].

3.3.2 *"Behavioural" Aspects in the Templates of the Residual HTemp Branches*

The Exist: Templates They can be classed in two main categories:

- Templates that represent specializations of Exist:BePresent, to be used to denote *situations where a given entity, human or not, is present at a given location.* They *all require as mandatory* the presence, in the derived occurrences, of the *location of the SUBJ(ect).* Moreover, OBJ(ect) is normally *forbidden* in these occurrences.
- Templates that represent specializations of Exist:OriginOrDeath and that are used to model *the 'birth' or the 'final end' or a given entity,* human or not. They can then be employed, in particular, to represent *the creation or the dismantling* of a social body, company, political party, university etc.

The Experience: Templates These templates are mainly used to represent events where a given entity, human or not, is exposed to some sort of 'experience' (illness, richness, economical growth, starvation, success, racism, violence...) that can be either 'positive' or 'negative'. The specific experience undergone is represented, in the derived predicative occurrences, by the filler of the OBJ(ect) role: this role is then *mandatory* for all the Experience: templates. On the contrary, given that all the experiences are considered as 'personal', the BEN(e)F(iciary) role is forbidden, +(BENF). An example of template pertaining to the Experience:GenericSituation sub-hierarchy is Experi-

Table 3.7 A simple example of use of the Move:ForcedChangeOfState template

```
aa11.c5)    MOVE    SUBJ       HOME_CONTROL_SYSTEM_1
                          OBJ     (SPECIF lighting_apparatus
BATHROOM_1): (switch_off, switch_on)
                    date-1:   11- april-2011/9:16
                    date-2:
Move:ForcedChangeOfState (4.12)
```

On April 4, 2011, at 9h16, the system has turned the bathroom lights on.

---

ence:BeAged while, e.g., Experience:HumanBeingInjuring pertains to the Experience:NegativeHumanSocial branch.

The Move: Templates These templates are distributed into four branches, Move:TransferToSomeone, Move:ForcedChange, Move:TransmitInformation and Move:AutonomousDisplacement. These templates present some interesting *syntactic variants* linked, at least partly, *to the different possible arrangements of the "location" items*—which are associated with the arguments of the predicate making use of the (external) operator ":". For example, an 'autonomous movement' of the SUBJ (Move:Autonomous Displacement templates) is always interpreted as: "The SUBJ moves herself/himself/itself as an OBJ": the location associated with the filler of the SUBJ role, l_1, is then interpreted as the 'initial location', and the location associated with the OBJ, l_2, as the 'arrival location', $l_1 \neq l_2$. As already stated, locations are represented in general as lists. For example, the Move:ForcedChange templates are used whenever an agent (SUBJ)—which is located in its proper, known or unknown, location—moves an entity (OBJ = physical object, animate entity, process, state...) from the 'initial' location l_1 to l_2. In this case, l_1 and l_2 correspond, respectively, to the initial and final term of the *location list* (l_1, l_2) associated with the filler of the OBJ role. A simple example of use of the Move:ForcedChangeOfState template—corresponding to an ongoing application on in the "assisted living" domain—is represented then in Table 3.7.

The Own: Templates They are mainly used to represent the different nuances of the notion of '*possessing some sort of entity*'. Moreover, under the Own:Property form, they are employed *to specify the 'properties' of NKRL inanimate entities* making use of the TOPIC role. Note that the 'properties' of human beings and social bodies must be described making use of the Behave: templates.

The Produce: Templates This class of templates is particularly rich: the Produce: templates can be used, in fact, to represent *a large number of conceptual domains* that are all of interest from a 'behavioural' point of view—see, for the details, [3, 169–174]. The meaning of the Produce:Entity templates (e.g., Produce:Hardware) is self-evident. Examples of the Produce:PerformTask/Activity templates are, e.g., Produce:Buy and

Table 3.8 An example of `Receive:DesiredAdvice` predicative occurrence

```
skin1.c8)    RECEIVE    SUBJ       CRYSTAL_EYES: (KANATA_ON)
                        OBJ        advice_
                        SOURCE     BEAUTYNET_COMMUNITY
                        TOPIC      (ALTERN    (SPECIF
                                                use_NOVA_UNDEREYE_CREAM)
                                              (SPECIF use_ baby_oil))
                        CONTEXT    (SPECIF therapy_ SPECIF
                                   undereye_dry_skin CRYSTAL_EYES))
                        { wish }
                        date-1:    28-december-2001
                        date-2:
Receive:DesiredAdvice (7.311)
```

On December 28, 2001, Crystal Eyes wishes to receive an advice from the Beauty Net community about the utilization of baby oil or of a product like Nova Undereye Cream in the context of her eye dry skin problem.

`Produce:Sell` (in the predicative occurrences derived from these two templates, the filler of the OBJ role is necessarily `purchase_` or `sale_`, or specialisations/instances of these concepts). Other `Produce:PerformTask/Activity` templates are, e.g., `Produce:Violence`, which involves several specializations, or `Produce:Acceptance/Refusal`. A particularly important specialization of `Produce:RelationInvolvement`, very significant from a 'behaviour' point of view, is `Produce:MutualCommittment`, to be used to represent *all forms of 'agreement' among several participants*. They are mentioned in a COORD(ination) list that fills the SUBJ(ect) role and that is duplicated as filler of the BEN(e)F(iciary) role. `Produce:PositiveCondition/Result` is an example of `Produce:CreateCondition/Result`: it can be used to express, e.g., an official approval with respect to a given action/situation. A predicative occurrence derived from the `Produce:Growth` specialisation of `Produce:Increment/Decrement` can be employed to represent the increase/acceleration/intensification/amplification etc. of a given process/action.

The `Receive:` Templates   Important `Receive:` templates are those in the style of `Receive:GetMoney` and, from a 'behavioural' point of view, those like `Receive:DesiredAdvice`. In the occurrences derived from this last template, the use of the modal modulator wish is *mandatory*. An example, pertaining to an NKRL application in the "beauty care" domain, is supplied in Table 3.8.

3.4 The Query/Inference Aspects

A detailed knowledge representation of "behaviour-like" phenomena would be of scarce utility without some means of *automatically exploiting its power*: in this

Section, we will supply some (basic) information about the query/inference aspects of NKRL, referring the interested readers to [10], [3, 183–243].

Reasoning in NKRL ranges from the *direct questioning* of a knowledge base of narratives in NKRL format—by means of *search patterns* p_i (formal queries) that unify information in the base thanks to a *Filtering Unification Module* (*Fum*)—to *high-level inference procedures*. Making use of a powerful *InferenceEngine*, these last utilise *mainly* two classes of rules, "transformations" and "hypotheses".

Transformation rules try to '*adapt*', from a *semantic* point of view, a search pattern p_i that '*failed*' (that was unable to find a unification within the knowledge base) to the *real contents* of this base making use of a sort of *analogical reasoning*. They attempt then to *automatically 'transform'* p_i into one or more *different* $p_1, p_2, \ldots, p_n$ that *are not strictly 'equivalent' but only 'semantically close'* (analogical reasoning) to the original one. A transformation rule can then be conceived as made up of a *left-hand side*, the "*antecedent*"—i.e. the formulation, in search pattern format, of the 'query' to be transformed—and of one or more *right-hand sides*, the "*consequent(s)*"—the representation(s) of one or more search patterns to be substituted for the given one. Denoting with A the antecedent and with Cs_i all the possible consequents, these rules can be expressed as:

$$A(var_i) \Rightarrow Cs_i(var_j), \quad var_i \subseteq var_j \tag{3.3}$$

The restriction $var_i \subseteq var_j$—all the variables declared in the antecedent A *must also appear* in Cs_i—assures the logical congruence of the rules.

Let us consider a concrete example, which concerns a recent NKRL application about the 'intelligent' management of "storyboards" in the oil/gas industry, see [11]. We want then ask whether, in a knowledge base where are stored all the possible *elementary and complex events* (narratives) related to the activation of a gas turbine, we can retrieve the information that a given oil extractor is running. In the absence of a direct answer we can reply by supplying, thanks to a transformation rule like that (*t11*) of Table 3.9, other related events stored in the knowledge base, e.g., *an information stating that the site leader has heard the working noise of the oil extractor*. Expressed in natural language, this result could be paraphrased as: "The system cannot assert that the oil extractor is running, but it can certify that the site leader has heard the working noise of this extractor".

With respect now to the *hypothesis rules*, these allow us to build up automatically a sort of '*causal explanation*' for an elementary event retrieved within a NKRL knowledge base. These rules can be expressed as *biconditionals* of the type:

$$X \text{ iff } Y_1 \text{ and } Y_2 \ldots \text{ and } Y_n, \tag{3.4}$$

where the 'head' X of the rule corresponds to a predicative occurrence c_j and the 'reasoning steps' Y_i—called 'condition schemata' in a hypothesis context—*must all be satisfied*. This means that, for each of them, at least one '*successful*' search patterns p_i must be derived: this p_i should then be able to find, using the standard *Fum* module (see above), a *successful unification* with some predicative occurrences of the base. *In this case, the set of* $c_1, c_2, \ldots, c_n$ *predicative occurrences retrieved by the* Y_i *thanks to their conversion into* p_i *can be interpreted as a context/causal explanation of the original occurrence* $c_j(X)$.

Table 3.9 An example of NKRL transformation rule

t11: "working noise/condition" transformation

antecedent:

```
OWN    SUBJ    var1
       OBJ     property_
       TOPIC   running_
var1 = consumer_electronics, hardware_, surgical_tool,
diagnostic_tool/system, small_portable_equipment,
technical/industrial_tool
```

first consequent schema (conseq1):

```
EXPERIENCE    SUBJ                      var2
              OBJ                       evidence_
              TOPIC                     (SPECIF var3 var1)
  var2 = individual_person
  var3 = working_noise, working_condition
```

second consequent schema (conseq2):

```
BEHAVE    SUBJ    var2
          MODAL   industrial_site_operator
```

Being unable to demonstrate directly that an industrial apparatus is running, the fact that an operator hears its working noise or notes its working aspect can be a proof of its running status.

To mention a well-known NKRL example, let us suppose we have directly re-trieved, in a querying-answering mode, information like: "Pharmacopeia, a USA biotechnology company, has received 64,000,000 dollars from the German com-pany Schering in connection with an R&D activity" that corresponds then to $c_j(X)$. We can then be able to automatically construct, using a "hypothesis" rule, a sort of 'causal explanation' of this event by retrieving in the knowledge base information like: (i) "Pharmacopeia and Schering have signed an agreement concerning the pro-duction by Pharmacopeia of a new compound", $c_1(Y_1)$ and (ii) "in the framework of this agreement, Pharmacopeia has actually produced the new compound", $c_2(Y_2)$.

An interesting, recent development of NKRL concerns the possibility of making use of the two above modalities of inference *in an 'integrated' way*, see [3, 216–234]. More exactly, *it is possible to make use of "transformations" when working within the "hypothesis" inference environment*. This means that, whenever a search pattern p_i is derived from a condition schema Y_i of a hypothesis to implement one of the reasoning steps, we can use it *'as it is'*—i.e., in conformity with its 'father' condition schema as this last has been originally encoded—but also in *a 'trans-formed' form if the appropriate transformation rules exist within the system*. The advantages are essentially of two types:

- From a 'practical' point of view, a hypothesis that was deemed to fail because of the impossibility of deriving 'successful' p_i from one of its condition schemata

Table 3.10 An example of gas/oil hypothesis in the presence of transformations

(**premise**)	An individual has carried out an "isolation procedure" in the context of an industrial accident.
(**cond1**)	A different individual had carried out earlier a (milder) "corrective maintenance" procedure.
(**cond2**)	This second individual has experienced a failure in this corrective maintenance context.
(**cond3**)	The first individual was a control room operator.
(**cond4**)	The second individual was a field operator.
(**cond5**)	The industrial accident is considered as a serious one.
	– (**Rule t6, consequent**) *The leakage has a gas cloud shape …*
	– (**Rule t8, consequent**) *A growth of the risk level has been discovered …*
	– (**Rule t9, conseq1**) *An alarm situation has been validated,* **and**
	– (**Rule t8, conseq2**) *the level of this alarm is 30% LEL.*

can now continue if a transformed p_i can find a unification within the knowledge base, getting then new values for the hypothesis variables.

- From a more general point of view, this strategy allows us to systematically explore all the possible *implicit* relationships among the data in the base. A modality of the integrated strategy allows us, in fact, *to transform all the p_i derived from the condition schemata of a hypothesis also in case of their successful unification with information in the base*. This permits, e.g., to confirm in many different ways the existence of relationships between people/entities.

Space limitations do not allow us to enter into details about the integrated strategy; the informal example of Table 3.10 refers again to the 'oil/gas industry' application. With this hypothesis, we should want to explain, see the *"premise"*, why an operator has activated a (particularly severe) "piping segment isolation procedure" in the context of, e.g., a gas leakage. The explication proposed is that (i) a previous 'milder' maintenance procedure has been executed (*cond1*), but this was unsuccessful (*cond2*); (ii) the accident is a very serious one (*cond5*). In the absence of predicative occurrences corresponding exactly to this last fact, the p_i derived from *cond5 can be transformed* to obtain *indirect* confirmations of the gravity of the accident, in the style of, e.g. "The gas leakage has a gas cloud shape" or "An alarm situation has been validated (*conseq1*) and the level of this alarm is 30% LEL, Low Explosion Level (*conseq2*)", etc.

3.5 Related Work

Because of their "limited expressiveness" from a *modelling* point of view, the so-called (*binary*) W3C languages like RDF(S), OWL, OWL 2 are not very useful for managing difficult representational problems like complex narratives, human intentions and behaviours, multi-modal interactions, spatio-temporal information, and any sort of events and complex events. An RDF [12] spin-off is the "*Friend of a Friend*" (FOAF) project, see [13], which aims at creating a Web of machine-readable pages describing people, the links between them and some aspects of the things they do. Unfortunately, given its RDF underpinning, FOAF cannot go beyond the expression of '*static*' information about people like publications, employment details group memberships and other sort of elementary personal information. More complex tools, making *necessarily* use of n-ary kinds of representations—e.g., Conceptual Graphs, Topic Maps or CYC—are required to describe and manage the '*dynamic*', '*syntactically complex*' and '*semantically rich*' *narrative/behavioural situations*.

Conceptual Graphs (CGs), see [14], are based on a powerful graph-based representation scheme that can be used to represent (n-ary) complex behaviours and relationships. A conceptual graph is a finite, connected, bipartite graph that makes use of two kinds of nodes, i.e., "concepts" and "conceptual relations"; every arc of a graph must link a conceptual relation to a concept. "Bipartite" means that every arc of a graph associates one concept with a conceptual relation: it is not possible to have arcs that link concepts with concepts or relations with relations. A main aspect of CGs concerns the "nested graphs" extensions that allow them to deal with "contexts" and that represent the CGs' solution to the "connectivity phenomena" problem.

A remark often raised about CGs concern the use of their "*canonical graphs*". Canonical graphs are general conceptual structures, similar to the NKRL templates, which could be used in principle for describing complex, dynamic and semantically-rich phenomena like narratives/behaviours. However, at the difference of what has been done for the NKRL templates, an exhaustive and authoritative list of "canonical graphs"—equivalent then to the HTemp hierarchy of Fig. 3.2—does not exist and its construction seems never have been planned. The practical consequence of this state of affairs could be the need, whenever a concrete application of CGs must be implemented, of defining anew for this a distinct list of canonical graphs. This contrasts with a fundamental characteristic of NKRL, where its catalogue of 'basic templates' (HTemp) is part and parcel of the definition of the language.

With respect now to Topic Maps (TMs), their key notions, see [15], concern "topics", "occurrences" and "associations". A topic is used to represent *any possible specific notion that could be interesting to speak about*, like the play Hamlet, the playwright William Shakespeare, or the "authorship" relationship: there is then no restriction on what can be represented as a topic. Topics can have names, and each individual topic is an instance of one or more classes of topics ("*topic types*") that may or may not be indicated explicitly. They can also have "occurrences", that is, information resources—e.g., links to external resources—that are considered to be relevant in some way to the subjects the topic reify. Links from topics to occurrences can be, e.g., discussed-in, mentioned-in or depicted-in. Finally, topics can participate in relationships with other topics, called "associations": an association consists in a number of "association roles" each of which has a topic attached as a "role player". The association role defines how its role player topic takes part in the association.

Given their (at least apparent) simplicity, Topic Maps have been sometimes considered—quite unfairly—as a sort of *downgraded version* of other conceptual proposals, like Semantic Networks, CGs, or NKRL.

CYC, [16], concerns one of the most controversial endeavours in the history of Artificial Intelligence. Started in the early '80, the project ended about 15 years later with the set up of a huge knowledge base containing about a million of hand-entered "logical assertions" including both simple statements of facts and rules about what conclusions can be inferred if certain statements of facts are satisfied. The "upper level" of the CYC ontology is now freely accessible on the Web, see http://www.cyc.com/cyc/opencyc. It is evident that CYC is, at least in principle, a really amazing achievement. It has also been, however, severely criticized. When examining the CYC's collection of entities, people are often struck by its complexity (i.e., the number of branches in the CYC's ontology and their inter-relationships) and wonder whether this complexity is really necessary. From a more 'technical' point of view, a criticism addressed to CycL concerns its uniform use of the same representation model (substantially, a frame system rewritten in logical form) to represent phenomena conceptually very different (the "uniqueness syndrome").

A system of Computational Linguistics origin that, at least in principle, could be used to represent complex narratives/behaviours is the "Text Meaning Representation" model (TRM), see [18].

3.6 Conclusion

"Behaviour computing" aims at investigating methods and tools for *human behaviour representation and processing*. In this Chapter, we have focused our attention on a fundamental aspect of behaviour computing, i.e., *behaviour modelling* and its *translation into computer-usable tools*. We have then supplied an extremely condensed overview of a knowledge representation language (and a wholly implemented computer system), NKRL, which deals with the behaviour of (human and non-human) characters when this is expressed under the form of "narratives"—i.e., logically and temporally structured sequences of elementary events.

After having supplied the basic syntactic and semantic principles of this language, we have presented in some depth the NKRL's templates—the Behave: templates in particular—that represent the main interest of this language from a behavioural point of view. Some information about the inference procedures that can be used in an NKRL context and a short analysis of few conceptual approaches that share some of the NKRL's ambitions and techniques conclude the chapter.

References

1. Inkpen, D., Strapparava, C. (eds.): Proceedings of the NAACL HLT Workshop on Computational Approaches to Analysis and Generation of Emotion in Text. The Association for Computational Linguistics (ACL), Stroudsburg (2010)

2. Ahn, H.I.: Modeling and analysis of affective influence on human experience, prediction, decision making, and behavior. Ph.D. thesis, MIT, Cambridge, MA (2010)
3. Zarri, G.P.: Representation and Management of Narrative Information, Theoretical Principles and Implementation. Springer, London (2009)
4. Finlayson, M.A., Gervás, P., Mueller, E., Narayanan, S., Winston, P., eds.: Computational Models of Narratives. Papers from the AAAI Fall Symposium. Technical Report (FS-10-04), AAAI Press, Menlo Park, CA (2010)
5. Jahn, M.: Narratology: A Guide to the Theory of Narrative (version 1.8). English Department of the Cologne University, http://www.uni-koeln.de/~ame02/pppn.htm (2005)
6. Noy, F.N., Fergerson, R.W., Musen, M.A.: The knowledge model of Protégé-2000: combining interoperability and flexibility. In: Proceedings of EKAW'2000. LNCS, vol. 1937, pp. 17–32. Springer, Heidelberg (2000)
7. Bechhofer, S., van Harmelen, F., Hendler, J., Horrocks, I., McGuinness, D.L., Patel-Schneider, P.F., Stein, L.A., eds.: OWL Web Ontology Language Reference, W3C Recommendation 10 February 2004. W3C, http://www.w3.org/TR/owl-ref/ (2004)
8. Hitzler, P., Krötzsch, M., Parsia, B., Patel-Schneider, P.F., Rudolph, S., eds.: OWL 2 Web Ontology Language Primer, W3C Recommendation 27 October 2009. W3C, http://www.w3.org/TR/owl2-primer/ (2009)
9. Zarri, G.P.: Differentiating between "functional" and "semantic" roles in a high-level conceptual data modeling language. In: Proceedings of the 24th International Florida AI Research Society Conference, FLAIRS-24, pp. 75–80. AAAI Press, Menlo Park (2011)
10. Zarri, G.P.: Integrating the two main inference modes of NKRL, transformations and hypotheses. J. Data Semant. 4, 304–340 (2005)
11. Zarri, G.P.: Knowledge representation and inference techniques to improve the management of gas and oil facilities. Knowl.-Based Syst. 24, 989–1003 (2011)
12. Manola, F., Miller, E.: RDF Primer, W3C Recommendation 10 February 2004. W3C, http://www.w3.org/TR/rdf-primer/ (2004)
13. Brickley, D., Miller, L.: FOAF Vocabulary Specification 0.98, Namespace Document 9 August 2010—Marco Polo Edition. http://xmlns.com/foaf/spec/ (2010)
14. Sowa, J.F.: Knowledge Representation—Logical, Philosophical, and Computational Foundations. Brooks/Cole, Pacific Grove (2000)
15. Rath, H.H.: The Topic Maps Handbook. White Paper, empolis GmbH, Gütersloh. http://www.empolis.com/downloads/empolis_TopicMaps_Whitepaper20030206.pdf (2003)
16. Lenat, D.B., Guha, R.V.: Building Large Knowledge Based Systems. Addison-Wesley, Reading (1990)
17. Cao, L.: In-depth behavior understanding and use: the behavior informatics approach. Inf. Sci. 180, 3067–3085 (2010)
18. Nirenburg, S., Raskin, V.: Ontological Semantics. MIT Press, Cambridge (2004)

Chapter 4
Semi-Markovian Representation of User Behavior in Software Packages

Prateeti Mohapatra and Howard Michel

Abstract Semi-Markov models have been used in the recent past to model user navigation behavior for personalization in the field of Web-based applications. However, research on its application to incorporate personalization in generalized software packages is rare. In this paper, we use a semi-Markov model to dynamically display personalized information in the form of high-utility software functions (states) of a software package to a user. We develop a demo package of ActiveX Servers and Controls as a test-bed.

4.1 Introduction

User behavior modeling [4] is one of the most important and interesting problems that need to be solved when developing and exploiting modern software systems [19]. User models usually derive the interests of the current and the past users from an analysis of their navigational movements. They enhance both *user control* (by providing the ability to filter content based upon personal interests and tailoring it in a suitable form) and *user experience* (by providing additional content in the form of recommendations derived from an analysis of user history and the interests of the user community).

There are numerous applications where the concept of user behavior modeling has been used—a notable application is in the field of personalization. Personalization is a process of building user loyalty by understanding the needs of each individual in a given context [20]. It is a process that changes the functionality, interface, information context, or distinctiveness of a software program to increase its relevance to an individual [8].

Personalization has found wide applications in web-based information services [2, 11, 14]. But its application to personal computing and generalized software pack-

P. Mohapatra (✉)
The Flash Center for Computational Science, University of Chicago, Chicago IL, 60637, USA
e-mail: mohapatra.prateeti@gmail.com

H. Michel
University of Massachusetts, Dartmouth MA, USA
e-mail: hmichel@umassd.edu

L. Cao, P.S. Yu (eds.), *Behavior Computing*,
DOI 10.1007/978-1-4471-2969-1_4, © Springer-Verlag London 2012

ages is slow despite its early use in personal computing and in the design of operating systems [21]. Today, a large number of generalized software packages are available in the areas of optimization, simulation, mathematical operations, design, and various domain-specific applications (finance, accounting, etc.). They provide multiple functionalities and have a large number of users across the world, and thus provide great opportunities for personalization. However, the use of personalization is not widespread in the design of these packages. Our work in this paper can be viewed as belonging to the broad field of intelligent user interfaces where the final objective of the work is to design dynamic display of the software features on a personalized basis.

There are many probabilistic methods to model user behavior [3, 10, 12, 16, 19]. Many soft computing techniques have also been used [6, 7, 17, 18, 22]. In the recent past, Markov models and its extensions have found quite a few applications for modeling users' behavior. Chen and Cooper [2] have used continuous-time semi-Markov models to derive user state transition patterns (both in rates and in probabilities) in a Web based information system. Manavoglu et al. [14] investigated the use of a maximum entropy based approach and a first order Markov mixture model to visualize individual behavior models for CiteSeer users. Lane [13] investigated behavioral user model with the help of Hidden Markov Models.

In this paper we use a semi-Markov model to depict software users' navigational behavior. The user behavior model considers the past users' navigation pattern and the current user's interest. A demo package of ActiveX Servers and Controls is used as the test bed. A discrete-event simulation model of user behavior is also developed. The test and the simulation results confirm the use of personalization features in generalized software packages.

4.1.1 Objectives of the Present Work

The objectives of this paper are to (1) Capture the interests of past and current users as they navigate through various functions of a software package (2) Develop a stochastic model that depicts user behavior in terms of the navigational pattern of the users using a semi-Markov model (3) Develop a simulation model of the user behavior (4) Analyze the test and the simulation results.

To achieve these objectives, (1) A relational database was designed to store the users' navigational pattern and the results of the analysis (2) A semi-Markov model was developed to depict the user behavior (3) A discrete-event simulation model of user behavior in the software package was developed in the Matlab language (4) Test and simulation results were statistically analyzed.

4.2 In Defence of Semi-Markovian Representation of User Behavior

Treating the behavior of a software user as semi-Markovian, this section shows how a software user's navigational behavior can be depicted as a semi-Markov process. To behave as a semi-Markov process, (1) a system should change state with time and the number of states are finite, (2) it should follow Markovian assumption (Memoryless property), and (3) time for stay at any state should be a random variable (holding time). All the three requirements are satisfied in case of a software user.

A software product has a finite number of functions (states). A user moves from one state to another depending on her requirements that vary from one encounter with the software to another and also from one user to another. A user of a software product normally follows a logical pattern of movement from state to state depending on her requirements in a particular encounter with the software product. This pattern will, however, change from the first encounter to subsequent encounters with the same set of requirements and with her requirements changing with time. A simple way to model her movement is to assume that the transition from the current state depends on the current state itself. Following this assumption made by many researchers in various fields [1, 2, 5, 11] the current work makes the Markovian assumption to model the movement of a user of a software product. The user occupies a state for some time before making a transition to the next state. Arguing in a fashion similar to the above two points, we can say that the residence time of a user in any state before making a transition to another state is probabilistic.

4.3 Modeling User Behavior

Personalization assumes that a current user's immediate move from a state (function) depends on the consideration of (1) Past users' interest, and (2) Current user's interest [11]. The past users' interest is derived from an analysis of the navigational data created during the past usage of the software and captured as an aggregate user profile. The analysis of the navigational data takes the form of a *transition probability matrix*.

A user might be more interested in a particular state than in other states or she may be more interested in one category of states than another. The current user's interest is stored as a *matrix of holding times*, each element in the matrix indicating the length of time she resides in one state before changing over to another state.

4.3.1 Computing User Behavioral Statistics

Holding Times   Holding times are computed on the basis of behavior of all users and are used for finding the interest of the current user. We define τ_{ij} as the holding time in state i before making a transition to state j in a session.

$$\tau_{ij} = entry\_time(j) - entry\_time(i). \qquad (4.1)$$

where, *entry_time*(j) is the time at which a user enters state j and *entry_time*(i) is the time at which a user enters state i.

Let K be the number of users and L be the total number of sessions. We define τ_{ijkl} as the time a user k spends in her session l in state i before making transition to state j, for all $i = 1, \ldots, N$; $j = 1, \ldots, N$; $k = 1, \ldots, K$; $l = 1, \ldots, L$. The time a user k spends in state i before making a transition to state j in all her sessions is given by:

$$\tau_{ijkl} = \sum_{l=1}^{L} \tau_{ijkl}, \quad i, j = 1, \ldots, N; k = 1, \ldots, K. \tag{4.2}$$

The average time a user spends in state i before making a transition to state j is:

$$\tau_{ij} = \frac{1}{K} \sum_{k=1}^{K} \tau_{ijk}, \quad i, j = 1, \ldots, N. \tag{4.3}$$

The average time a user spends in state i is given as:

$$\tau_i = \frac{1}{N} \sum_{j=1}^{N} \tau_{ij}, \quad i = 1, \ldots, N. \tag{4.4}$$

Let T denote the total time a user spends on average in the system in all sessions given by

$$T = \sum \tau_i. \tag{4.5}$$

We denote pf_i as the fraction of time a user spends in state i while using the system:

$$pf_i = \frac{\tau_i}{T}, \quad i = 1, \ldots, N. \tag{4.6}$$

We denote $t_i = \sum_j \tau_{ijkl}$ as the total time the current user k in her session l has spent in state i. Let T_{kl} denote the time a user k has spent in a session when she is making a transition from state i to state j.

$$T_{kl} = \sum_i \sum_j \tau_{ijkl}. \tag{4.7}$$

The fraction of the time the current user has spent her time while using the system is given by:

$$f_i = \frac{t_i}{T_{kl}}, \quad i = 1, \ldots, N. \tag{4.8}$$

The algorithm for obtaining the holding time τ_{ij} is given below.

function *holding_time* returns τ_{ij}
input: *previous_state*, *current_state*, *entry_time* in current state, *entry_time* in previous state

begin
      $i := previous\_state$;
      $j := current\_state$;
      $\tau_{ij} = entry\_time(j) - entry\_time(i)$
      return τ_{ij}
end

Transition Probabilities Transition probabilities are computed on the basis of behavior of all users including the current user. We define w_i as the number of times the state i is visited by all users, and define w_{ij} as the number of times they visit state j immediately after visiting state i. We count the number of transitions taking place from state i to state j to determine w_{ij}. Using these counts, we can estimate the transition probabilities between any two states. The probability of a transition p_{ij} from state i to state j is estimated by computing the ratio of the number of transitions from state i to state j to the total number of transitions taking place from state i by all users in all sessions:

$$p_{ij} = \frac{w_{ij}}{\sum_{k=2}^{N+2} w_{ij}}, \quad i = 1, \ldots, N+1; i = 2, \ldots, N+2. \tag{4.9}$$

Semi-Markov process allows transition from a state to the same state. We model this by assuming that if a user continues in a state for more than a specified maximum time she is considered to be making a transition to the same state that she is occupying. This maximum time is taken as the minimum of the holding times of a sample of past users.

The algorithm for obtaining the transition probabilities is given below.

function *transition_probability* returns p_{ij}
input: *previous_state, current_state, threshold_time*
Initialize w_{ij} to 0 for all i and j for the first time use of the software package.
begin
      $i := previous\_state$;
      $j := current\_state$;
      $\tau_{ii} = holding\_time$ in i before making a transition to j
      if $\tau_{ij} > threshold\_time$ then $w_{ii} = w_{ji} + 1$
      $w_{ij} = w_{ij} + 1$
      $w_i = \sum_k w_{ik}$
      $p_{ij} = \frac{w_{ij}}{w_i}$
      return p_{ij}
end;

4.4 ActiveX Servers and Controls—A Demo Software Package

A demo software for ActiveX Servers and Controls was developed at a company reputed for its generalized software package that has a large number of functions to support control, optimization, modeling and simulation, statistical, and image processing applications, to name a few. The demo package was to be used by individuals who wish to know how they can incorporate ActiveX Servers and Controls in their Matlab programs. The user requirements of the demo software had been defined after intense interaction with the Usability and Marketing Teams of the Company. The design details of the demo software package are given elsewhere [15].

The personalization features were implemented in the software as follows. A user was uniquely identified by her user-id. She can have many sessions. Users and their sessions were uniquely identified by their session keys. An agent tracked a user through the software product, collected the navigational data (generated dynamically) during tracking, and stored it in a database. The agent generated the users' aggregate profile, created the transition probability matrix, and updated it with every new session. It also generated and stored the holding time matrix for each user session. Using the afore-mentioned knowledge, a set of personalized navigational links to functions were generated and displayed to the user [15]. As the user continued to move from state to state, the displayed links changed dynamically. The *transition_probability* and the *holding_time* algorithms were presented in Sect. 4.3. Each function (state) needed to call these algorithms to generate the transition probability matrix and the holding time matrix.

4.5 Testing

4.5.1 *Organization of the User Testing of the Package*

The demo package for ActiveX Servers and Controls was used by a number of graduate students of UMass Dartmouth, USA over a period of two months. Student users were combined into groups. A one-page write-up on the package was given to each member of both the groups. The researcher explained various features of (i) ActiveX Servers and Controls, and (ii) the software package, gave a demonstration of the package, and outlined the steps that the users needed to follow to use the package. The users were requested to observe the system operations during the demonstration of the package (Debriefing Session), work with the software, and learn how to use it (Learning Session), use the software with and without the personalization features (Test Session) and give their feedback, either verbal or written (Feedback Session). Each member used the package individually. The user navigational data stored during the learning session were excluded from the data base for user behavior analysis.

4.5.2 *The User Behavior Database*

User navigational data and other information were stored in a database in the form of (1) Personal information of each user, (2) Number of transitions from each state to

Fig. 4.1 Metric values

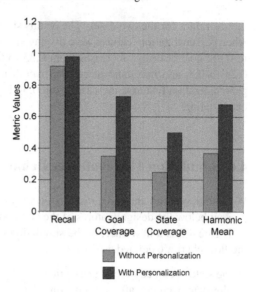

other states, accumulated over all users, (3) Holding time in each state before making a transition for each user, and (4) Recommended states sorted in the decreasing order of their utility values.

Following the procedure given in Sect. 4.3, the transition probability matrix was computed and updated with every move of the user. At the end of all the user sessions, the average holding times, averaged over all the users, was computed.

4.5.3 Evaluation Metrics for Personalization

Following Srivihok and Sukonmanee [23], we have adopted the following evaluation metrics for personalization:

$$Recall = \frac{total\_no\_of\_clicks\_on\_the\_recommended\_states}{total\_no\_of\_clicks\_byall\_users} \quad (4.10)$$

$$Goal\_Coverage = \frac{total\_no\_of\_clicks\_on\_the\_goal\_states}{total\_no\_of\_clicks\_byall\_users} \quad (4.11)$$

$$State\_Coverage = \frac{total\_no\_of\_clicks\_on\_max\_utility\_states}{total\_no\_of\_clicks\_byall\_users} \quad (4.12)$$

$$Harmonic\_Mean = \frac{3*(a*b*c)}{bc+ac+ab} \quad (4.13)$$

where, $a = Recall$, $b = Goal\ Coverage$, and $c = State\ Coverage$, *goal states* are those software functions that the developer wants all the users to be aware of and use, and *max utility states* are those software functions having the highest utility.

We compute these metrics when the users test the package without and with the display of personalization features. Figure 4.1 shows, in a bar chart form, the metric values for the two cases. The results show that all the metric values increase when personalization features were displayed. The system recall increases from 0.92 to 0.98, goal coverage increases from 0.35 to 0.73, state coverage increases from 0.25 to 0.5, and harmonic mean, the overall indicator of the effectiveness of the personalization features, increases from 0.37 to 0.68. These results are indeed very encouraging.

4.6 Simulation Model of User Behavior

We also developed and ran a *discrete-event simulation model* of user behavior. We present below the design features of the simulation model.

Figure 4.2 is a flow chart of the simulation model. The various symbols used in the flow chart are defined as under:

i: the state occupied before a transition
j: the state occupied after a transition
k: the cumulative number of user sessions
t_{ij}:   time of transition from state i to state j
t_i: holding time in state i
n_i: frequency of visit to state i
Clock: simulation time

As evident from Fig. 4.2, the sequence of operations for a user session in the model is the following:

1. From any start node i find the next state j using the transition probability matrix.
2. Generate holding time t_{ij} using the holding time matrix.
3. Update the frequency of visits to state i, the holding time in state i, and the clock.
4. Stop a user session when the exit state 13 is reached or when clock equals 950 seconds. The maximum time was taken equal to 950 seconds—the maximum session length during testing of the demo package by the dummy users.
5. Stop simulation after a fixed number of user sessions.

4.6.1 Data for the Simulation Model

The final transition probability values, obtained for both without and with personalization when the dummy users tested the package, were used as initial transition probabilities in the simulation model to generate the states to which transitions are made from given states. It was observed that the transition probability matrix contained a large number of zero entries for elements that do not involve either the entry or the exit state. This was caused, presumably, due to a small number of user sessions that does not reflect the true situation. We have

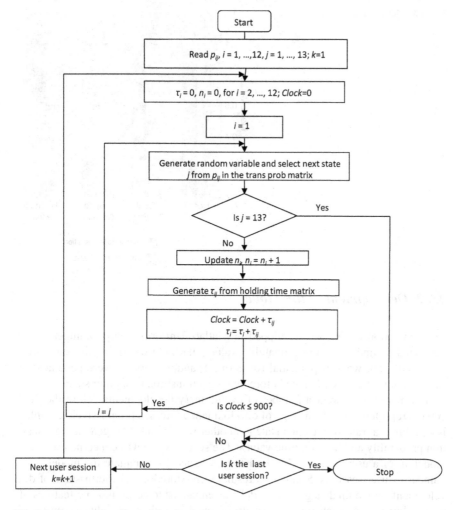

Fig. 4.2 Program flow chart of the simulation model

deliberately replaced these zeros by small probability values while maintaining the row sums to 1. As simulation proceeded, the transition probabilities were updated.

Exponential and normal distributions were fitted, separately for without- and with-personalization cases, to the holding time values, which were obtained during user testing of the demo package. Matlab statistical tools were used for the purpose. Small number of holding time values often did not permit fitting a distribution. In such cases, we took exponential—the continuous analog of geometric distribution, the one that has been assumed by Howard [9] in his monumental work on semi-Markov process.

Fig. 4.3 Metric values

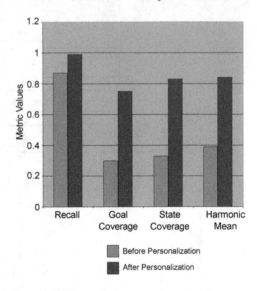

4.6.2 Development of the Model

The simulation model was developed in Matlab. Uniform random numbers were generated to find transition probability values that help determine the state after each transition, while exponential (or normal) random variates were generated to find holding times in each state before making a transition to another state.

The model underwent a number of evolutionary developments before the final version reported here. Each stage of development addressed progressively complex issues. First, a transition from a state to another was modeled by generating transition probability and holding time values. The second stage was concerned with the modeling of a user session. The third stage was used to simulate the model for a number of user sessions. Simulation results were stored in the fourth stage of development. In the final stage, the model was enhanced to accommodate features of personalization. At each stage of development, the results were edited online and verified, by manually analyzing the system state and the input conditions.

4.6.3 Analysis of the Simulation Results

Evaluation metrics—Recall, Goal Coverage, State Coverage, and Harmonic mean of Recall, Goal Coverage and State Coverage—are plotted in Fig. 4.3. The figure clearly shows that these values are higher when the personalization features were present compared to those when the features were absent.

It may be indicated here that we had made an attempt at the initial stage of model building to use the personalization algorithms for generating the personalized states in the simulation model. The attempt was successful. But we did not have enough information on the basis of which the users were to select the next state from a given state. We thereafter decided to rely, instead, on the test data and used them for modeling the without- and with-personalization situations.

4.7 Conclusions

This paper captured the user's behavior from a semi-Markov model of user navigation in the software. Frequency of visit to a state and the holding time in the state together determined the utility of the state.

A demo package of ActiveX Servers and Controls was built to test the personalization features based on the semi-Markov model of user behavior. The package was tested by graduate students. The analysis of the test results indicated that the users utilized the information provided in the personalized software functions to a large extent

A discrete-event simulation model of user behavior was developed on the basis of data generated from the test results cited above. The simulation results confirmed the use of personalization features in generalized software packages.

The following three features in this paper are quite novel:

1. Many reported algorithms (e.g., [11]) take stationary transition probabilities for depicting past users' interests. In contrast, the algorithm, used in the paper, uses a transition probability matrix that gets updated after every user session.
2. The current user's interest is judged by not only the frequency of user visits to a state (a commonly used criterion), but also the user residence time in that state compared to the total time she has spent on the system.
3. The transition probability and holding time matrices, obtained during the tests, were used as input to develop a discrete-event simulation model which was then tested to firmly establish the effectiveness of personalization.

Acknowledgements Many thanks to The Mathworks Inc., Natick, to provide the first author with a research assistantship and the facility for carrying out a major part of this work. Thanks are also due to Mr. Dave Foti and Mr. Fazil Peermohammed of The Mathworks Inc. for their many helpful comments during the progress of the work.

References

1. Anderson, C., Domingos, P., Weld, D.: Relational Markov models and their applications to web navigation. In: 8th ACM SIGKDD International Conference on Knowledge Discovery and Mining (2002)
2. Chen, H., Cooper, M.: Stochastic modeling of usage patterns in a web-based information system. J. Am. Soc. Inf. Sci. Technol. **53**, 536–548 (2002)
3. D'Ambrosio, B., Altendorf, E., Jorgensen, J.: Probabilistic relational models of on-line user behavior early explorations. In: 5th WEBKDD Workshop, pp. 9–16 (2003)
4. Cao, L.: In-depth behavior understanding and use: the behavior informatics approach. Inf. Sci. **180**, 3067–3085 (2010)

5. Eirinaki, W., Vazirgiannis, M., Kapogiannis, D.: Web path recommendations based on page ranking and Markov models. In: 7th Annual ACM International Workshop on Web Information and Data Management (2005)
6. Frias-Martinez, E., Magoulas, G.D., Chen, S.Y., Macredie, R.D.: Modeling human behavior in user-adaptive systems. In: Recent Advances Using Soft Computing Technique. Expert Systems with Applications, vol. 29, pp. 104–114 (2005)
7. Goren-Bar, D., Kuflik, T., Lev, D., Shoval, P.: Automatic Personal Categorization Using Artificial Neural Networks. In: Proceedings of the 8th International Conference on User Modeling 2001. Lecture Notes in Artificial Intelligence, vol. 2109, pp. 188–198 (2001)
8. Hiltunen, M., Heng, L., Helgesen, L.: Personalized electronic banking services. In: Karat, C.-M. (ed.) Designing Personalized User Experiences in eCommerce. Kluwer Academic, Dordrecht (2004)
9. Howard, R.: Dynamic Probabilistic Systems. Semi-Markov and Decision Processes, vol. II. Wiley, New York (1971)
10. Iwata, T., Saiti, K., Yamada, T.: Modeling user behavior in recommender systems based on maximum entropy. In: 16th International Conference on World Wide Web, pp. 1281–1282 (2007)
11. Jenamani, M., Mohapatra, P., Ghosh, S.: Online customized index synthesis in commercial websites. IEEE Intell. Syst. **17**, 20–26 (2002)
12. Jin, X., Zhou, Y., Mobasher, B.: Task-oriented web user modeling for recommendation. In: 10th International Conference on User Modeling (2005)
13. Lane, T.: Hidden Markov models for human/computer interface modeling. In: Proceedings of the IJCAI-99 Workshop on Learning about Users, pp. 35–44 (1999)
14. Manavoglu, E., Pavlov, D., Giles, C.E.: Probabilistic user behavior models. In: 3rd International Conference on Data Mining, pp. 203–210 (2003)
15. Mohapatra, P.: Dynamic function personalization in generalized software packages. Master's thesis, University of Massachusetts, Dartmouth, USA (2006)
16. Mustapha, N., Jalali, M., Jalali, M.: Expectation maximization clustering algorithm for user modeling in web usage mining systems. Eur. J. Sci. Res. **32**, 467–476 (2009)
17. Nandedkar, A.V., Biswas, P.K.: A fuzzy min-max neural network classifier with compensatory neuron architecture. IEEE Trans. Neural Netw. **18**, 42–54 (2007)
18. Nasraoui, O., Krishnapuram, R.: Extracting Web user profiles using relational competitive fuzzy clustering. Int. J. Artif. Intell. Tools **9**, 509–526 (2000)
19. Petrovskiy, M.: A data mining approach to learning probabilistic user behavior models from database access log. In: Filipe, J., Shishkov, B., Helfert, M. (eds.) ICSOFT 2006. LNCS, CCIS, vol. 10, pp. 323–332. Springer, Heidelberg (2006)
20. Rieken, G., Tuzhilin, A.: Personalization technologies: a process-oriented prospective. Commun. ACM **43**, 83–90 (2005)
21. Singh, M.: Practical Handbook of Internet Computing. Chapman and Hall/CRC Press, London/Boca Raton (2004)
22. Stathacopoulou, R., Grigoriadou, M., Magoulas, G.D.: A neurofuzzy approach in student modeling. In: Proceeding of the 9th Int. Conf. on User Modeling. Lecture Notes in Artificial Intelligence, vol. 2702, pp. 337–342 (2003)
23. Sukonmanee, P., Srivihok, A.: E-commerce intelligent agent: personalization travel support using Q-learning. In: 7th International Conference on Electronic Commerce, pp. 287–292 (2005)

Part II
Behavior Analysis

Chapter 5
P-SERS: Personalized Social Event Recommender System

Yun-Hui Hung, Jen-Wei Huang, and Ming-Syan Chen

Abstract As the increasing popularity of social networking functions, people interact with others in social events everyday. However, people are easily overwhelmed by hundreds of social events. In this work, we propose P-SERS, a Personalized Social Event Recommender System, which consists of three phases: (1) Candidate selection, (2) Social measurement and (3) Recommendation. Among these, potential candidate events are selected based on user preference and the social network. In our opinion, every social event is composed of three critical elements: (1) the initiator, (2) the participants and (3) the target item. These elements possess different types of influential power on a social event. Therefore, we design algorithms to compute three social measures, i.e., initiator score, participant score and target score, which model expertise of the initiator, group influence of participants and global popularity of the target item respectively. P-SERS evaluates each candidate social event by these social measures and produces a recommendation list. In addition, explanations and the grouping function are provided to improve the recommendation. Finally, we examine P-SERS by recommending group buying events in a real world online group buying website. The experimental results show the superiority of P-SERS over conventional social recommendation methods.

5.1 Introduction

In recent years, lots of websites introduce social networking functions. For example, on Facebook [8], users share their interests or chat with people after adding them as friends. In addition, users are able to join clubs they are interested in and talk

Y.-H. Hung (✉) · M.-S. Chen
National Taiwan University, No. 1, Sec. 4, Roosevelt Road, Taipei, 10617, Taiwan
e-mail: yhhung@arbor.ee.ntu.edu.tw

M.-S. Chen
e-mail: mschen@cc.ee.ntu.edu.tw

J.-W. Huang
Yuan Ze University, 135 Yuan-Tung Road, Chung-Li, 32003, Taiwan
e-mail: jwhuang@saturn.yzu.edu.tw

L. Cao, P.S. Yu (eds.), *Behavior Computing*,
DOI 10.1007/978-1-4471-2969-1_5, © Springer-Verlag London 2012

about items with other club members. It is easier for users to construct their own social networks in online communities according to their friends in real lives or their interests. In addition, social networks are also built by discovering the relationships between items and users. The online social network is critical for every user when he/she chooses what activity to do and what information to know.

Furthermore, social networks help users form groups with ease when they want to pursue a mutual goal. We regard a social event as an activity in which people form a group to achieve a common goal or share information. A social event is initiated by a user, targeting on a specific item and is participated by a group of people. For example, Alice posts a personal note talking about a popular video on Youtube [18], and then her friends, Bob and Chris, comment on the note. In this social event, Alice is the initiator, and the video is the target item. Bob and Chris are the participants. Another example is that people buy products together in online group buying communities to get discounts from retailers. A user initiates a group buying event in a target store. Then other people join the event and get products together.

Hundreds to thousands of social events are initiated and joined by people everyday. Users are easily overwhelmed by a significant number of events which are shared by others. As a result, people give up looking for interesting events and miss the items they probably like. It is important to address the challenges by providing personalized social event recommendations and give the most valuable and relevant information to users. The significant amount of social information opens great opportunities to filter out irrelevant items to users. Many previous researches, such as [9, 11, 17], studied the social information filtering which aims to provide recommendation by users' similarity or familiarity in social networks. With the technique in [11], social software items are able to be recommended by considering social relations. However, there is no technology focusing on social event recommendation in the past.

In our opinion, every social event consists of three major elements, (1) the initiator, (2) the participants, and (3) the target item. All of these elements influence a social event in different views and should be modeled and measured by different techniques. First, good initiators usually bring better experiences to participants and easily gather people together to hold events successfully. Therefore, we consider the expertise of the initiator, that is, the ability of holding successful social events. The ability can be the popularity or the experiences of holding the social events. Second, an individual might become interested in an event if many people who have relationships with him/her have joined the event. We model the group influence of a social event by social networks of the target user. At last, an item becomes popular if many people successively talk about the item. The item is more popular if more people recently adopt the item or people give good comments to the item. We also measure the popularity of the target item. In addition to three major components, the location is important to every social event, which is considered as a constraint in this work. It is reasonable because in our daily lives, we may not join an event if the place is too far from us.

Our goal is to design a Personalized Social Event Recommender System, abbreviated as P-SERS. The system uses the social information in a community to model

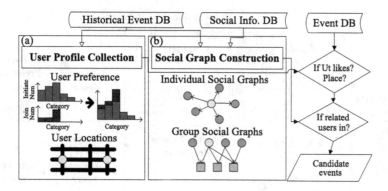

Fig. 5.1 Candidate selection phase

three critical components and give personalized social event recommendations to the target user. There are three phases in our system, i.e., (1) Candidate selection, (2) Group measurement and (3) Recommendation.

In order to provide personalized recommendation, P-SERS selects possible recommended events by constructing the user profile and the social graphs in the first phase, Candidate selection, as shown in Fig. 5.1. The user profile describes the preference categories and locations where the target user usually appears. User preference consists of how and which categories of social events the user have initiated and joined in the historical data. User locations are also collected from historical data or user settings. In addition to user preference, the social graph is used when P-SERS selects possible candidate events because people are usually interested in friends' events as modeled in [6]. P-SERS collects similar users who involved in the historical events with the target user and familiar users, such as friends and club members of the target user. Two types of social graphs are constructed by individual-level social behavior, such as adding someone as a friend, and group-level social behavior, such as co-joining a club with others. As a result, a personalized candidate set is chosen according to user preference, locations and his/her social network.

As mentioned earlier, three critical components, the initiator, participants and the target item, influence every social event in different aspects. Therefore, we aim to measure every candidate social event by three different measures. As illustrated in Fig. 5.2, we propose three algorithms to calculate three measures including the initiator score, the participant score, and the target score for every event in the candidate set. The final relevance score is computed by summing three scores and adjusting them according to the user preference.

First, we propose Social-HITS to compute the initiator score.Social-HITS models the relative expertise of the initiator to the target user in the community. Social-HITS takes both user preference and social network of the target user into consideration to discover initiators who are experts in the target user's favorite categories and are recommended by related users to the target user. As shown in Fig. 5.2(a), an expertise graph is built by extracting users and their initiating and joining behavior which are related to the target user in the community. The local information of

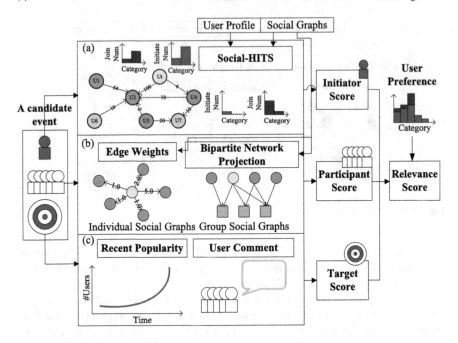

Fig. 5.2 Social measurement phase

every node is the user experience of initiating and joining events in different categories. Every edge implies which initiator a user recommends and the weight of the edge represents how the user recommends the initiator. Social-HITS automates the process of recommending initiators by propagating scores through edges. By considering both user experience and automating the recommendation, the initiator score related to the target user is calculated by Social-HITS. As for the participant score, P-SERS models the group influence of a social event to the target user. As mentioned in [5], whether a user joins a community is affected by his/her friends who have already joined the community. Therefore, the group influence of a social event is measured by individual influence of every participant to the target user as depicted in Fig. 5.2(b) The influence between every user and the target user is represented by an edge weight in individual social graphs, while in group social graphs, the influence is modeled by bipartite network projection similar to [20]. Finally, the target score is calculated by a global popularity measure, which takes the recent popularity and comments of the target item into consideration as shown in Fig. 5.2(c). The recent popularity reflects how the item has gained attention recently in the community. In addition, people tend to consult others' opinions before they adopt an item. Therefore, the target score includes the historical comments of the target item. With three measures, the final relevance score is computed accordingly.

After Social measurement phase, P-SERS is able to provide a recommendation list by ranking candidate events according to their relevance scores. In addition to the ranking list, we design two important functions, explanations and grouping, as

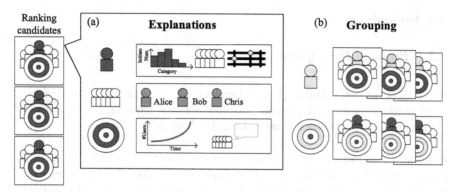

Fig. 5.3 Recommendation phase

depicted in Fig. 5.3. A good recommender system is able to give the reasons why an event is recommended to the target user. Therefore, explanations about the initiator, participants and the target item are given to help users understand every recommended event better. Moreover, the explanations of the initiator and the participants enhance the probability that the target user gets familiar with other users and expands his/her own social network. Next, the grouping function is provided to avoid repetitive recommendations. By the grouping function, the events which are initiated by the same initiator or target on the same item are clustered into groups. In addition, the grouping function gives more space for other initiators or items in the list and makes the recommended list consist of a diversity of recommended events.

For the experiments, we apply P-SERS to group buying events. We collect the real world data from the biggest group buying website, IHERGO [12], in Taiwan. There are more than 600,000 members on IHERGO. We give real recommendations to 3,000 users and evaluate the recommendation satisfaction and the effectiveness of explanations. In the satisfaction study, we compare P-SERS with three methods, Random, Collaborative Filtering (CF) and Social Filtering (SF). In the satisfaction results, P-SERS outperforms other methods and is able to provide recommendation which users like and would like to join. Moreover, there is a trend that the number of interested events increases when every score gets higher, especially the initiator score. As for explanation study, every user is given a recommendation list in which every event is accompanied with explanations. 700 users are randomly chosen to study the helpfulness of the explanations. The results show that explanations of initiator expertise and target popularity are helpful for over 73% of events. However, explanations of participants are marked as helpful by only about 44% of events. In sum, the experimental results show the superiority of P-SERS over other methods.

The rest of this work is organized as follows. In Sect. 5.2, we define our problem and survey some related works. Then, our system model is presented in Sect. 5.3 with detailed descriptions of three phases. In addition, running examples for illustrating algorithms in the system are given. The experiments on real world data and the results are discussed in Sect. 5.4. Finally, we conclude this work in Sect. 5.5.

5.2 Preliminaries

5.2.1 Problem Description

In this work, we aim to build a personalized recommender system to provide social event recommendations, which contain the most relevant and attractive social events to the target user U_t. A social event is defined as follows.

Definition 1 A social event E_i is held by an initiator I_{E_i} with the target item T_{E_i} and joined by the participants P_{E_i}. E_i belongs to a category C_{E_i} according to its target item T_{E_i}, such as sports, news, etc. The event is held at location $Loc(E_i)$ if the location is specified by the initiator.

For example, a group buying event E_1 is initiated by Alice (I_{E_1}) to buy the bread of Bread Bakery store (T_{E_1}) at Taipei Main Station ($Loc(E_1)$). Bob and Chris, who are the participants (P_{E_1}), join the event and buy bread with Alice together. In this example, the category of this event is food ($C_{E_1} = Food$).

In the following section, we discuss about group buying communities and how online social networks bring possibilities of recommendations using social information filtering.

5.2.2 Related Works

Group Buying Communities   Virtual communities and online communities have been well studied by many researches in social science, [4, 13, 16]. People share common interests through a specific media and form virtual communities across spatial and temporal boundaries. In online communities, people exchange contents through online services and construct bonds with others during interactions. In recent days, many online communities focus on group activities. In online group buying communities, people seek for others who are able to form groups and buy products or services together. Some online group buying websites provide users social networking functions, such as adding friends or setting up clubs. Users are allowed to join clubs and easily exchange their information with club members. The social relations become important to users because users receive information of group buying events by the relations.

Social Information Filtering   As social networking functions become prevalent, online social networks are more visible and useful to find out contents which users are interested in. As mentioned in [17], the process of word of mouth occurs in our daily lives. It is useful and effective to make personalized recommendations according to the target user's social network. Social networks can be built by aggregating social information, such as SONAR [10] and SocialScope [3]. There are two main categories of the related techniques, (1) Collaborative Filtering and (2) Social Filtering. The former is based on similarity of users in social networks and the later is based on familiarity of users in social networks to the target user.

Collaborative Filtering (CF) makes recommendations based on other users who have similar tastes. By analyzing the relations between users and objects, the similarity between users can be computed. Xerox PARC launches Information Tapestry project [9] to help document search by users' previous comments. GroupLens [15] also uses CF to recommend netnews. CF is broadly used in taste-related recommendations, such as book, music and movie recommender systems. For example, Ringo [17] and MovieLens [7] give music and movie recommendations based on the historical ratings of users. Furthermore, expert-CF [1] is based on expert opinions because experts tend to identify high quality items and give reliable ratings.

Social Filtering (SF) uses explicit familiarity relations to provide items which are related to the user. In [11], social information is aggregated by SONAR [10] and the authors find that familiar network has better results than the similarity network when recommending social software items, such as bookmarks and blog articles. In [5], the authors show the high probability of joining a community when many friends are already members.

In this work, we compare P-SERS with recommendation by CF and SF. For CF, the similarity network are users who have involved in the same event with the target user in the past. The candidates are selected from the events which are initiated or joined by similar users and the recommendation score of every event is measured by the number of similar users who have participated in the event. Similarly, SF chooses candidates by the friends and club members of the target user, measures each event by the number of familiar users, and produces the list by ranking the numbers.

5.3 System Model of P-SERS

P-SERS consists of three phases, (1) Candidate selection, (2) Social measurement, and (3) Recommendation. At first, P-SERS selects candidate events by analyzing U_t's behavior, which includes user profile collection and social graph construction. By the user profile, P-SERS is able to know the preferences of U_t among all categories and the U_t's locations. In addition, the candidate events are chosen from the social graphs of U_t. Second, P-SERS computes relevance values of the candidate events to U_t by three social measures which are based on three major components of an event. We propose three algorithms to calculate the initiator score, the participant score and the target score respectively. Finally, after social scores of candidate events are acquired, P-SERS generates a top-k recommendation event list. In addition, the explanations of why each event is recommended to the target user are reported and the grouping function is provided to avoid repetitive recommendations in Recommendation phase. In the following sections, we will explain how these phases work in detail.

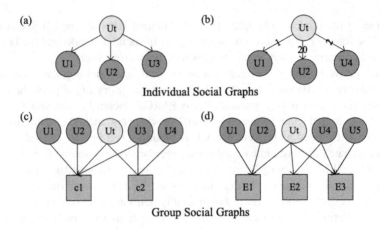

Fig. 5.4 (a) Friend graph, (b) Join graph, (c) Co-club graph, (d) Co-join graph

5.3.1 Candidate Selection

To give personalized recommendation to a target user U_t, P-SERS chooses the potential events by user profile collection and social graph construction (Fig. 5.1).

User Profile Collection   P-SERS collects related information from U_t's past behavior to analyze U_t's preference. The following three data which are *InitNum*, *JoinNum*, and *Place* are selected to reflect U_t's favors and U_t's locations. *InitNum*(C_j, U_t) and *JoinNum*(C_j, U_t) are the number of times U_t initiated and participated in an event of category C_j in the past. For example, *InitNum*$(C_j, U_t) = 10$ represents that U_t has held 10 events of category C_j. *Place* describes the locations where U_t usually appears. Locations in *Place* are coordinates of longitude and latitude.

In addition, U_t's preference vector, denoted as *Pref*, is built by the first two data, *InitNum* and *JoinNum*.

$$Pref(C_j, U_t) = InitNum(C_j, U_t) + JoinNum(C_j, U_t). \tag{5.1}$$

For instance, $Pref(Food, U_t) = 15$ where $InitNum(Food, U_t) = 10$ and $JoinNum(Food, U_t) = 5$. Consequently, the system is able to know which categories U_t is interested in and the relative strength to U_t.

Social Graph Construction   In order to provide events which are joined by U_t's related users in the community, P-SERS builds two types of egocentric social graphs of U_t according to U_t's individual-level and group-level behavior. As shown in Fig. 5.4, individual social graphs include (a) a friend graph and (b) a join graph and group social graphs include (c) a co-club graph and (d) a co-join graph. Co-club and co-join mean that the users join the same club or events. The friend graph and the co-club graph are built according to the explicit relationships of U_t, and the join graph and the co-join graph are constructed by the historical event-joining behavior of U_t. Take Fig. 5.4(b) for example, the edge $(U_t, U_1) = 1$ means that U_t has joined the events initiated by U_1 for once. The co-join graph specifies the co-participation, i.e., $U_1, U_2, U_t \in P_{E_1}$. The users in Fig. 5.4(a), (c), say $U_1 \sim U_4$, are familiar users of U_t. On the other hand, the users in the join and co-join graphs, i.e., U_1, U_2, U_4, U_5 in Fig. 5.4(b), (d), share the similarity of choosing events with U_t. All users who appear in the egocentric social graphs of U_t are thought of as related users, denoted as U_{rel}, to U_t. For example, $U_{rel} = \{U_j | 1 \le j \le 5\}$ in Fig. 5.4.

Every candidate event is an ongoing event which is participated or initiated by U_{rel} with the constraints $Pref(C_{E_i}, U_t) > 0$, $dist(Loc(E_i), Place) < Th$. The distance is the minimum distance which is calculated by the coordinates of $Place$ and $Loc(E_i)$.

5.3.2 Social Measurement

To calculate the relevance value of each candidate event to U_t, we propose three algorithms to compute three social measures as illustrated in Fig. 5.2. For every candidate event E_i, P-SERS computes the initiator score $IS(I_{E_i}, U_t)$, the participant score $PS(P_{E_i} + I_{E_i}, U_t)$ and the target score $TS(T_{E_i})$ respectively. Then the relevance score RS_{it} of an event E_i, to a target user U_t, is calculated as follows.

$$RS_{it} = Pref(C_{E_i}, U_t)$$
$$* \left[w_{IS} * IS(I_{E_i}, U_t) + w_{PS} * PS(P_{E_i} + I_{E_i}, U_t) + w_{TS} * TS(T_{E_i}) \right], \quad (5.2)$$

where w_{IS}, w_{PS}, and w_{TS} are the importance weights of IS, PS and TS respectively. The weights can be set by a prior user study or with the historical transactions. In this work, we set the weights according to the user study.

Initiator Score The initiator score, $IS(I_{E_i}, U_t)$, quantifies I_{E_i}'s expertise of holding an event in the favorite categories of U_t. As in [19], P-SERS builds an expertise graph based on the event-joining behavior in the community as shown in Fig. 5.5(a). Each edge is weighted by the number of joining behavior from the participant to the initiator and specified with a category. We propose Social-HITS which is similar to [14] to compute the relative initiator score $IS(U_j, U_t)$ of each user U_j to U_t in the expertise graph. Next, we illustrate Social-HITS with the pseudo code in Figs. 5.6 and 5.7.

Social-HITS consists of four stages, (1) relevant set retrieval, (2) subgraph growth, (3) initial score assignment, and (4) score propagation. At first, U_{rel} is assigned as the relevant set and U_t is also included into the relevant set (lines 7, 8). Then, Social-HITS grows an expertise subgraph by the relevant set and user preference in procedure *growSubGraph*. Any edge in the expertise graph including endpoints in the relevant set will be aggregated into the subgraph with the weight of

(a)   (b)

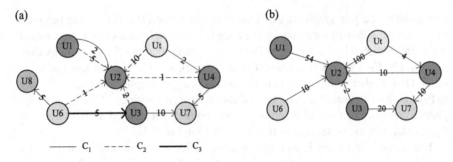

Fig. 5.5 (a) An example of an expertise graph and (b) the subgraph after the procedure *growSubGraph* in Social-HITS

$Pref(C_{edge}, U_t)$ according to the category of the edge as in Fig. 5.5(b). The endpoints of the edge are included in the subgraph as well. Therefore, the resultant subgraph indicates which initiator U_{rel} usually seek for when they want to join the events of U_t's favorite categories. After the subgraph growth, the initial scores of the nodes in subgraph are calculated by user profiles, *InitNum* and *JoinNum* (lines 10 to 17). $InitNum(C_i, U_j)$ describes how U_j is good at holding events of category C_i. If $JoinNum(C_i, U_j)$ is large, U_j has higher probability to recommend good initiators from their historical experience. The initiating expertise is treated as the authority score, $AutScore(U_j)$ and the joining expertise is seen as the hub score, $HubScore(U_j)$. In order to find the experts who are most related to U_t, $InitNum(C_i, U_j)$ and $JoinNum(C_i, U_j)$ are weighted by $Pref(C_i, U_t)$ and then summed up to be the initial authority and hub scores, $iniAutScore(U_j)$ and $iniHubScore(U_j)$. Initial scores are normalized to make the maximal score equal to 1.0. With the subgraph and the initial scores, users propagate their scores through the edges in the final stage for *iter* iterations (line 25 to 33). Score propagation is composed of two directions of propagation, hub propagation and authority propagation as in procedure *propagate* in Fig. 5.7. After *iter* iterations of propagation, Social-HITS outputs the final *AutScore* which quantifies the relative expertise of other users to U_t. As a result, the initiator score $IS(U_j, U_t)$ equals $AutScore(U_j)$. Then *IS* of all candidate events will be normalized to make maximal *IS* equal to 1.0.

Participant Score For each candidate event E_i, we measure its group influential power to U_t as the participant score $PS(PE_i + I_{E_i}, U_t)$. The group influence of an event can be calculated by the relationships between U_t and all participants, including the initiator. For individual social graphs, the influential power of U_j to U_t, denoted as $p(U_j, U_t)$, is the weight on the edge from U_t to U_j, i.e., $p(U_2, U_t) = 2$ in Fig. 5.4(b). For the unweighted friend graph, all edge weights equal to 1. For group social graphs, we revise bipartite network projection [20], to compute the influential power of all users in the U_t's group social network. Take Fig. 5.4(c) for example, U_t is influenced by c_1 and c_2 with equal weights 0.5 so $f'(U_t) = \frac{1}{2}f(c_1) + \frac{1}{2}f(c_2)$, where $f(c_j)$ is defined as the influential power of c_j and $f'(U_t)$ is the resultant influence to the target user U_t. In the next

---

Algorithm Social-HITS(*Ut, iter, alpha, Graph*)

1. var *relSet*; // used to store the relevant set of *Ut*
2. var *subGraph*; // graph for subgraph growth
3. var *iniAutScore*; // initial authority score
4. var *iniHubScore*; // initial hub score
5. var *AutScore*; // authority score
6. var *HubScore*; // hub score
7. *relSet* = the *related users* of *Ut*; // get relevant set
8. add *Ut* into *relSet*;
9. **growSubGraph**(*relSet, Graph*); // grow subGraph
10. **FOR** each user *Uj* in *subGraph* // assign initial score
11.   *iniAutScore*(*Uj*) = 0;
12.   *iniHubScore*(*Uj*) = 0;
13.   **FOR** each category *Ci* in *Pref*
14.   *iniAutScore*(*Uj*)+ = *Pref*(*Ci, Ut*) * *InitNum*(*Ci, Uj*);
15.   *iniHubScore*(*Uj*)+ = *Pref*(*Ci, Ut*) * *JoinNum*(*Ci, Uj*);
16.   **END FOR**
17. **END FOR**
18. normalize *iniAutScore* to max(*iniAutScore*(*Uj*)) = 1
19. normalize *iniHubScore* to max(*iniHubScore*(*Uj*)) = 1
20. **FOR** each vertex *Uj* in *subGraph* // initialize in-weight and out-weight
21.   **SET** $Uj.w_{out}$ to the sum of *edge.w* of all out-edges of *Uj*;
22.   **SET** $Uj.w_{in}$ to the sum of *edge.w* of all in-edges of *Uj*;
23. **END FOR**
24. **SET** *HubScore* = *iniHubScore*;
25. **FOR** *i* = 1 to *iter* // score propagation for *iter* iterations
26.   **propagate**(*subGraph, AutScore, HubScore*); //propagation
27.   normalize *AutScore* to max(*AutScore*(*Uj*)) = 1;
28.   normalize *HubScore* to max(*HubScore*(*Uj*)) = 1;
29.   **FOR** each vertex *Uj* in *subGraph*
30.   *AutScore*(*Uj*) = (1 − *alpha*) * *AutScore*(*Uj*) + *alpha* * *iniAutScore*(*Uj*);
31.   *HubScore*(*Uj*) = (1 − *alpha*) * *HubScore*(*Uj*) + *alpha* * *iniHubScore*(*Uj*);
32.   **END FOR**
33. **END FOR**
34. normalize *AutScore* to max(*AutScore*(*Uj*)) = 1;
35. **OUTPUT** *AutScore*;

End

Fig. 5.6 The pseudo code of Social-HITS algorithm

step, each club node is contributed by the users with equal weights. They can be represented as $f(c_1) = \frac{1}{4}f(U_1) + \frac{1}{4}f(U_2) + \frac{1}{4}f(U_t) + \frac{1}{4}f(U_3)$, and $f(c_2) = \frac{1}{3}f(U_t) + \frac{1}{3}f(U_3) + \frac{1}{3}f(U_4)$. Therefore, the resultant influential power $f'(U_t)$ can be rewritten as $f'(U_t) = \frac{1}{8}f(U_1) + \frac{1}{8}f(U_2) + \frac{7}{24}f(U_3) + \frac{1}{6}f(U_4) + \frac{7}{24}f(U_t)$. In sum, the influential power $f'(U_t)$ can be represented as the following formula:

$$f'(U_t) = \sum_{G_i \in N(U_t)} \sum_{U_j \in N(G_i)} \frac{f(U_j)}{degree(G_i) * degree(U_t)}, \tag{5.3}$$

Procedure growSubGraph(*relSet,Graph*)

1. **INIT** *subGraph* to a empty directed graph;
2. **FOR** each *edge* from n_1 to n_2 in *Graph*
3.   **IF** (*relSet* contains n_1 or n_2) and *Pref*(C_{edge}, *Ut*) > 0 **THEN**
4.   **IF** *subGraph* contains *edge$_2$* from n_1 to n_2 **THEN**
5.   add *edge$_2$.w* with *edge.w* * *Pref*(C_{edge}, *Ut*);
6.   **ELSE**
7.   *edge.w* = *edge.w* * *Pref*(C_{edge}, *Ut*);
8.   add *edge*, n_1 and n_2 to *subGraph*;
9.   **END IF**
10.   **END IF**
11. **END FOR**

END

Procedure propagate(*subGraph, AutScore, HubScore*)

1. **FOR** each vertex *Uj* in *subGraph*
2.   **If** *Uj* has in-neighbors **THEN**
3.   *AutScore*(*Uj*) = 0;
4.   **FOR** each in-edge *e* from *Un* to *Uj*
5.   *AutScore*(*Uj*)+ = (*e.w*/*Un.w_{out}*) * *HubScore*(*Un*);
6.   **END FOR**
7.   **END IF**
8. **END FOR**
9. **FOR** each vertex *Uj* in *subGraph*
10.   **If** *Uj* has out-neighbors **THEN**
11.   *HubScore*(*Uj*) = 0;
12.   **FOR** each out-edge *e* from *Uj* to *Un*
13.   *HubScore*(*Uj*)+ = (*e.w*/*Un.w_{in}*) * *AutScore*(*Un*);
14.   **END FOR**
15.   **END IF**
16. **END FOR**

END

Fig. 5.7 Procedure *growSubGraph* and *propagate* of Social-HITS

where G_i is a club node which is the neighbor of U_t, U_j is a user node which is the neighbor of G_i and *degree*(G_i) and *degree*(U_j) are the indegree of G_i and the outdegree of U_j. Every user U_j influences U_t with power $w_j * f(U_j)$. We assume $f(U_j) = 1$, so the influential power to U_t owned by U_j becomes w_j, that is, $p(U_j, U_t) = w_j$. Therefore, P-SERS is able to model how much U_t is influenced by others from the projection weights. For instance, $p(U_3, U_t) = \frac{7}{24}$, $p(U_2, U_t) = \frac{1}{8}$, $p(U_4, U_t) = \frac{1}{6}$. Finally, the influential power in every social graph is normalized to the maximal $p(U_j, U_t) = 1$, $U_j \neq U_t$ and then summed up for each user as the final influential power $p(U_j, U_t)$, which is formulated as follows.

$$p(U_j, U_t) = \hat{p}_{friend}(U_j, U_t) + \hat{p}_{join}(U_j, U_t)$$
$$+ \hat{p}_{co\text{-}club}(U_j, U_t) + \hat{p}_{co\text{-}join}(U_j, U_t). \quad (5.4)$$

For example, $p(U_1, U_t) = 1 + \frac{1}{20} + \frac{3}{7} + \frac{2}{5} = 1.88$. P-SERS measures the group influence of an event E_i by summing up individual influence of the participants, P_{E_i}, and the initiator I_{E_i} as follows.

$$PS(P_{E_i} + I_{E_i}, U_t) = p(I_{E_i}, U_t) + \sum_{U_j \in P_{E_i}} p(U_j, U_t). \qquad (5.5)$$

PS of all candidate events will be normalized to make maximal PS equal to 1.0.

Target Score   The target score $TS(T_{E_i})$ captures the global popularity of T_{E_i} in the community. We utilize two attributes of the target item, Num, the number of people who have adopted the item recently, and $avgScore$, the average comment score which is computed by the historical comments by other users who have adopted the item in the community. $TS(T_{E_i})$ is computed by the following formula,

$$TS(T_{E_i}) = Num(T_{E_i}) * \left(1 + avgScore(T_{E_i})\right), \qquad (5.6)$$

where $avgScore \in [0, 1]$. With the global target measure, the events with popular items will be recommended to U_t. All TS are also normalized to make the maximal TS equal to 1.0.

5.3.3 Recommendation

In the above two phases, P-SERS selects candidate social events and calculates their corresponding relevance scores (E_i, R_{it}) to the target user U_t. Then, in Recommendation phase, P-SERS generates a top-k relevance list by ranking the events according to the relevance scores as depicted in Fig. 5.3. In light of [2], explanations and the grouping are important functions to a recommender system. Explanations help users understand recommendations better and the grouping function can avoid the repetitive recommendations. We illustrate explanations and grouping function in details as follows.

Explanations   In addition to representing the list to the target user, P-SERS adds explanations accompanied with every event in order to enhance the influential power of the three elements. For every recommended event E_i, we add three reasons to the initiator I_{E_i}, the participants P_{E_i} and the target item T_{E_i} respectively.

The reason of the initiator is to show (1) the popularity, how many people have joined the events held by I_{E_i} recently, (2) the experience, how many times I_{E_i} hold events in the categories which U_t is interested in and (3) the locations where the initiator usually appears. We make the target user believe that the initiator can be trusted or reliable. Next, we list people who have close relationships with U_t to be the reason of the participants. The social influence makes U_t to become interested in the events if their good friends have joined those recommended events. Finally, the target reason is to show the popularity of T_{E_i} by $Num(T_{E_i})$ and $avgScore(T_{E_i})$

to attract U_t. With the above three reasons, users will not only understand our recommendations better but also have chances to get familiar with other users in the community.

Grouping The goal of the grouping function is to avoid recommending a repetitive event list to U_t. There are two cases leading to the repetitive situations, an active initiator and a popular target item. Active initiators initiate many social events everyday, while others might initiate a few in several days. The number of the events initiated by an active user is much larger and those active users usually bring higher *IS* to their events. Consequently, there might be many events from the same active initiator in the recommendation list. Furthermore, a popular target item also results in the repetition of events. An interesting item usually gains lots of popularity when it comes into the community, i.e., a funny Youtube video or a product with good discounts. As a result, there are many events talking about the same item around us. P-SERS provides the grouping function to cluster the events of the same initiator or the same target item together in order to give spaces for other interesting events and get users rid of repetitive events.

5.4 Experiments

In our experiment, we evaluate our system P-SERS with real world data in the online group buying website, IHERGO [12]. In order to understand users' needs, we conduct an user study with online questionnaires to 5,000 active users in IHERGO. We also examine user satisfaction with real recommendation. More details about the prior survey and the recommendation are presented in the following sections.

5.4.1 Dataset and User Understanding

IHERGO [12] is the biggest group buying website in Taiwan and there are more than 600,000 members. We collect the event data of the food category from IHERGO website which happened in 2010/01/01~2010/04/30. The total number of events in this period is 460,000. The total number of users in the period is 124,000. We randomly sample 5,000 active users in IHERGO to conduct a prior user study by online questionnaires. Several findings prove that it is effective to use social information to filter potential events for users. For example, 87% users browse events of their friends and clubs and 83% of them would like to join events held by their familiar initiators. Filtering by user preference is also essential because 91% choose events of their preference categories. In addition, location constraints of P-SERS are necessary because 96% users only choose events around their locations. Furthermore, we investigate how users are affected by the critical elements, say the initiator, the participants and the target store of a group buying event. 66% of them take the experience of the initiator into consideration. Moreover, 61% might join the events if

Fig. 5.8 A recommended
group buying event

Event: Noodle Group Buying Event
Initiator: Alice
Participants: Bob, Chris, ...
Store: Boss Q Noodles
Place: Taipei Main Station
Are you interested in this event? YES, NO.

the initiator is recommended by others. However, the influence of the participants is
relatively weak in group buying events. Only about 40% people will become interested in an event if their similar or familiar users have joined the event. As for the
target store, 73% will pay attention to its recent popularity and 91% take account of
the historical comments. Therefore, we set $w_{IS} = 0.6$, $w_{PS} = 0.4$, and $w_{TS} = 0.7$ by
the above findings to recommend food group buying events in IHERGO.

5.4.2 Recommendation Satisfaction

We compare P-SERS with three conventional algorithms, say Random, Collaborative Filtering (CF) and Social Filtering (SF). The score calculation of CF and SF
has been presented in Sect. 5.2. Furthermore, three methods are applied the place
constraint which is the same as P-SERS's. The methods become Place, Place + CF
and Place + SF. We invite 3,000 active users and assign the users into 4 groups to be
recommended by Place, Place + CF, Place + SF and P-SERS. Of 3000 users, 671
submitted their results of recommendation. There are 117, 78, 121 and 178 from
Place, Place + CF, Place + SF and P-SERS. Every user is given a ranked list which
contains 15 group buying events. As depicted as Fig. 5.8, every recommended event
shows the information of the event with a question which studies user satisfaction.

We measure the recommendation results by the top-k satisfaction, denoted as
Satisfaction(k, method), which shows how users are pleased with the top-k recommendation by *method*. When users are satisfied with more recommendation with
higher rank, the satisfaction score is higher. *Satisfaction(k, method)* is computed as
follows.

$$Satisfaction(k, method) = \frac{\sum_{U_j \in method} \sum_{l=1}^{k} (ans_l(U_j) \cdot \frac{1}{l})}{|U_j \in method| \sum_{l=1}^{k} \frac{1}{l}}, \quad (5.7)$$

where $ans_l(U_j) = 1$ if U_j's answer of the lth recommendation is YES while
$ans_l(U_j) = 0$ if U_j responses NO. $|U_j \in method|$ is the total number of users who
are recommended by *method*. According to Fig. 5.9, P-SERS outperforms than other

Fig. 5.9 User satisfaction in four methods

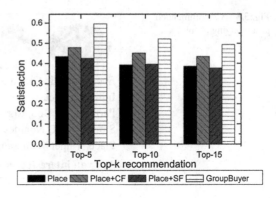

Fig. 5.10 The relations between the score and satisfaction count of three components

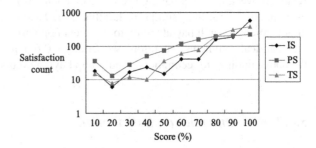

methods in top-5, top-10 and top-15 satisfaction, especially in the top-5 recommendation. Place + CF shows that similar users also bring interesting events to target users. However, most events recommended by Place + CF are initiated by familiar initiators of target users. As for Place + SF, familiar users dominate the events which belong to categories which do not appeal to target users. As a result, Place + SF also fails in this experiment and it is even worse than Place as shown in Fig. 5.9. In sum, P-SERS has higher recommendation power than other three methods.

Furthermore, we examine the importance of each component in our model. In Fig. 5.10, the *Score* is *IS, PS* and *TS* respectively and the satisfaction count is the total number of the events marked as *like*. We can observe that *IS* is crucial to the recommendation because the users are more interested in the events held by the initiators whose $IS = 1.0$. The experts make group formation successful and attract people to join their events. The users like popular items so they prefer the events with high *TS* ($0.8 < TS < 1.0$). However, the flat trend of *PS* infers that the participants are not so important to the target user in group buying events because the users join the events to get desired commodities from a reliable initiator but not for social purposes.

Fig. 5.11 Group buying
event with explanations

> **Initiator Score:** 100%
> Recent popularity: 198,
> Specialized in holding events: cookie: 5, dessert: 69, etc.
> **Participant Score:** 82%
> Number of related users: 5, Bob, Chris, etc.
> **Target Score:** 82%
> Recent popularity: 137, Average comment: 72%
>
> Do you think the explanation of the initiator helpful? YES, NO.
> Do you think the explanation of participants helpful? YES, NO.
> Do you think the explanation of the store helpful? YES, NO.

Fig. 5.12 The effectiveness
of three types of explanations

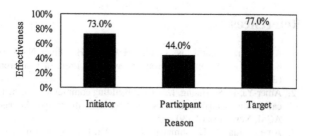

5.4.3 Explanation Effectiveness

As described previously, we investigate whether the explanations help users understand recommendations better with three questions as shown in Fig. 5.11. Of 700 users, 168 users respond to the explanation survey. The effectiveness is computed by

$$effectiveness(reason) = \frac{\sum_{U_j} \sum_{l=1}^{k} ans_l(U_j, reason)}{|U_j| \cdot k}, \qquad (5.8)$$

where *reason* can be reasons of *initaitor, participant* or *target* and $k = 15$. $ans_l(U_j, reason) = 1$ if U_j feels helped by *reason* of the lth recommendation. $|U_j|$ is the total number of users in this experiment so $|U_j| = 168$ in this experiment. The effectiveness of three reasons is plotted in Fig. 5.12. The results display good effectiveness of initiator reasons and target reasons and reveal that reliable initiators and popular targets attract users' attention. However, the usefulness of participant reasons is relative low. It is because participants have weak influence to the recommendation or there are many events with no participants. In our survey, the participant explanation becomes helpful if the number of related users is large, i.e., the effectiveness reaches 78% if over 40 related users participate an event. However, there are usually less than 20 participants in a group buying event.

5.5 Conclusions

In this work, we design P-SERS which uses social information to provide personalized group buying event recommendation. First, P-SERS selects potential candidate events by user's preference and social network. Second, three social scores model the different influence of three elements respectively in P-SERS. In our experiments, we conduct user study to support our model and examine P-SERS with real group buying event recommendation. The experimental results show that P-SERS brings good satisfaction and we discover that a reliable initiator and a popular target commodity are crucial when recommending group buying events in our analysis.

References

1. Amatriain, X., Lathia, N., Pujol, J.M., Kwak, H., Oliver, N.: A collaborative filtering approach based on expert opinions from the web. In: Proc. of Conference on Information Retrieval (2009)
2. Amer-Yahia, S., Huang, J., Yu, C.: Building community-centric information exploration applications on social content sites. In: Proc. of Conference on Management Of Data, pp. 947–952. ACM, New York (2009)
3. Amer-Yahia, S., Lakshmanan, L.V.S., Yu, C.: Socialscope: Enabling information discovery on social content sites. In: Proc. of Conference on Innovative Data Systems Research (2009). arXiv:0909.2058
4. Anderson, B.R.O.: Imagined communities: reflections on the origin and spread of nationalism (1983)
5. Backstrom, L., Huttenlocher, D., Kleinberg, J., Lan, X.: Group formation in large social networks: Membership, growth, and evolution. In: Proc. of Conference on Knowledge Discovery and Data Mining (2006)
6. Crandall, D., Cosley, D., Huttenlocher, D.P., Kleinberg, J.M., Suri, S.: Feedback effects between similarity and social influence in online communities. In: Proc. of Conference on Knowledge Discovery and Data Mining, pp. 160–168. ACM, New York (2008)
7. Dahlen, B.J., Konstan, J.A., Herlocker, J.L., Good, N., Borchers, A., Riedl, J.: Jump-starting movieLens: User benefits of starting a collaborative filtering system with "dead-data". In: University of Minnesota TR 98-017 (1998)
8. Facebook: http://www.facebook.com
9. Goldberg, D., Nichols, D., Oki, B.M., Terry, D.: Using collaborative filtering to weave an information tapestry. Commun. ACM 35(12), 61–70 (1992)
10. Guy, I., Jacovi, M., Shahar, E., Meshulam, N., Soroka, V., Farrell, S.: Harvesting with sonar: the value of aggregating social network information. In: Proc. of Conference of ACM SIGCHI, April (2008)
11. Guy, I., Zwerdling, N., Carmel, D., Ronen, I., Uziel, E., Yogev, S., Ofek-Koifman, S.: Personalized recommendation of social software items based on social relations. In: Proc. of Conference on Recommender System, October, pp. 53–60 (2009)
12. IHERGO: http://www.ihergo.com
13. Kim, A.J.: Community Building on the Web: Secret Strategies for Successful Online Communities. Addison-Wesley, Boston (2000)
14. Kleinberg, J.M.: Authoritative sources in a hyperlinked environment. J. ACM 46, 604–632 (1999)
15. Resnick, P., Iacovou, N., Sushak, M., Bergstrom, P., Riedl, J.: Grouplens: An open architecture for collaborative filtering of netnews. In: Proc. of Conference on Computer Supported Cooperative Work, October (1994)

16. Rheingold, H.: The Virtual Community. Addison-Wesley, Boston (1993)
17. Shardanand, U., Maes, P.: Social information filtering: Algorithms for automating word of mouth. In: Proc. of Conference of ACM SIGCHI, vol. 1, pp. 210–217 (1995)
18. Youtube: http://www.youtube.com
19. Zhang, J., Ackerman, M.S., Adamic, L.: Expertise networks in online communities: structure and algorithms. In: Proc. of the 16th International Conference on World Wide Web (2007)
20. Zhou, T., Ren, J., Medo, M.c.v., Zhang, Y.-C.: Bipartite network projection and personal recommendation. Phys. Rev. E **76**(4), 046115 (2007)

Chapter 6
Simultaneously Modeling Reply Networks and Contents to Generate User's Profiles on Web Forum

Zhao Zhang, Weining Qian, and Aoying Zhou

Abstract Capturing individual profiles is one of the key tasks in behavior computing. Now, web forum has been one of the main platforms to exchange information. In this paper, we focus on get extension profiles for web forum users. Extension profiles are the types and areas of that forum users concern about ("term-profile") plus a description of user's collaboration networks (neighborhood-profile). We define and implement the tasks of automatically determining an extension profiles of a forum user from a web forum corpus. We propose the tripartite graph model to effectively capture the user's profiles. This tripartite graph integrate forum user's posts and reply networks in web forums. Furthermore, we discuss how to implement an application by following the model. And the efficiency of the application is discussed on the basis of an experimental study using a real data set of online forum.

6.1 Introduction

Behavior computing include modeling, analysis, mining of human behavior within virtual and physical organizations. Find people's profiles is one of the key tasks in behavior computing. In this paper, we talk about how to find extension profiles of web forum users. With the rapid development of Internet, the popularity of online forums has been increasing dramatically over the last couple of years. CNNIC Internet Report showed [3] that the size of registered users on the online forum has been more than 148 millions in the end of 2010 in china. People can talk about anything they like on web forum. General topics are from experiences (travelling,

This work is partially supported by National Science Foundation of China under grant number 61070051, and National Basic Research (973 program) under grant number 2010CB731402.

Z. Zhang (✉) · W. Qian · A. Zhou
Institute of Massive Computing, East China Normal University, Shanghai 200062, P.R. China
e-mail: zhzhang@sei.ecnu.edu.cn

W. Qian
e-mail: wnqian@sei.ecnu.edu.cn

A. Zhou
e-mail: ayzhou@sei.ecnu.edu.cn

L. Cao, P.S. Yu (eds.), *Behavior Computing*,
DOI 10.1007/978-1-4471-2969-1_6, © Springer-Verlag London 2012

Wedding, looking after the baby, decorating, etc.), opinions (about products, films, events, business, etc.) to information technology. So, online forum has been one of main platforms to exchange information. On the one hand, forum users contribute their experiences and knowledges; on the other hand, forum users build their collaboration networks by posts and comments. For example, Alice belong to an young mother virtual community and concerns about children's goods by mining her accessing history in web forum. In this paper, we called the mining results as Alice's extension profiles.

Actually, Web forums provide such an easy and convenient way of communicating that every day thousands and thousands people share their information on web forums. Consequently, it demands immense efforts to find one's needed information from the ocean of information contributed by the uncountable web forum users. To relieve this itch, applications, such as customized advertisements, product recommendations, and personalized information discovery, are created and get popular. Therefore, it is critical for using the applications to get the users' preferences and concerns. Our research focuses on how to build a user's extension profile by capturing and analyzing his or her accessing history, an important analytical task in behavior informatics [1]. Considering the fact that posting message is the only action taken by a user on web forum, and a posting carries two notes: one is the posted messages, and the other is people who receive or response to the posting, we invented two terms to represent a user's information. The neighborhood-profile is for most users who often communicate with the user. The term-profile is for the contents concerned by the user. Working together, these two terms will cover a user's all information.

We identify the following the research challenges and opportunities on generating extension profiles for web forum user.

- *Building suitable data model for web forum users.* Forum users express their views by writing down texts on forum. Forum users are associated by submitting their own posts and commenting on other user's posts. This association between users means a kind of social networks. Usually, User's posts is very short, and some pieces, such as 're' and 'support' , are absolutely meaningless. So, it's difficult to capture user's extension profiles by pure analyzing the web forum contents or pure analyzing web forum social network. So, how to integrate web contents with social network to capture extension profiles for users is one of challenges.
- *Finding the all links among users.* Links among the users are too complex to be discerned. Due to the exiting social networks in web forums,the relevancies among the users can be generated in various ways. For instance, the links can be generated by communicating directly or indirectly, or just by focusing on the same topics.
- *Finding a uniform measure function to measure correlation among different type of object.* There are four type objects in web forum, such as user, topic and keyword. In order to find extension profiles, we must construct relationships among users by topic and keywords. And, we must construct relationships between users and keywords by topics. Furthermore, Constructing these relationships need cross

Fig. 6.1 An example of
transitive similarity

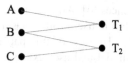

different type objects. So, finding a uniform measure function between heterogeneous objects is one of challenges.

We will address the challenges listed above and propose our solutions. The main ideas of this paper are as follows.

- *We present a tripartite graph model in order to integrate contents with social information on web forum.* There are two relationships in web forum: (1) Co-author relationship, that is between users involved in the same topic. This relationship can be extracted directly by forum crawler. (2) Containing relationship, that is between topics and keywords. This relationship can be obtained using Natural-language processing tools. We construct a tripartite graph based on users, topics and keywords. We unify similarity measurement between users and correlation measurement between users and keywords into link analysis on this tripartite graph.
- *We define user's extension profiles based on contents and his collaborative networks.* In order to understand a user's intention, we use two kinds users profiles to describe this user's preference. They are neighborhood-profile and term-profile. Neighborhood-profile is used to illustrate who is closely related to this user, term-profile is used to illustrate what this user concerned about. For Example, a java programmer who belongs to a java language community concerns about java programming language book. But, an expected mother who belongs to a young mother group focuses on products for baby. Finally, we provide personalization information discovery services for users based on their two kinds of profiles.
- *We propose an effective algorithm to measure correlation among nodes in tripartite graph. And the algorithm is implemented and its effectiveness is tested on a real forum dataset.* According to the previous description, we just need measure similarity between nodes on tripartite graph in order to find user's extension profile. In fact, the similarity or correlation between nodes on tripartite graph is transitive. For example, as Fig. 6.1 shows, user A, B reply to the same topic T_1, user B, C reply to the same topic T_2, It means user A is similar to user B, user B is similar to user C, so user A may be similar to user C. RWR (Random walk with restart) succeed in many applications on measuring the relevance between two nodes on graph [12, 15]. And, it can measure transitive similarity between nodes on graph. In this paper, we use RWR algorithm to obtain user's extension profiles.

The majority of research done on user profiles in web sites deal with content and collaborative filtering separately. In references [6], keywords with various weights and concepts are used alternatively to represent a user's profile. It is recommended by Google News that using collaborative filtering alone to describe a user's preferences [4]. After reviewing relevant literatures, we believe that the research on web forum user profiles has not been launched until now.

The rest of this paper is organized as follows: We introduce web forum data model and user extension profiles which includes neighborhood profile and term-profile in Sect. 6.2; In Sect. 6.3 we give two approaches to generate extension profiles of user based on web forum graph model; We review related work in Sect. 6.4; We report the experimental results on the dataset collected from a real popular web forum in china in Sect. 6.5. Section 6.6 is conclusions and future work.

6.2 Problem Statement

The extension profiles of a user is a record of the types and areas of that user concern about ("term profile") plus a description of her collaboration networks ("neighborhood profile"). Web forum allows users to contribute, share and communicate their ideas and experiences easily. A user, therefore, is able to participate in discussions on the varieties of topics. To gain a user's concerns, it is essential to plot the history which reveals the contents and reply network generated by the user toward one topic. However, an application, which can be used to grasp the contents and reply network simultaneously, has not been available until now. Our goal is to develop a model to discern the complexities of the posting history so that an effective algorithm can be found to solve the problem.

Table 6.1 lists the main symbols we use throughout the paper.

6.2.1 Concepts Used in the Proposed Model

An online forum can be viewed as the combination of four types of entity. They are authors, posts, topics and keywords. Each forum contains several topics, and each topic contains a variety of posts published by authors. The three types of objects in web forum are presented in Fig. 6.2. In our research, a user's posting is treated as a vote, and limited to text. The meanings of the terms are listed as the following.

Author A registered user who left message.
Post A text which published by a user.
Topic A set of posts under the same topic.
Keyword The word extracted from a topic by using NLP (Natural-Language Processing) tools.

Two relationships are invented to denote the links among author, topic and keyword.

Co-author relationship The relationships between authors who make posts on the same topic. They can be extracted directly using forum crawler.
Containing relationship The relationships between the topic and keywords. They can be obtained using NLP (Natural-language Processing) tools.

Table 6.1 Main notations used in this paper

Symbols	Definition and description
G	Web forum tripartite graph
G_c	Content-based bipartite graph
G_{cf}	Collaborative-filtering-based bipartite graph
W	Matrix of tripartite graph G
W_1	Matrix of tripartite graph G_{cf}
W_2	Matrix of tripartite graph G_c
V_i	The ith component of vertices of the tripartite graph. $i = 1, 2, 3$
E_i	The ith component of edges of the tripartite graph. $i = 1, 2$
$|V_i|$	The number of vertices in set $|V_i|$
$sim(u_i, u_j)$	A similarity function between users
$rel(u_i, t_j)$	A relevance function for between user and term
$np(u_i)$	Neighborhood-profile of u_i
$tp(u_i)$	Term-profile of u_i
w	A transition matrix which is column normalized
$(1 - c)$	Random particle that starts from node i
$\vec{e}_i$	A vector that the i-th element is 1 and other elements all are 0
$\vec{r}_i$	An vector which has $|V_1| + |V_2| + |V_3|$ components

6.2.2 Web Forum Tripartite Graph Model

To denote the entities and relationships on web forum, we invented the tripartite graph model. The entities are the vertex, and the relationships are edges on the graph. A tripartite graph model is graphically presented in Fig. 6.3.

The tripartite graph grasps the complexities of contents and reply networks related to a author, topic, and keywords simultaneously. Its virtue will be discussed in the rest part of the paper on the contrast to the traditional bipartite graph based on contents or reply network.The definition of tripartite graph and the traditional bipartite graph are listed as the following.

Definition 6.1 (Web forum tripartite graph G) Integrated contents and social network on web forum, the web forum is viewed as a tripartite graph $G = \{V_1, V_2, V_3, E_1, E_2\}$, where $V_1 = \{a_i | 1 \le i \le n\}$, $V_2 = \{t_i | 1 \le i \le m\}$, $V_3 = \{k_i | 1 \le i \le p\}$, $E_1 \subset V_1 \times V_2$, $E_2 \subset V_2 \times V_3$. In this tripartite graph, a vertex of V_1 represents an author, a vertex of V_2 represents a topic, a vertex of V_3 represents a term that user concerned about. The terms can be extracted from posts using NLP (Natural-language Processing) tools. One topic contains some posts of users. Some terms comprise topic.

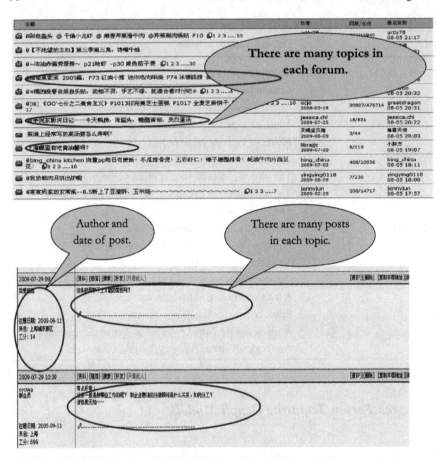
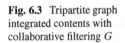
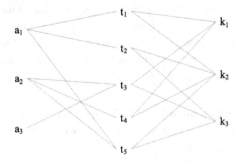

Fig. 6.2 An example of online forum

Fig. 6.3 Tripartite graph
integrated contents with
collaborative filtering G

Definition 6.2 (Web forum content-based bipartite graph G_c) This bipartite graph
represents the relationship between users and keywords. $G_c = \{V_1, V_3, E_2\}$, where
$V_1 = \{a_i | 1 \leq i \leq n\}$, $V_3 = \{k_i | 1 \leq i \leq m\}$, $E_2 \subset V_1 \times V_2$.

Fig. 6.4 Reply-networks
based bipartite graph G_{cf}

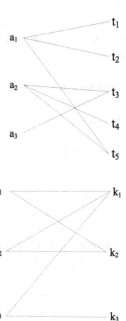

Fig. 6.5 Contents-based
bipartite graph G_c

In this bipartite graph, a vertex of V_1 represents an author, a vertex of V_3 represents a keyword that user concerned about. The keywords can be extracted from posts using NLP (Natural-language Processing) tools.

Definition 6.3 (Web forum collaborative-filtering-based graph G_{cf}) Collaborative-filtering-based graph represents relationship between users and topics. $G_{cf} = \{V_1, V_2, E_1\}$, where $V_1 = \{a_i | 1 \leq i \leq n\}$, $V_2 = \{t_i | 1 \leq i \leq m\}$ $E_1 \subset V_1 \times V_2$.

In this bipartite graph, a vertex of V_1 represents an author, a vertex of V_2 represents a topic, one topic contains several posts of users.

6.2.3 Extension Profiles of Users

Traditional user profiles include demographic information, such as name, age, country, and occupation. They are not sufficient to describe the authors' behaviors. The extension profiles, therefore, invented to denote and measure the authors' activities. The extension profiles include the term-profile and the neighborhood profile. The term-profile is used to describe the contents which are concerned by the authors. The neighborhood-profile is used to depict the group which the author belongs to and the relationships among peers. In another word, the extension profiles of a user is a record of the types and areas of that user concern about ("term profile") plus a description of her collaboration networks ("neighbor-hood profile").

Following is the definitions of neighborhood-profile and term-profile. In the definitions, For forum user u_i, neighborhood-profile is denoted by $np(u_i)$, term-profile is denoted by $tp(u_i)$.

Definition 6.4 (ε-neighborhood of a user) The ε-neighborhood of a user $u_i \in V_1$, denoted by $\varepsilon - N(u_i)$, is defined by $\varepsilon - N(u_i) = \{u_j | u_j \in V_1, i \neq j, sim(u_i, u_j) > \varepsilon\}$

Definition 6.5 (ε-term of a user) The ε-term of a user $u_i \in V_1$, denoted by $\varepsilon - T(u_i)$, is defined by $\varepsilon - T(u_i) = \{t_j | t_j \in V_3, rel(u_i, t_j) > \varepsilon, rel(u_i, t_j) \in [0, 1]\}$.

Definition 6.6 Neighborhood profile (k nearest neighbor of u_i): $np(u_i) = \{u_j | u_j \in \varepsilon - N(u_i), |\varepsilon - N(u_i)| = k, \forall \varepsilon'(\varepsilon' < \varepsilon, \varepsilon'' > \varepsilon | \varepsilon' - N(u_i)| < k, |\varepsilon'' - N(u_i) \geq k)\}$.

Definition 6.7 Term-profile (k nearest term of u_i): $tp(u_i) = \{t_j | t_j \in \varepsilon - T(u_i), |\varepsilon - T(u_i)| = k, \forall \varepsilon'(\varepsilon' < \varepsilon, \varepsilon'' > \varepsilon | \varepsilon' - T(u_i)| < k, |\varepsilon'' - T(u_i)| \geq k)\}$

$sim(u_i, u_j)$ is a similarity function between users. $rel(u_i, t_j)$ is a relevance function between user and term. The measure function $sim(u_i, u_j)$ and $rel(u_i, t_j)$ can be replaced by any existing measure function. For example, it can be cosine distance measurement, L_p norms measurement, or simrank [8].

6.3 Approaches of Generating Extension Profiles

Bipartite graph G_c describe relationships between users and keywords. And an example of G_c is shown in Fig. 6.5. Bipartite graph G_{cf} describe relationships between users and topics. And Fig. 6.4 is an example of G_{cf}. However, tripartite graph G simultaneously describe relationships between users, topics and keywords. Furthermore, considering contents of posts and reply networks in web forum, we can capture neighborhood-profile and term-profile for forum users by two approaches: (1) *Based on bipartite graph* Neighborhood-profile can be obtained based on bipartite graph G_c and G_{cf}. Term-profile can be obtained based on bipartite graph G_c. (2) *Based on tripartite graph* Neighborhood-profile and term-profile can all be capture based on tripartite graph G at the same time.

We believe tripartite graph based approach is more reliable than bipartite graph based approach. The reasons will be discussed in the following Sect. 6.3.3.

6.3.1 Random Walk with Restart (RWR) on Web Forum Graph

There are several algorithms computing proximities among vertex on a graph. RWR (Random Walk with Restart) is an effective algorithm for making proximity scores since it gives the lengths of all the possible paths between two nodes on a graph [5].

	a_1	a_2	a_3	t_1	t_2	t_3	t_4	t_5	k_1	k_2	k_3
a_1	0	0	0	1/3	1/3	0	0	1/4	0	0	0
a_2	0	0	0	0	0	1/4	1/3	1/4	0	0	0
a_3	0	0	0	0	0	1/4	0	0	0	0	0
t_1	1/3	0	0	0	0	0	0	0	1/3	1/4	0
t_2	1/3	0	0	0	0	0	0	0	0	1/4	1/3
t_3	0	1/3	0	0	0	0	0	0	1/3	0	1/3
t_4	0	1/3	0	0	0	0	0	0	1/3	1/4	0
t_5	1/3	1/3	1	0	0	0	0	0	0	1/4	1/3
k_1	0	0	0	1/3	0	1/4	1/3	0	0	0	0
k_2	0	0	0	1/3	1/3	0	1/3	1/4	0	0	0
k_3	0	0	0	0	1/3	1/4	0	1/4	0	0	0

Fig. 6.6 transition matrix of Fig. 6.3

In other words, RWR can reflect transitive relationships among nodes on the graph. Furthermore, RWR has been successfully used in many applications to measure the proximities between two nodes in a graph [12, 15]. Therefore we implement RWR to compute an author's neighbor-hood profile in our research.

As in references [12], Random walk with restart is defined as equation (6.1). In equation (6.1), $(1 - c)$ is a random particle that starts from node i. Matrix w is a transition matrix which is column normalized, and the values in the cells in one column sum up to 1. $\vec{e}_i$ is a vector in which the i-th element is 1 and other elements are 0. Equation (6.1) is proved convergent in reference [14].

If we need capture extension profiles of user a_2 in Fig. 6.3, firstly we construct transition matrix w for Fig. 6.3, which is shown in Fig. 6.6. Secondly, we construct initial vector $\vec{e}_i$ for user a_2, where $\vec{e}_i = [0, 1, 0, 0, 0, 0, 0, 0, 0, 0, 0]^T$

The following algorithms are iterative implementation of equation (6.1) on bipartite graph G_{cf}, G_c, and tripartite graph G. Iterating the equation (6.1) until term-profile or neighborhood-profile turns convergent.

$$\vec{r}_{i+1} = cw\vec{r}_i + (1 - c)\vec{e}_i \qquad (6.1)$$

6.3.2 Algorithms of Generating Extension Profiles

Based on the above analysis about correlation measurement, we know random walk with restart is a good correlation measurement between nodes in graph. In this paper, we focus on two types of graph model, tripartite graph model and bipartite graph model to get extension profiles of forum user by using the same RWR algorithm. Furthermore, we use RWR algorithm on bipartite graph G_{cf}, G_c and tripartite graph G to obtain term-profile and neighborhood-profile. Bipartite graph G_{cf} describes the co-author relationship between users involved the same topic. Bipartite graph G_c describes the relationship between users shared the same keyword. Algorithm 6.1 generates neighborhood-profile for user based on bipartite graph G_{cf} and G_c. For user u_i top k nearest neighborhood, we get top $\lceil k/2 \rceil$ nearest neighbors based on

---

Algorithm 6.1: Get neighborhood-profile of user u_i based on bipartite graph $G_{cf} = \{V_1, V_2, E_1\}$ and $G_c = \{V_1, V_3, E_2\}$

Input: $u_i \in V_1$, matrix W_1 of bipartite graph G_{cf}, matrix W_2 of bipartite graph G_c, restarting probability c, tolerant threshold ε

Output: Neighborhood-profile $np(u_i)$

1   Initialize $\overrightarrow{e}_i = 0$ except the i-th element is 1, $\overrightarrow{r}_i = 0$ except the i-th element is 1;

2   Construct adjacent M_1 and transition matrix $w_1 = col\_norm(M_1)$ based on matrix W_1;

3   **repeat**

4      $np1_i(u_i) = \{$top $\lceil k/2 \rceil$ elements with the highest value of $\overrightarrow{r}_i(1 : |V_1|)\}$;

5      $\overrightarrow{r}_{i+1} = cw\overrightarrow{r}_i + (1 - c)\overrightarrow{e}_i$;

6      $np1_{i+1}(u_i) = \{$top $\lceil k/2 \rceil$ elements with the highest value of $\overrightarrow{r}(i + 1)(1 : |V_1|)\}$;

7   **until** $|np1_{i+1}(u_i) - np1_i(u_i)|/k < \varepsilon$;

8   Construct adjacent M_2 and transition matrix $w_2 = col_norm(M_2)$ based on matrix W_2;

9   **repeat**

10      $np2_i(u_i) = \{$top $\lfloor k/2 \rfloor$ elements with the highest value of $\overrightarrow{r}_i(1 : |V_1|)\}$;

11      $\overrightarrow{r}_{i+1} = cw\overrightarrow{r}_i + (1 - c)\overrightarrow{e}_i$;

12      $np2_{i+1}(u_i) = \{$top $\lfloor k/2 \rfloor$ elements with the highest value of $\overrightarrow{r}(i + 1)(1 : |V_1|)\}$;

13   **until** $|np2_{i+1}(u_i) - np2_i(u_i)|/k < \varepsilon$;

14   $np(u_i) = np1(u_i) \cup np2(u_i)$;

15   return $np(u_i)$

---

collaborative filtering based bipartite graph G_{cf}. Top $\lfloor k/2 \rfloor$ k nearest neighbors is obtained by analyzing content based bipartite graph G_c.

Algorithm 6.2 generates term-profile for a forum user based on bipartite graph G_c which describes the relationships between users and keywords.

Algorithm 6.3 can generate both neighbor-hood profile and term-profile at the same time, due to tripartite graph combined collaborative filtering information with contents.

6.3.3 Comparisons of Two Approaches of Generating Extension Profiles

Intuitively, the tripartite graph integrates reply networks with contents. This model can build associations among users, due to replying the same topic or writing down

---

Algorithm 6.2: Get term-profile of user u_i based on bipartite graph $G_c = \{V_1, V_3, E_2\}$

Input: $u_i \in V_1$, matrix W_2 of bipartite graph G_c, restarting probability c, tolerant threshold ε

Output: term-profile $tp(u_i)$

1 Initialize $\vec{e}_i = 0$ except the i-th element is 1, $\vec{r}_i = 0$ except the i-th element is 1;

2 Construct adjacent M_2 and transition matrix $w_2 = col\_norm(M_2)$ based on matrix W_2;

3 **repeat**

4 $\quad$ $tp_i(u_i) = \{$top k elements with the highest value of $\vec{r}_i(|V_1| + 1 : |V_1| + |V_3|)\}$;

5 $\quad$ $\vec{r}_{i+1} = cw\vec{r}_i + (1 - c)\vec{e}_i$;

6 $\quad$ $tp_{i+1}(u_i) = \{k$ elements with the highest value of $\vec{r}_{i+1}(|V_1| + 1 : |V_1| + |V_3|)\}$;

7 **until** $|tp_{i+1}(u_i) - tp_i(u_i)|/k < \varepsilon$;

8 **return** $tp(u_i)$;

Algorithm 6.3: Get extension profiles of user u_i on tripartite graph $G = \{V_1, V_2, V_3, E_1, E_2\}$

Input: $u_i \in V_1$, matrix W of tripartite graph G, restarting probability c, tolerant threshold ε

Output: Neighborhood-profile $np(u_i)$, Term-profile $tp(u_i)$

1 Initialize $\vec{e}_i = 0$ except the i-th element is 1, $\vec{r}_i = 0$ except the i-th element is 1;

2 Construct adjacent M and transition matrix $w = col\_norm(M)$ based on matrix W;

3 **repeat**

4 $\quad$ $tp_i(u_i) = \{$top k elements with the highest value of $\vec{r}_i(|V_1| + |V_2| + 1 : |V_1| + |V_2| + |V_3|)\}$;

5 $\quad$ $np_i(u_i) = \{$top k elements with the highest value of $\vec{r}_i(1 : |V_1|)\}$;

6 $\quad$ $\vec{r}_{i+1} = cw\vec{r}_i + (1 - c)\vec{e}_i$;

7 $\quad$ $np_{i+1}(u_i) = \{$top k elements with the highest value of $\vec{r}_{i+1}(1 : |V_1|)\}$;

8 $\quad$ $tp_{i+1}(u_i) = \{$top k elements with the highest value of $\vec{r}_{i+1}(|V_1| + |V_2| + 1 : |V_1| + |V_2| + |V_3|)\}$;

9 **until** $|np_{i+1}(u_i) - np_i(u_i)|/k < \varepsilon$ and $|tp_{i+1}(u_i) - tp_i(u_i)|/k < \varepsilon$;

10 **return** $tp_i(u_i), np_i(u_i)$;

the same keywords on web forum. It makes our problem simpler. we can generate both neighborhood-profile and term-profile on the tripartite graph at the same time. The result of Algorithm 6.3 based on tripartite graph is more accurate than the results of other algorithms based on bipartite graph.

Firstly, we evaluate effectiveness of algorithms on generating neighborhood-profiles. Compared with Algorithm 6.3, Algorithm 6.1 only captures the relationship in which authors share same keywords or the relationships in which authors response to the same topic. However, Algorithm 6.3 captures two kinds of relationships at the time. For example, if user u_j and user u_i have not only strong co-author relationship but also similar interests in the same content. When two factors are taken into account, similarity score between u_j and u_i is fully measure. The result from Algorithm 6.3 is, therefore, more reliable than the one from Algorithm 6.1.

Secondly, we justify effectiveness of algorithms on generating term-profiles. The graphs referred in this paper are 'undigraph without weight'. However, the tripartite graph G implies the weight by its middle layer topic, compared with bipartite graph G_c that has no way to represent the weight. For example, a user u_i participate in two topics, and these two topics respectively contain same keyword T. In this case, user u_i has two paths to reach the keyword T. However, in such a case, bipartite graph G_c can only represent one path. Consequently, the results of Algorithm 6.3 is more reliable than the results of Algorithm 6.2.

In summary, we think tripartite graph model is superior to bipartite graph in generating extension profiles. This conclusion is supported by the experimental result.

6.4 Experiments

In this section, we verify the advantages of the proposed tripartite graph model on a real-world web forum dataset.

6.4.1 Data Collection

Crawling a travelling board from a popular web forum (bbs.liba.com) in China from March 25 to April 15 2010, we got 493 topics, and extracted names of the users who involved in these topics. Segmenting texts of topics using NLP tools and selecting the noun words, we construct a tripartite graph G, a bipartite graph G_c and a bipartite graph G_{cf} based on the data set. See Table 6.2 for more details.

6.4.2 Competitive Approaches

We choose 300 forum users at random from 1519 users in Table 6.2, capturing their neighborhood-profiles and term-profiles based on bipartite graph and tripartite graph

Table 6.2 Dataset

Board	Travelling
Topics	493
Users	1519
Noun terms	5120
Time span	March 25 to April 15 2010

respectively. In the rest of experiments, we will compare these two approaches from iteration times and computation results.

Comparing Iteration Times of Two Approaches   Based on our real data distribution, we set parameters $c = 0.9$, threshold $\varepsilon = 0.01$ and $k = 1, \ldots, 100$ in Algorithm 6.1, Algorithm 6.3 and Algorithm 6.2.

In Figs. 6.7, 6.8, Horizontal axis denotes parameter top-k, longitudinal axis denotes average iteration times what algorithms need for all selected users. Figure 6.7 is experimental results about generating neighborhood-profile, and Fig. 6.8 is experimental results about generating term-profile. Figures 6.7 and 6.8 both indicate the iteration times of the algorithm based tripartite graph is obviously fewer than the iteration times of the algorithm based bipartite graph during generating neighborhood-profile or term-profile. So, we believe approach based tripartite graph is more efficient than approach based bipartite graphs.

Comparing Results of Two Approaches   Now, we compare the outputs of tripartite based approach and bipartite based graph. By comparison with bipartite graph, tripartite graph can build more relationships between users or between users and keywords. This will lead the tripartite graph based approach is more accurate than bipartite graph based approach to capture user's profiles.

For each user in test, we respectively compute this user's neighborhood profile by Algorithm 6.1 and Algorithm 6.3 and this user's term profile by Algorithm 6.2 and Algorithm 6.3. In Fig. 6.6, horizontal axis denotes the parameter top-k in Algorithms 6.1 and 6.3, longitudinal axis denotes the number of overlap algorithm 6.1 and 6.3. Similarly, in Fig.10, horizontal axis denotes the parameter top-k in algorithm 6.2 and 6.3, longitudinal axis denotes the number of overlap Algorithms 6.2 and 6.3.

Figure 6.9 shows that the intersection of two neighborhood-profiles computed by two approaches is empty when $k \leq 30$. Figure 6.10 shows that the intersection of two term-profiles computed by two approaches is also empty when $k < 20$. With k increasing, the results of these two approaches become overlapped. In other words, top 30-nearest neighbors and top 20-nearest terms of a forum user are generated due to combination of contents and reply networks simultaneously. we can not capture these neighbors and terms, if considering contents and reply networks separately. So, we think tripartite graph based approach is more reliable.

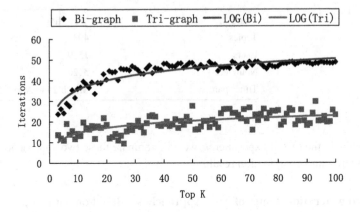

Fig. 6.7 Iterations of neighborhood profile based on bipartite graph and tripartite graph

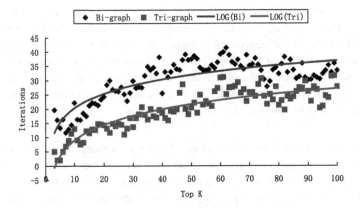

Fig. 6.8 Iterations of term profile based on bipartite graph and tripartite graph

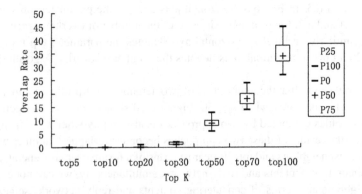

Fig. 6.9 Overlap of neighborhood profile based on two approaches

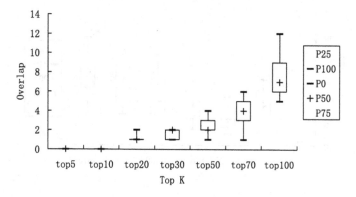

Fig. 6.10   Overlap of term profile based on two approaches

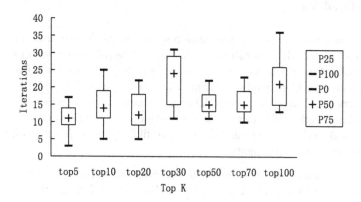

Fig. 6.11   Iterations of neighborhood profile based on tripartite graph

6.4.3   Our Approaches to Generate Extension Profiles for Forum Users

Above experimental results show the approach based tripartite graph is more reasonable than the approach based bipartite graph. Therefore, we capture neighborhood and term profiles for forum users based on tripartite graph. The following experimental results indicate that it is also reasonable using RWR method to generate neighborhood and term profiles for users on tripartite graph. Usually, the disadvantage of the RWR is large iterations. However, it can be seen from the Figs. 6.11, 6.12 that iterations is only about 12 times for top k (where is $k \leq 10$) nearest neighbors or terms for most of forum users. In real applications about user's profile, $0 \leq k \leq 10$ is a reasonable range to capture user's preference accurately. So, we believe the efficiency of the RWR method is acceptable in our problem.

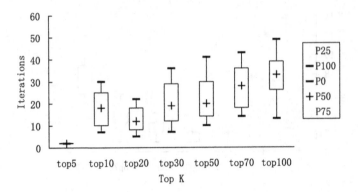

Fig. 6.12 Iterations of term profile based on tripartite graph

6.5 Related Work

In information retrieval or recommendation system research field, user profiles is a foundation to provide personalized information access [6].Recommender systems or personalized information discovery can be broadly categorized into two types: Content based and Collaborative filtering. Our work is generating profiles for web forum users based on content and collaborative filtering. In this section, we briefly review related work, which can be categorized into three groups: (1) Data analysis and statistical work on web forum; (2) user's profile; (3) Random walk with restart related methods and applications.

- *Data analysis and statistics on web forum.* Some researches have been carried out on user behaviors and characterizes web forum. Reference [9] reveals the structure of the reply networks. Reference [7, 9] focus on analyzing the social network on online forum. Reference [11] studies the patterns of user participation behaviors, and the feature factors that influence such behaviors on web forum. These papers are concerned about social behaviors and social networks on web forum.
- *User profiles.* User profiles include explicit profiles and implicit profiles [2]. Explicit profile is user's personal information, such as age, gender and occupation. It is easy to get explicit profile from the user's primitive register information on web forum. Implicit profile is extracted from a user's history behaviors and it is generally transparent to the user. In this paper, we focus on implicit profiles, called extension profiles. Extension profiles are generally represented as sets of weighted keywords, semantic networks, or weighted concepts [6]. Keywords profiles, semantic networks, and weighted concepts are based on contents. Now, applications about profiles mainly focus on generally e-commerce web site or news web site [4, 10]. Google news personalization [4] focuses on collaborative filtering. Extension profiles in our research integrate contents and collaborative filtering, such as a user's extension profiles is organized into two types, called term-profile and neighborhood profile. To the best of our knowledge, we are the first able to cap-

ture web forum user's profile with collaborative filtering information and contents at the same time.
- *Random walk with restart related methods and applications.* We use RWR in our research to measure proximities between nodes in tripartite graph. Faloutsos et al. consider RWR is a good means to score proximities between nodes in a graph [5]. Hanghang Tong and others present several good applications using RWR. Reference [12] presents a anomaly detection method in Bipartite, reference [13] presents a center-piece subgraph solution, reference [15] gaves a network analysis method using RWR.

6.6 Conclusions

In this paper, we study how to obtain extension profiles for forum users. we integrate contents-based and collaborative filtering information to obtain extension profiles for forum user. The major contributions of the paper include:

- *Propose the definitions of the extension profiles of the forum user.* Extension profiles include neighborhood-profile and term-profile. Neighborhood-profile represents a group of people who share characteristics of a user. Term-profile represents a user's preference for contents.
- *Propose the tripartite web forum graph model.* This graph model integrates social networks and contents in a web forum. It help us to capture a user's all behaviors and characteristics easily.
- *Present a similarity measure between users and a correlation measure between users and keywords by link analysis on a tripartite or bipartite graph.* We use RWR on tripartite graph to measure correlation measure between users and keywords. The algorithm is implemented and its effectiveness is tested on a real data set.

Our further research direction is to use suitable machine learning techniques to determine how to combine neighborhood-profile and term-profile for providing personalized information service on web forum. Another effort is to make our approaches scalable by using MapReduce framework in Hadoop.

References

1. Cao, L.: In-depth behavior understanding and use: The behavior informatics approach. Inf. Sci. **180**(17), 3067–3085 (2010)
2. Chu, W., Park, S.-T.: Personalized recommendation on dynamic content using predictive bilinear models. In: Quemada, J., León, G., Maarek, Y.S., Nejdl, W. (eds.) WWW, pp. 691–700. ACM, New York (2009)
3. CNNIC: 24th China Internet development research report. Website (2009), http://research.cnnic.cn

4. Das, A., Datar, M., Garg, A., Rajaram, S.: Google news personalization: scalable online collaborative filtering. In: Williamson, C.L., Zurko, M.E., Patel-Schneider, P.F., Shenoy, P.J. (eds.) WWW, pp. 271–280. ACM, New York (2007)
5. Faloutsos, C., Tong, H.: Large graph mining: patterns, tools and case studies tutorial proposal for ICDE 2009, Shanghai, China. Website (2009), http://www.cs.cmu.edu/~htong/tut/icde2009/icde_tutorial.html
6. Gauch, S., Speretta, M., Chandramouli, A., Micarelli, A.: User profiles for personalized information access. In: Brusilovsky, P., Kobsa, A., Nejdl, W. (eds.) The Adaptive Web. Lecture Notes in Computer Science, vol. 4321, pp. 54–89. Springer, Berlin (2007)
7. Gómez, V., Kaltenbrunner, A., López, V.: Statistical analysis of the social network and discussion threads in slashdot. In: Huai, J., Chen, R., Hon, H.-W., Liu, Y., Ma, W.-Y., Tomkins, A., Zhang, X. (eds.) WWW, pp. 645–654. ACM, New York (2008)
8. Jeh, G., Widom, J.: Simrank: a measure of structural-context similarity. In: KDD, pp. 538–543. ACM, New York (2002)
9. Kou, Z., Tao, B., Zhang, C.: Discovery of relationships between interests from bulletin board system by dissimilarity reconstruction. In: Grieser, G., Tanaka, Y., Yamamoto, A. (eds.) Discovery Science. Lecture Notes in Computer Science, vol. 2843, pp. 328–335. Springer, Berlin (2003)
10. Li, L., Chu, W., Langford, J., Schapire, R.E.: A contextual-bandit approach to personalized news article recommendation. In: Rappa, M., Jones, P., Freire, J., Chakrabarti, S. (eds.) WWW, pp. 661–670. ACM, New York (2010)
11. Shi, X., Zhu, J., Cai, R., Zhang, L.: User grouping behavior in online forums. In: Eder, J.F. IV, Fogelman-Soulié, F., Flach, P.A., Zaki, M.J. (eds.) KDD, pp. 777–786. ACM, New York (2009)
12. Sun, J., Qu, H., Chakrabarti, D., Faloutsos, C.: Neighborhood formation and anomaly detection in bipartite graphs. In: ICDM, pp. 418–425. IEEE Comput. Soc., Los Alamitos (2005)
13. Tong, H., Faloutsos, C.: Center-piece subgraphs: problem definition and fast solutions. In: Eliassi-Rad, T., Ungar, L.H., Craven, M., Gunopulos, D. (eds.) KDD, pp. 404–413. ACM, New York (2006)
14. Tong, H., Papadimitriou, S., Yu, P.S., Faloutsos, C.: Proximity tracking on time-evolving bipartite graphs. In: SDM, pp. 704–715. SIAM, Philadelphia (2008)
15. Xia, J., Caragea, D., Hsu, W.H.: Bi-relational network analysis using a fast random walk with restart. In: Wang, W., Kargupta, H., Ranka, S., Yu, P.S., Wu, X. (eds.) ICDM, pp. 1052–1057. IEEE Comput. Soc., Los Alamitos (2009)

Chapter 7
Information Searching Behavior Mining Based on Reinforcement Learning Models

Liren Gan, Yonghua Cen, and Chen Bai

Abstract Mining the behavioral characteristics and the adaptive learning mechanism of users during their information searching is meaningful for the academic database providers to improve their service and build e-learning platform to help users manipulate their search products more effectively The paper comprises four main parts: the first, Related work, makes a literature review and declares the concerns of our research; the second, Theories and models, explains the basic idea of reinforcement learning behavior, and introduces three representative reinforcement learning models, i.e. BM model, BS model and CR model; the third, Experiments and analysis, experimentally observes the characteristics of the academic users' reinforcement learning in the process of search tasks performing, further quantitatively simulates their reinforcement learning behavior in information seeking using the three learning models, and gives extensive discussions about these models; and the fourth, Conclusions, makes some suggestions for the academic database providers efficiently. Based on the theories and models of reinforcement learning behavior, this research takes the freshmen and senior students from universities as user samples, experimentally observes the explicit behavioral and implicit psychological characteristics of their learning behavior in the process of search tasks performing, and further quantitatively simulates their reinforcement learning behavior in information seeking using the Bush-Mosteller model, Borgers-Sarinare model and Cross model. Finally, the paper makes some extensive discussions about these models and gives some advices to the database providers.

L. Gan (✉) · Y. Cen · C. Bai
School of Economics and Management, Nanjing University of Science and Technology, 210094 Jiangsu, China
e-mail: gan5707@vip.sina.com

Y. Cen
e-mail: justin.cen@gmail.com

C. Bai
e-mail: flyluo77@sina.com

L. Cao, P.S. Yu (eds.), *Behavior Computing*,
DOI 10.1007/978-1-4471-2969-1_7, © Springer-Verlag London 2012

7.1 Introduction

As a primary method for researchers and university users to obtain necessary information, academic search plays an important part in scientific research and knowledge learning. To facilitate access to high quality literatures, many academic database providers have invested a great deal to build comprehensive electronic content collections and offer a variety of search products/functionalities, including traditional simple search (which provides single textbox for quick search), advanced search (which provides multiple textboxes to support more controlled and pinpoint search), expert search (which allow users to form complex queries combined with keywords, field codes, Boolean operators, wildcard or truncations), as well as other special purpose products such as number search, image search and academic trend search. Maximizing the return on these huge investments is the key for the database companies or even the nonprofit digital libraries to survive and develop. To achieve that, the academic database providers have to constantly update the content and improve the service based on their insight into users' characteristics and requirements so as to enable the users to maximize the effectiveness of their knowledge discovery process. Also the database providers should understand whether the electronic resources and search capabilities offered are sufficiently and productively utilized by academic users. In other word, it is of great significance to investigate whether the users can choose the most appropriate product and correctly use it to search the information desired.

Unfortunately, according to our previous research, many academic users (including undergraduate and graduate students, as well as experienced researchers) could not control the elaborately designed search functions effectively. We have organized searching experiments in which some university users with different academic backgrounds were asked to perform some predefined search tasks through a famous academic database system in China. To finish these tasks, the participants might have to use the multi-textbox search function provided by the system. However, our study proved that most participants preferred initially to choose simple search and input keywords in a single textbox to finish searching. Obviously, the functionalities offered by the database websites are not fully understood by the end users. Thus, it is necessary for the academic users to learn the skills for more effective information seeking. At the same time, it is necessary for the academic database providers to develop an environment more conducive to users' adaptive learning of search skills.

In response to these challenges, this paper focuses on mining the characteristics of users' information search behaviors, which is an important task in behavior computing [3], and the learning mechanism by which the users adjust their strategies during information searching. Probe to these problems is meaningful for the academic database providers to develop online learning platforms and offline services to guide, help or train the users to manipulate the search products more effectively and efficiently.

The paper comprises four main parts: the first, Related work, makes a literature review and declares the concerns of our research; the second, Theories and models, explains the basic idea of reinforcement learning behavior, and introduces three representative reinforcement learning models, i.e. BM model, BS model and CR model; the third, Experiments and analysis, experimentally observes the characteristics of the academic users' reinforcement learning in the process of search tasks performing, further quantitatively simulates their reinforcement learning behavior in information seeking using the three learning models, and gives extensive discussions about these models; and the fourth, Conclusions, makes some suggestions for the academic database providers.

7.2 Related Work

The nature of learning is a process that an organism adjusts its behavior according to the reaction of the environment to its preceding behavior so that to obtain an optimized output expectation. Research on learning behavior began from a century ago in psychology and education. Later, based on the developed learning theories, many quantitative models from different perspectives were established through different methodologies to probe and solve the learning issues in different decision-making contexts. Nowadays, theories and models of learning were introduced and applied extensively in engineering, economics and management [1]. A significant viewpoint is what Drew Fudenberg & David K. Levine proposed in their book, The Theory of Learning in Games [2], that the equilibrium state in game theory is not completely a result of the process that an rational player try to find an optimized solution, and that learning models may found a new theoretical basis for solving the equilibrium problem. Brian applied learning models to simulate and predict the behavior strategy selection of drivers on highways in New York in order to improve the management of traffics, accidents and safety on highways [4]. Wang Xun & Chen Jinxian employed learning theories and models to validate the marketing performance of a certain operational decision made by an enterprise [5]. Han Ling investigated how an investor changes his/her stock-holding action according to the rule of profit maximization [6]. Wang Tao resorted to reinforcement learning models and agent-simulation to explore the interconnection between macro polices, marketing, farmers and their inter-influence, as well as the soil fertility so that to discuss the behavior adjustment issues of farm utilization [7]. Machine learning is a recently highlighted topic, in which so called machine is taken advantage to replace or assist human decision-making through an iterative process of try and error to seek the action satisfying specified requirement based on the environmental feedback and under some reinforcement learning rules. For example, Wang Wenxi & Xiao Weide et al. designed a routine programming agent based on reinforcement learning [8]. In biology, Ma Qi & Zhang Liming established reinforcement learning models to discover the relationship between the visual information of fruit flies and their behavior selection [9].

In the field of information seeking, many studies about learning behavior involve with the aspects of objectives, contents, manners, influential factors and experimental methodologies.

As for the objectives and contents of learning, most research emphases users' learning or understanding of the search tasks and information needs, as well as the search strategies. Charles Cole et al. conducted a field study to examine how domain novice users learn to represent the topic space of the search tasks [10]. Diane Kelly & Xin Fu explored the techniques to elicit more robust information need descriptions from users of information systems [11]. In the Internet age, learning of the search functionalities on the web interfaces and the search strategies is also an important concern. For example, Hitomi Saitoa & Kazuhisa Miwab proposed an instructional design that supports reflective activities by presenting learners' problem-solving processes in information seeking on the Web and conducted experiments to evaluate its educational effects [12]. Wu He et al. examined the effects of two different training approaches, referred to conceptual description and search practice, on users' learning and understanding of using a case-based reasoning retrieval system [13]. The system relies on unstructured user queries and ranks search results according to their probability of relevance to the query. Besides, Kai Halttunen investigated students' conceptions of IR know-how and analyzed its implications for designing learning environments for IR [14].

Substantially, we can categorize individual learning behavior into two forms, i.e. internally self-assisted explorative learning and external instruction assisted learning. In terms of learning theories, the former concurs with reinforcement learning [15]. The latter refers to social learning or training based learning, which involves external intervention. Accordingly, in information searching, most previous work about learning behavior highlights the influences of individual personality characteristics (including individual experience, knowledge, cognitive style, learning style and so on) and external intervention, instruction or stimulation on the learning process.

Comparative studies about learning behavior of users with different knowledge background or cognitive capacity (such as novice users and experts, children and adults) during information seeking can be considered to be related to the individual characteristics [16–18]. Recently, Bernard J. Jansen et al. examined the learning characteristics of users with different cognitive style during completing information search tasks of different complexities [19]. Yan Zhang explored the effects of undergraduate students' mental models of the Web on their online searching behavior and concluded from experimental studies that the subjects' satisfaction with their own performances is significantly correlated with the time to complete the task and the familiarity of the task to subjects had a major effect on their ways to start interaction, query construction, and search patterns [20]. Efforts of Bernard J. Jansen et al. [19] indicated that a learning theory may better describe the information searching process than more commonly used paradigms of decision making or problem solving, and the learning style of the searcher does have some moderating effect on exhibited searching characteristics.

In recent years, social learning or training based learning are much stressed and external intervention influence is extensively analyzed, which may be driven by the difficulties and confusion that users face in information seeking. Charles Cole et al. claimed that some instructive intervention could help the novice users to align their mental models with the thesaurus's hierarchical representation of the search topic area and therefore facilitated their learning [10]. According to studies of Diane Kelly, Xin Fua [11] and Hitomi Saitoa & Kazuhisa Miwab [12], when provided with analogous searching information such as keywords description of similar search topics, information about other participants' search process, a participant may greatly improve his/her search effectiveness by learning. Max L. Wilson & M.C. Schraefel quantified the strengths and weaknesses of the advanced search interfaces in supporting user tactics and varying user conditions by combining some established models of users, user needs, and user behaviors [21]. Kai Halttunen focused on an integration of IR instruction and constructive learning environments, presented five qualitatively different conceptions of IR know-how, explored their connections with learning style, enrollment type, and major subject, and furthermore discussed the implications for learning environment [15]. Studies of Peter Gerjets & Tina Hellenthal-Schorr [22] showed that a user-oriented (instead of a technically oriented) web training based on a conceptual analysis in the fields of media literacy research and information retrieval research, as well as a task analysis of information problems on the web improves pupils' declarative knowledge of the web as well as their search performance and thereby outperforming the conventional Internet training.

As mentioned in above reviews, field experiments and comparative analysis is the main treatment for researchers to gain insights of learning behavior in information seeking. One kind of comparative experimental analysis is by controlled group experiments [18, 22, 23], which observes the characteristic difference of learning behavior of users with different knowledge background or personal traits (like novice users and experts) in information seeking in the same test environment and same time period. Another kind is consecutive observation experiments, in which a group of participants are asked to complete a sequence of homogeneous or evolutionary search tasks during a relatively long period and the process of their behavior adjustments and knowledge structure changes are monitored and analyzed. These adjustments or changes may be catalyzed by the participants' learning effects. Sometimes, the participants are provided with some external intervention, such as instructions or training of knowledge about the information need description [24], information about other participants' search strategies [11, 12], textual materials for understanding the knowledge related to the search tasks [25], and so forth.

In our research, we don't pay much attention to users' learning or understanding of the search tasks and information needs. Actually, we are greatly concerned about the effects of personal traits (academic background) on users' learning of search strategy selection in information seeking through field experiments and comparative analysis. Moreover, contrasted to previous work, we are not motivated to only analysis the influential mechanism by some cause-effect induction, but rather to mine the mechanism and rules underlying the user's leaning, behavioral adjustments or knowledge evolution in search strategy choosing during executing consecutive homogeneous search tasks by quantitatively fitting the entire process of two groups of participants at different cognitive level with some mathematical models introduced from economics and game theories, i.e. reinforcement learning models.

We combine the ideas of controlled group experiments and consecutive observation experiments mentioned above in our methodology design. Undoubtedly, the reinforcement learning models from other domains endowed us with mathematical insight to simulate the whole process of multiple search tasks performing of heterogeneous user groups, and our efforts of experimental comparison and model fitting may provide a complement to the research in information seeking behavior analysis.

7.3 Theories and Models

7.3.1 Reinforcement Learning Behavior

According to the learning theory of cognitive behaviorism [26], learning is a process that animals and humans could obtain experience and change their behavior to gain a better feedback. And humans share with other animals a simple way of learning, which is usually called reinforcement learning. This reinforcement learning seems to be biologically inherent. If an action leads to a negative outcome, i.e. a punishment, this action will be avoided in the future. Otherwise, if an action leads to a positive outcome, i.e. a reward, it will reoccur [15].

During the process of search knowledge learning, a user manifests similar reinforcement characteristics: when the user completes an information search task by a certain strategy, he/she will evaluate this process in terms of time cost, quantity of relevant results achieved, and form a tendency to or not to adopt this strategy or switch to other strategies in next tasks.

Here, a search strategy refers to the action that an individual participant selects one kind of search functionalities (such as simple or basic search, advanced search, expert search and so forth) from the required search system and formulate a search query when carrying on each of the search tasks. In order to construct a search query, a participant can input several keywords in one search textbox, or simply input several keywords with each keyword in one search box, or input several keywords with each keyword in one box and at the same time use Boolean operators to connect the keywords into a meaningful query form.

Consider the evolution of an individual user's information seeking behavior. He/she may experience a process from not knowing how to use, never using, or being not familiar with the information retrieval system to well controlling the system. This sequence of behavior adjustments is a dynamic learning process according to learning theories. The strategy formulation process of information seeking is a process of dynamic alignment, in which a user's strategy in each turn may not be consistent, and the adjustment of behavior is reached from the user's database using experience under the influence of external stimulations (tips on the system interface, training or communication) and based on user's knowledge about the available strategy set. This process tallies with the description of learning behavior in learning theories.

7.3.2 Models of Reinforcement Learning Behavior

Enlightened by the basic idea of reinforcement learning, a variety of quantitative models were proposed to fit different learning processes in different situations. Considering the characteristics of information seeking process, three classic models are introduced into our research for quantitative analysis: Bush-Mosteller model (BM model) [27], Borgers-Sarinare model (BS model) [28] and Cross model (CR model) [29]. All these models take the reward or punishment as the basis for the adjustment of strategy attraction of next period. Nevertheless, to the rules of this adjustment, they are not the same.

BM Model In 1955, psychologist Bush and Mosteller started a quantitative research of learning process and set up the BM model, one of the oldest and most famous learning models. In this model, the attraction of a strategy is defined as a probability variable $P(\cdot)$, the strategy which is chosen by a user in period t is defined as $d(t)$, and the reward or punishment fed back to the user in period t is $\pi(t)$ (a nonnegative $\pi(t)$ means the user gets a reward, otherwise a punishment). Suppose in period t, a user chooses the jth strategy from the strategy set, i.e. $j = d(t)$. Then for this user, the attraction of strategy j is updated under the following rules:

$$P(j, t+1) = \begin{cases} P(j,t) + \alpha^{BM} \cdot (1 - P(j,t)) & j = d(t) \wedge \pi(t) \geq 0, \\ P(j,t) - \beta^{BM} \cdot P(j,t) & j = d(t) \wedge \pi(t) < 0. \end{cases} \tag{7.1}$$

For each strategy k other than j (i.e. strategies not chosen by the user), the attraction value is updated according to the following rules:

$$P(k, t+1) = \begin{cases} P(k,t) - \alpha^{BM} \cdot P(k,t) & k \neq d(t) \wedge \pi(t) \geq 0, \\ P(k,t) + \beta^{BM} \cdot (1 - P(k,t)) & k \neq d(t) \wedge \pi(t) < 0. \end{cases} \tag{7.2}$$

In the above adjusting rules, α^{BM} and β^{BM} are two parameters to be estimated. $\alpha^{BM} \in [0, 1]$ is the weight factor assigned to a nonnegative payoff, while $\beta^{BM} \in [0, 1]$ the weight factor to a negative payoff. A smaller α^{BM} means that a nonnegative payoff plays a less part in the strategy selection, while a smaller β^{BM} means that a negative payoff plays a less part in the strategy selection.

More intuitively, the learning rules that BM model describes can be interpreted: when a certain strategy leads to a positive payoff, then the probability that this strategy will be chosen again increases and the probability it will be avoided decreases. Otherwise, the probability that the strategy will be selected decreases and the probability it will not be selected increases.

BS Model Based on BM model, Borgers and Sarin proposed another model to explain reinforcement learning behavior. Compared to BM model, BS model details the information for evaluating the payoff of a strategy adoption under the supposition that the evaluation of a strategy choice may not directly rely on the absolute value of actual payoff, but on the difference between the actual payoff and the expected one. Let $P(\cdot)$, $d(t)$ and $\pi(t)$ be the variables similar to those in BM model.

Besides, let $A(t) \in [0, 1]$ denote the payoff expectation for a user after employing a strategy in period t and $A(1)$ be the initial expectation for the user before decision-making.

If $\pi(t) \geq A(t)$, the attraction values of strategies after period t are updated:

$$P(j, t+1) = \begin{cases} P(j, t) + \alpha^{BS} \cdot (1 - P(j, t)) & j = d(t), \\ P(j, t) - \alpha^{BS} \cdot P(j, t) & j \neq d(t), \end{cases} \tag{7.3}$$

$$A(t+1) = \left(1 - \beta^{BS}\right) \cdot A(t) + \beta^{BS} \cdot \pi(t). \tag{7.4}$$

Else if $\pi(t) < A(t)$, the attraction values are updated as following:

$$P(j, t+1) = \begin{cases} P(j, t) - \alpha^{BS} \cdot P(j, t) & j = d(t), \\ P(j, t) + \alpha^{BS} \cdot (1 - P(j, t)) & j \neq d(t), \end{cases} \tag{7.5}$$

$$A(t+1) = \left(1 - \beta^{BS}\right) \cdot A(t) + \beta^{BS} \cdot \pi(t). \tag{7.6}$$

The parameter α^{BS} is regarded as the reinforcement strength, whose value is the absolute difference between the actual payoff and the expected one, i.e. $\alpha^{BS} = |\pi(t) - A(t)|$. The parameter β^{BS} is set fixed, which stands for the adjustment speed of payoff expectation. The bigger β^{BS} is, the more greatly the current payoff influences further strategy selection.

Similarly, BS model can be intuitively summarized as: if the actual payoff exceeds the expectation of an individual after a strategy is settled, then the probability that this strategy will be further selected increases. Contrarily, if the actual payoff is less than the expectation, the probability that it will be adopted in future periods decreases. Obviously, the expected payoff is dynamically changing according to the actual payoff of the previous strategy adoption.

CR Model   CR model was proposed by Cross in 1973. As a modification to BM model, it is one of the most acknowledged reinforcement learning models. Let $R(\pi(t))$ be the reinforcement strength, which is a monotonic function of the payoff $\pi(t)$. Then, in CR model, the attraction values of strategies and the reinforcement strength after period t are updated under the following rules:

$$P(j, t+1) = \begin{cases} P(j, t) + R(\pi(t)) \cdot (1 - P(j, t)) & j = d(t), \\ P(j, t) - R(\pi(t)) \cdot P(j, t) & j \neq d(t), \end{cases} \tag{7.7}$$

$$R(\pi(t)) = \alpha^{CR} \cdot \pi(t) + \beta^{CR}. \tag{7.8}$$

In the above rules, α^{CR} and β^{CR} are two parameters that control the updating mechanism of the attraction of each strategy.

CR model modifies BM model in the following two points: (i) In CR model, the attraction of a strategy is defined as a linear function of the payoff by configuring the reinforcement strength as a correlated variable to the payoff, while in BM model, the reinforcement strength factors, α^{BM} and β^{BM}, are fixed and independent of payoff. (ii) BM model treats the nonnegative and negative payoff differently, namely assigns them different reinforcement strength, i.e. α^{BM} (to a nonnegative payoff) and β^{BM} (to a negative payoff). Comparatively, in CR model, the reinforcement strength assigned to a nonnegative payoff or a negative payoff is the same. According to Cross [29], it is meaningless to differentiate between the nonnegative payoff and the negative payoff, since the payoff doesn't obey the von Neumann-Morgenstern axioms. For example, 3 units of positive payoff for a user to choose the jth strategy and 3 units of negative payoff for him/her to choose another strategy influences his/her adopting probabilities of strategy j in next period in a same way, as reflected in CR model, the updating rule of reinforcement strength corresponding to the nonnegative and negative payoff is identical.

7.4 Experiments and Analysis

In January and June 2009, we organized experiments to fit the reinforcement learning models for explaining the learning behavior in information seeking.

7.4.1 Experimental Research Design

Purposes   The overall purpose of our experiments is to observe how the academic users, who have the experience of using general search engine and are used to input keywords in a single textbox to finish searching (hereinafter referred to as "single textbox search"), learn to use multiple textboxes and Boolean droplists so as to logically combine several keywords to finish searching (referred to as "multi-textbox search"), namely to observe how they learn search knowledge and adjust their search strategy from simple search to multi-textbox search to perform complex search. More specifically, the intents of our experiments are:

(i) To reveal the behavioral characteristics of the academic users with different backgrounds during their adaptive learning of search knowledge.
(ii) To observe the belief adjustment process of academic users during their adaptive learning of search knowledge.
(iii) To discover what kinds of behavior adjustment rules govern the process of academic users' adaptive learning, in other word, to find which reinforcement learning model can better fits users' learning behavior in information searching.
(iv) To verify whether the users with different academic backgrounds but sharing the same search preference show different explicit behavioral characteristics and implicit psychological mechanism in the process of search knowledge learning and search strategy adjusting.

Sampling   Two groups of academic users, i.e. freshmen and senior students from universities participated in our experiments. In their earlier Internet surfing, these participants are inclined to use single textbox search strategy as they do with general search engine. Each participant's searching behavior was recorded by a kind of screen video recording software and the data from the video records were quantitatively organized to fit the reinforcement learning models.

Search Task Design   The participants were all required to log in the search page of the CNKI academic database system (http://www.cnki.net) and perform 10 turns of search tasks. In each turn, each participant could only choose one search function provided by the system to finish the task. Then based on the feedback, the participant could adjust their search strategy in next turn. This process was iterated until all the search tasks were completed. In each turn of search task, the participants were provided with several optional keywords. To finish these tasks, the participants might have to use the multi-textbox search function offered by the system. We assigned each item in all the possible search results a standard evaluation criterion in advance, so as to reduce the inaccuracy of statistics and analysis caused by the bias and inconsistency existed in participants' subjective judgments.

Framework of Data Analysis   Analysis to the experiment data involves the following three levels:

(i) Explicit level. Analysis at this level is to mine users' explicit behavioral characteristics in the process of learning in which users choose and formulate a strategy, conduct trial by error, and finally obtain the information desired.

(ii) Implicit level. Analysis at this level is to mine users' implicit psychological characteristics that control the explicit behaviors and explain why a user chooses a strategy, why a user adjusts his/her previous strategy, and what kinds of psychological changes and correspondingly what kinds of behavior adjustments result in the success or failure of task performing. Two variables depict the implicit psychological characteristics, i.e. expectation and satisfaction. Expectation means the expected probability of success before a user adopts a strategy, while satisfaction is an evaluation that a user gives to the results obtained.

(iii) Rule level. This level is a nexus between the above levels. Mining at this level is to discover the quantitative rules of user behavior adjustment by which the implicit psychological effects control the explicit behavioral presentation.

7.4.2 Mining of Explicit Behavioral Characteristics of Reinforcement Learning

We use four indicators, i.e. *initial strategy, stabilized strategy, stabilized turn, search time cost* to characterize the explicit reinforcement learning behavior of academic users during their information searching.

Table 7.1 The statistics of academic users' strategy selection

	Freshmen	Senior
Percent of users with initial strategy as single textbox search	95%	82.4%
Percent of users with stabilized strategy as multi-textbox search	69%	91.2%
Average stabilized turn	5.61	3.94

Table 7.2 Results of statistical test to the search time cost

	Mean	Normality test		Wilcoxon rank sum test	
		W	$Pr < W$	Z	$Pr > \lvert Z \rvert$
Freshmen	163.011667	0.900662	<0.0001	−4.4359	<.0001
Senior students	144.085	0.810849	<0.0001		

Fig. 7.1 Search time cost comparison between freshmen and senior students

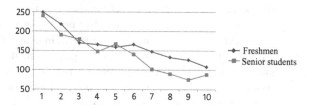

The *initial strategy* is the product (search function) which a user chooses as the strategy in the first turn of search task. To a great extent, this indicator is determined by users' habitual preferences. Analysis to this indicator can help us gain the insight of behavioral disposition and knowledge background of the users before their search tasks performing. We introduce stability to describe a user who chooses a certain strategy in one turn of search task and never changes the strategy in the succeeding turns. And the chosen and never changed strategy is referred to as *stabilized strategy*. This indicator can be examined to discover which search strategy is the most accepted by the users after learning and experiencing. Another related indicator, *stabilized turn*, is a number represents that after how many turns, a user never changes his/her strategy. To some extent, this indicator reflects the learning time cost or learning efficiency of a user from his/her original preferred strategy to the stabilized strategy. The *search time cost* is the total time for a user to finish all the 10 search tasks. Also this indicator can be used to measure users' learning speed.

Table 7.1 and Fig. 7.1 are comparisons of the above indicators between freshmen and senior students. Table 7.2 is the results of statistical test to the search time cost in which the Wilcoxon test indicates that the learning time cost of freshmen and that of senior students are significantly different with a power of 99.99%.

Several conclusions can be drawn from the above statistical data:

(i) The academic users' learning to use the search products are influenced greatly by their prior preference. For both the freshmen and the senior students, the simple search preference formed in their earlier Internet surfing experience affected them so deeply that when facing the new search environment, most users indeliberately employed the single textbox search strategy as they have done before. Although the percent of senior students who followed their earlier preference in the first turn is less than that of freshmen (see Table 7.1), it is not dispensable for senior students to learn academic search skills, despite the fact that they have already studied in the universities for almost four years.

(ii) Driven by the search tasks, the majority of academic users could finally choose the most appropriate search strategy. This conforms to the contingency theory of preference [30]. For an individual, when placed in different task contexts, his/her preference is not fixed. Instead, this preference is constructed dynamically. When the task context or the behavioral utility changes, the preference changes consequently [31]. In our research, this behavioral utility means to what degree the results fed back by the database system can satisfy the requirement of a user after adopting a certain search strategy. That is to say, when a preferred strategy always leads to failures, the individual will naturally give up this preference and learn to seek another solution.

(iii) Reinforcement learning is an adaptive process of trial and error, a process that users have to take time for. The statistics of indicator stabilized turn in Table 7.1 tell that after about 4 to 6 turns, users gave up their initially preferred strategy and learn to use multi-textbox search strategy. Obviously, the average learning time cost of freshmen is larger than that of senior students. From this point, it is important for the universities to impart the newly entered students knowledge and skills of academic literature searching. Besides, based on the statistics, it is still necessary to further train the senior students to more effectively retrieve academic literatures.

7.4.3 Mining of Implicit Psychological Characteristics of Reinforcement Learning

Before adopting a strategy, a user may form an expectation or a belief of the outcome that the strategy will produce. The embodiment of this expectation and its relation to the behavioral outcome is what we try to reveal in this section.

To observe the implicit psychological characteristics of reinforcement learning in academic information searching, we asked the participants to write down their certainty (confidence or belief) that a search strategy will bring about a satisfying feedback before they choose to adopt the strategy and to write down their evaluation of satisfaction to the results fed back after adopting the strategy.

Figure 7.2(a) draws the comparison between the confidence curve of freshmen and that of senior students, while Fig. 7.2(b) compares the satisfaction curves of this two user groups. Table 7.3 is the results of statistical test to the confidence and satisfaction of freshmen and senior students, from which, we can conclude that the average confidence and satisfaction of freshmen and that of senior students are significant different with a power more than 98.

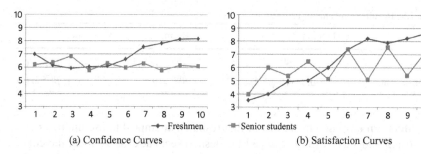

(a) Confidence Curves   (b) Satisfaction Curves

Fig. 7.2 Confidence and satisfaction comparison between freshmen and senior students

Table 7.3 Results of statistical test to the confidence and satisfaction

		Mean	Normality test		Wilcoxon rank sum test	
			W	$Pr < W$	Z	$Pr > |Z|$
Confidence	Freshmen	6.20588235	0.967095	<0.0001	3.8039	0.0001
	Senior students	6.97	0.930911	< 0.0001		
Satisfaction	Freshmen	5.99117647	0.866132	<0.0001	2.1804	0.0146
	Senior students	6.4	0.791905	<0.0001		

Table 7.4 Pearson correlation coefficient between confidence and satisfaction

	Pearson correlation coefficient between confidence and satisfaction
Freshmen	0.62925 (<.0001)
Senior students	0.41050 (<.0001)

Table 7.4 gives the correlation analysis of the confidence and satisfaction of freshmen and senior students, in which the Pearson coefficients indicate some connections between confidence and satisfaction especially for freshmen users.

Further analysis to the statistical data reveals:

(i) From Table 7.1, we have already known that after 4 to 6 turns, the users stabilized their strategy to multi-textbox search. That means in the first 3 to 5 turns, users' confidence can be attributed to their expectation of desired outcome brought about by the single textbox search strategy, while after the 5th or 6th turn by the multi-textbox search strategy. In the same way, in first 3 to 5 turns, users' satisfaction can be understood as their evaluation to the feedback from single textbox search, while after the 5th or 6th turn from multi-textbox search.

(ii) The confidence of freshmen in the first turn is higher than that of senior students. This may be induced by the strong belief of the freshmen towards their earlier preferred strategy, i.e. single textbox search. After the first turn, due to the unsatisfying search results, the confidence of freshmen declines till the

Table 7.5 Results of parameter estimation

	BM model	BS model	CR model
Freshmen	$\alpha^{BM} = 0.1; \beta^{BM} = 1$	$\beta^{BS} = 0.28$	$\alpha^{CR} = 0.1; \beta^{CR} = 0.1$
Senior students	$\alpha^{BM} = 0.50846; \beta^{BM} = 0.775$	$\beta^{BS} = 0.5$	$\alpha^{CR} = 0.1; \beta^{CR} = 0.1$

4th or 5th turn, in which most of the freshmen attempted to use multi-textbox search and successfully obtained the desired feedback. Since then, the confidence of freshmen increases linearly. These varying characteristics signal the remarkable connection between the satisfaction and belief, as shown in Table 7.4, the correlation coefficient between these two variables is 0.62925 for freshmen users with a probability of 0.999.

(iii) Interestingly, although the learning efficiency of senior students exceeded that of freshmen (as shown by the average stabilized turn in Table 7.1), they might not be so active to express their implicit feeling of confidence and satisfaction, and their confidence curve and satisfaction curve are almost flat without any significant trend.

7.4.4 Mining of User Behavior Adjustment Rules

In this section, we will apply BM model, BS model and CR model to fit the experimental data and mine the rules by which the users adjust their behavior. We used the first 70% turns of the experimental data for model inference (parameter estimation), and the remaining 30% for model verification and validation.

Estimation of Model Parameters α^{BM} and β^{BM} are the parameters to be estimated for BM model, while α^{BS} for BS model, and α^{CR} and β^{CR} for CR model. We applied maximum likelihood estimation for model inference to determine these parameters. The likelihood function is defined as:

$$LL(\Theta) = \prod_{n=1}^{N} \left(\prod_{t=1}^{T} P_n^k(t) \right). \tag{7.9}$$

where Θ denotes the parameters, T the total task turns, N the total participants, and $P_n^k(t)$ denotes the attraction of strategy k adopted by the nth user in period t. The results of parameter estimation are shown in Table 7.5.

In BM model, the parameters α^{BM} and β^{BM} reflect the mechanism by which the searching strategy adopted in a certain turn updates the attraction of this strategy for a user in the subsequent turns, in which α^{BM} means the weight assigned by the user to the positive payoff obtained in the latest period. For freshmen, $\alpha^{BM} = 0.1$, while for senior students $\alpha^{BM} = 0.50846$. This means comparatively, the freshmen concern more on the results obtained in all the finished periods (not only the latest one), while the senior students are more subjected to the outcome of the previous period.

Table 7.6 Results of model verification and validation

	Differences	BM model	BS model	CR model
Freshmen	Mean	0.040004	0.0039	0.005919
	Standard derivation	0.002137662	0.004219317	0.11219055
Senior students	Mean	0.010969	0.024872	0.00782
	Standard derivation	0.000420205	0.100978	0.0020750

Besides, freshmen are more sensitive to previous failures than senior students. This also verifies why the correlation between confidence and satisfaction of freshmen is stronger than that of senior students as discussed in Sect. 7.4.3.

In BS model, the parameter β^{BS} takes the roles of α^{BM} and β^{BM} in BM model. $\beta^{BS} = 0.28$ for freshmen users means that they assign a weight of 0.28 to the current search results, while 0.72 to all the previous experience. Contrastively, for senior students, $\beta^{BS} = 0.5$. More intuitively, freshmen are more strongly disposed to act according to the past experience, and for them, overcoming the attraction of prior strategies and switching to a new one may take a long time, even though the efficiency of the habitual behavior is low.

The parameters α^{CR} and β^{CR} in CR model for the two groups of users all equal 1. That means the influence of the search feedback in the previous period on the strategy selection behavior in the subsequent periods is not so remarkable.

Model Verification and Validation Replacing the update rules of strategy attraction with the estimated parameter values, we can obtain the models. Using the learned models, we can compute the strategy choosing probability for the remaining 30% experimental turns and predict the strategies the users may choose and adopt in those turns. Then, by computing the difference between the mean of the simulated strategies and that of the actual strategies as well as the difference between the standard derivation of the simulated strategies and that of the actual strategies for each model and each user group, we can verify the validity of those models in explaining the reinforcement learning behavior. The smaller the differences, the better the model.

Table 7.6 gives the results of model verification and validation, which indicate that BS model can best predict the learning behavior of freshmen users, while CR model best predicts the behavior of senior students. On the whole, taking all the users into consideration, CR model may be the best.

Discussions About the Models From the above research, we can conclude:

(i) As shown in Table 7.6, the differences between the mean and standard derivation of the simulated strategies and those of the actual strategies are all very small, which means that academic users manifest significant behavioral characteristics of reinforcement learning during their learning of information search. In other word, users can adjust their search strategy selection behavior according to the search results in an independent learning environment.

(ii) BS model can best fit the learning behavior of freshmen users, which implies that freshmen always adjust their behavior based on their prior knowledge and experience, their habitual preference halters their later behavior and therefore they may need more time to learn new search knowledge. This consists with the conclusions in Sects. 7.4.2 and 7.4.3. As novices, they have little academic background or poor literature retrieval skills, so they show a strong viscosity and inertia to the preference of single textbox search formed in their earlier Internet surfing. Besides, BS model implies that freshmen always adjust their search strategy according to the agreement between their evaluation to the feedback and their expectation, as analyzed in Sect. 7.4.3. A serious divergence between the confidence and satisfaction certainly leads to the strategy adjustment.

(iii) CR model best explains the learning behavior of senior students, which suggests that not the search result fed back in the previous period but the accumulated utilities from all the finished periods influence the senior students' strategy selection behavior in the subsequent periods.

Further comparison and examination to these models may also reveal:

(i) As for which model is the best, no absolute conclusion can be made. Different model gives different perspective to describe users' learning behavior.

(ii) As the oldest learning model, BM model was sometimes suspected because its strength parameters α^{BM} and β^{BM} are set to constants. However from the parameter estimation of our research, BM model is also useful, especially for explaining that freshmen are more ready to adjust their search strategy facing unsuccessful feedbacks and are more motivated to learn new search knowledge. Besides, from the results of model verification and validation, we can see that there is no remarkable difference between BM model and other two models.

(iii) Modification to BM model may better depict the detailed characteristics of users' learning behavior. Take BS model for example. We have known BS model best fits freshmen users. According to BS model, the freshmen will update their expectation $A(t)$ by $A(t+1) = (1 - \beta^{BS}) \cdot A(t) + \beta^{BS} \cdot \pi(t)$ (see (7.4) and (7.6)).For freshmen, $\beta^{BM} = 0.28$, which means they assign a small weight to the current search results ($\pi(t)$) and therefore the change of their expectation is not so active. Obviously, BS model can better explain this hidden mechanism.

(iv) BM model and BS model are mutually complementary and supportive. BM model shows that freshmen users are more sensitive to failures. BS model manifests that freshmen users adjust their search behavior by comparing the feedback and expectation. Since the expectation of freshmen users are relatively stable, we can infer that their behavior adjustments are mainly based on the feedback $\pi(t)$, which conforms with BM model.

There are still many problems deserving further study, such as experimental design, model analysis under different experimental conditions, initialization of variables and parameters and so on.

7.5 Conclusions

According to the behavioral and psychological characteristics and the adaptive learning mechanism of academic users during their information searching, as analyzed above, we propose some advices for academic database providers:

Firstly, it is of great significance to help users learn to use the search products offered. Due to the cognitive limitation, users may not thoroughly understand the functionalities and applicability of the offered search products. Undoubtedly, this cognitive incompleteness may lead to the misunderstanding or even misuse of the search products, therefore lead to the low economic efficiency and user turnover, with the huge investments idling. Since users have inherent potentiality of learning, appropriate and effective guidance to them is meaningful, especially for those novice users who have the behavioral inertia. This guidance will benefit the novices to switch their strategy to more effective search products.

Secondly, real-time, convenient and effective e-learning environment implanted into the academic database systems becomes necessary for literature search in the Internet age. On one hand, most academic users in universities rely heavily on the Internet for problem solving. They may not be willing to join in the traditional offline training. As proved by our previous investigations, those university students score only 5.88 towards the necessity of offline tanning (with 10 the max), and almost 60% to 70% of users prefer to adaptive learning. On the other hand, most students in universities to some extent lack the knowledge of academic literatures searching, which makes learning indispensable. Assertively, the most effective solution to this problem in the new age is to develop effective e-learning systems especially question-and-answer based learning systems.

Acknowledgements   This research was supported by the National Natural Science Foundation of China under contract No.70773054, No. 71001052 and No. 71003049, as well as the Graduate Student Innovation Project of Jiangsu, China under contract No. CX08B_203Z.

References

1. Bergemann, D., Valimaki, J.: Learning and strategic pricing. Econometrica **64**(5), 1125–1149 (1996)
2. Fudenberg, D., Levine, D.K.: The Theory of Learning in Games. MIT Press, Cambridge (1998)
3. Cao, L.: In-depth behavior understanding and use: the behavior informatics approach. Inf. Sci. **180**, 3067–3085 (2010)
4. Sallans, B.: Learning Factored Representations for Partially Observable Markov Decision Processes. Advances in Neural Information Processing Systems, vol. 12, pp. 1050–1056 (2000)
5. Xun, W., Jinxian, C.: The evolutionary analysis of supply chain cooperative relationship between reciprocal and contract mechanisms. Oper. Res. Manag. Sci. (China) **17**(5), 26–31 (2008)
6. HanLing: The influence of feedback information signalization on learning behavior: an experimental research. Master Thesis of Nanjing University of Science and Technology, China (2010)

7. Tao, W.: Mechanism & simulation research on farm utilization behavior adjustment decision-making from the perspective of farmer. Master Thesis of Northwest University, China (2009)
8. Wenxi, W., Shide, X., Xiangyin, M., et al.: Model and architecture of hierarchical reinforcement learning based on agent. J. Mech. Eng. **46**(2), 76–82 (2010)
9. Qi, M., Li-ming, Z.: A reinforcement learning model to simulate insect's choice behavior facing visual cues. Acta Biophys. Sin. **24**(3), 211–220 (2008)
10. Cole, C., Lin, Y., Leide, J.: A classification of mental models of undergraduates seeking information for a course essay in history and psychology: Preliminary investigations into aligning their mental models with online thesauri. J. Am. Soc. Inf. Sci. Technol. **58**(13), 2092–2104 (2007)
11. Kelly, D., Fua, X.: Eliciting better information need descriptions from users of information search systems. Inf. Process. Manag. **43**(1), 30–46 (2007)
12. Saitoa, H., Miwab, K.: Construction of a learning environment supporting learners' reflection: A case of information seeking on the Web. Comput. Educ. **49**(2), 214–229 (2007)
13. He, W., Erdelez, S., Wang, F.-K., Shyu, C.-R.: The effects of conceptual description and search practice on users' mental models and information seeking in a case-based reasoning retrieval system. Inf. Process. Manag. **44**(1), 294–309 (2008)
14. Halttunen, K.: Students' conceptions of information retrieval: Implications for the design of learning environments. Libr. Inf. Sci. Res. **25**(3), 307–332 (2003)
15. Brenner, T.: Agent learning representation: Advice in modeling economic learning. In: Tesfatsion, L., Judd, K.L. (eds.) Handbook of Computational Economics, vol. 2, pp. 895–947. Elsevier, Amsterdam (2006)
16. Bilal, D., Kirby, J.: Differences and similarities in information seeking: children and adults as Web users. Inf. Process. Manag. **38**(5), 649–670 (2002)
17. Tabatabai, D., Shore, B.M.: How experts and novices search the Web. Libr. Inf. Sci. Res. **27**(2), 222–248 (2005)
18. Thatcher, A.: Web search strategies: The influence of Web experience and task type. Inf. Process. Manag. **44**(3), 1308–1329 (2008)
19. Jansen, B.J., Booth, D., Smith, B.: Using the taxonomy of cognitive learning to model online searching. Inf. Process. Manag. **45**(6), 643–663 (2009)
20. Zhang, Y.: The influence of mental models on undergraduate students' searching behavior on the Web. Inf. Process. Manag. **44**(3), 1330–1345 (2008)
21. Wilson, M.L., Schraefel, M.C., White, R.W.: Evaluating advanced search interfaces using established information-seeking models. J. Am. Soc. Inf. Sci. Technol. **60**(7), 1407–1422 (2009)
22. Gerjets, P., Hellenthal-Schorr, T.: Competent information search in the World Wide Web: Development and evaluation of a web training for pupils. Comput. Hum. Behav. **24**(3), 693–715 (2008)
23. Tsai, C.-F., McGarry, K., Tait, J.: Qualitative evaluation of automatic assignment of keywords to images. Inf. Process. Manag. **42**(1), 136–154 (2006)
24. Tenopir, C., Wang, P., Zhang, Y.: Academic users' interactions with ScienceDirect in search tasks: Affective and cognitive behaviors. Inf. Process. Manag. **44**(1), 105–121 (2008)
25. Hsu, Y.-c.: The effects of metaphors on novice and expert learners' performance and mental-model development. Interact. Comput. **18**(4), 770–792 (2006)
26. Fudenberg, D., Levine, D.K.: The Theory of Learning in Games. MIT Press, Cambridge (1998)
27. Bush, R.R., Mosteller, F.: A stochastic model with applications to learning. Ann. Math. Stat. **24**(4), 559–585 (1953)
28. Borgers, T., Sarin, R.: Naive Reinforcement Learning with Endogenous Aspirations. Int. Econ. Rev. **41**(4), 921–950 (2000)
29. Cross, J.G.: A stochastic learning model of economic behavior. Q. J. Econ. **87**, 239–266 (1973)
30. Xiong, Y.: A comment on preference reversal and its theoretical explanation. J. Zhejiang Shuren Univ. **8**(5), 68–72 (2008)
31. Zhu, L.: Survey of preference logic study. J. Tianjin Univ. Commer. **28**(6), 27–31 (2008)

Chapter 8
Estimating Conceptual Similarities Using Distributed Representations and Extended Backpropagation

Peter Dreisiger, Wei Liu, and Cara MacNish

Abstract The ability to perceive similarities and group entities into meaningful hierarchies is central to the processes of learning and generalisation. In artificial intelligence and data mining, the similarity of symbolic data has been estimated by techniques ranging from feature-matching and correlation analysis to *Latent Semantic Analysis (LSA)*. One set of techniques that has received very little attention are those based upon cognitive models of similarity and concept formation. In this paper, we propose an extension to a neural network-based approach called *Forming Global Representations with Extended backPropagation (FGREP)*, and show that it can be used to form meaningful conceptual clusters from information about an entity's perceivable attributes or its usage and interactions. By examining these clusters, and their classification errors, we also show that the groupings identified by FGREP are more intuitive, and generalise better, than those formed using LSA.

8.1 Introduction

The ability to perceive similarities, and form meaningful, or 'natural', groups underlies some of our most important mental processes and behaviors [1]. In remembering, it allows us to go beyond superficial correspondences and identify precedents based upon structural, or deeper, similarities. In problem solving, it allows us to draw upon our past observations and experiences, and to find solutions in a timely

P. Dreisiger (✉)
Maritime Operations Division, Defence Science and Technology Organisation, Perth, Australia
e-mail: prd@csse.uwa.edu.au

P. Dreisiger · W. Liu · C. MacNish
School of Computer Science & Software Engineering, The University of Western Australia, Perth, Australia

W. Liu
e-mail: wei@csse.uwa.edu.au

C. MacNish
e-mail: cara@csse.uwa.edu.au

L. Cao, P.S. Yu (eds.), *Behavior Computing*,
DOI 10.1007/978-1-4471-2969-1_8, © Springer-Verlag London 2012

and context-appropriate manner. And in learning, these groupings determine the nature of the associations and the quality of our generalisations.

In the fields of artificial intelligence and data mining, the process of placing similar objects or observations into groups is called *cluster analysis*, and its goal is to find an arrangement that maximises both the *intra*-cluster similarities and the *inter*-cluster differences. While the most common estimates of these differences are also measures of distance, choosing the 'best' measure can be far from trivial—on the one hand, it depends upon the form and representation of our observations; on the other, it determines the make up of the clusters and the types of relationships they capture.

For symbolic textual data, such measures should capture salient differences in the terms' roles and features. Techniques such as feature matching and correlation analysis are commonly used to estimate the difference between feature vectors or sets of objects [2]. However, these approaches have several limitations: firstly, the vectors' high dimensionality, and their resulting sparseness, can be a problem for traditional measures of distance; secondly, they treat each variable, or dimension, as equally important—an assumption that is often incorrect; and thirdly, their focus on the presence or absence of terms makes them blind to the order of terms and, thus, their roles.

One way to provide a more realistic estimate of the differences is to manually weight each variable according to its importance, or salience. However, this requires the weights to be known beforehand, and it assumes that they are constant across situations and contexts. Another solution is to find a transformation that better captures, or accounts for, the variations in the raw data. The most common way of finding these transforms involves the use of *dimensional reduction*, and within data mining, one of the most powerful and widely used examples of this is *Latent Semantic Analysis (LSA)*.

LSA was developed by Landauer et al. to analyse and characterise written documents [3], and it uses dimensional reduction to minimise the effects of its inputs' sparseness and biases. Not only does this produce representations that better reflect the words' average meanings, but also the distances between them, which can be used by traditional clustering algorithms to identify groups of related terms. What it cannot do, however, is record word order or differentiate between roles.

In this paper, we will focus on a particularly promising model of sub-symbolic processing that was first described by Miikkulainen and Dyer [4]. Like LSA, their technique, called *Forming Global Representations with Extended backPropagation (FGREP)*, represents symbolic data as vectors, uses a form of dimensional reduction to emphasise deeper correlations, and produces representations whose relative distances reflect underlying similarities. Unlike LSA, it develops these representations along a relatively small set of opaque dimensions, and it takes order into account.

Through a series of experiments, we investigate FGREP's ability to produce meaningful clusters. We compare these groupings to the reference class hierarchies and those found using LSA, we examine how well the resulting clusters generalise and capture differences in the terms' roles, and we show that it can be used to provide a more intuitive estimate of similarity. We begin, in Sect. 8.2, with an overview of LSA and FGREP. In Sect. 8.3, we describe our variation of the FGREP model and the test data, and in Sect. 8.4, we present the results of two experiments that focus on perceptual data and term usage. We close, in Sect. 8.5, with a summary of our results and an outline of our future work.

8.2 Background

LSA is a well established statistical technique that accepts a matrix of term–passage frequencies, and uses dimensional reduction to compute their 'average meanings'. More specifically, it uses reduced-rank singular value decomposition to produce estimates of the average frequencies that are based upon the terms' shared neighbourhood, and their distribution across passages. Not only do these estimates reflect the words' meanings more closely than the raw frequencies, but the distances between the resulting row and column vectors can be used by traditional clustering algorithms to identify groups of related terms.

LSA is fundamentally deterministic, it can scale to relatively large corpora, and studies have shown that its estimates of inter-document similarity are comparable to those produced by human subjects [3]. LSA does, however, have some limitations: it assumes that the inputs can, in fact, be divided into natural bundles of information such as paragraphs or documents, and it is blind to the ordering of words within a document [5]. While LSA is well suited to the analysis of *textual* documents, this latter fact, in particular, affects both its ability to form clusters from declarative knowledge, and its ability to arrange them into meaningful hierarchies.

FGREP is a cognitive model that was introduced by Miikkulainen and Dyer [4], and explored further by Miikkulainen [6]. It is built around a multi-layer perceptron, and the essence of this model is that: (1) every term is represented by a numerical vector, called a *distributed representation*; (2) these representations change, from initially random values, to ones that capture patterns in the terms' usage; and (3) they are developed *automatically* by propagating the error signals back through the hidden layer to the network's inputs. Put another way, this extended form of backpropagation changes both the network's weights *and* its inputs, and FGREP uses these additional degrees of freedom to further improve the network's accuracy.

Structurally, FGREP consists of three components: a distributed representation store, a 'routing' network, and a three-layer perceptron. The store keeps track of the terms' current distributed representations; the purpose of the routing network is to convert tuples of terms into input and target vectors, and vice versa. At the beginning of each training cycle, the routing network retrieves the appropriate representations from the store, concatenates their current patterns of activation, and presents the resulting vector to the neural network's input and target layers (see Fig. 8.1a). The errors are then calculated, and once the inputs have been updated, the routing network reverses the process by extracting the terms' new representations and placing them back into the store (Fig. 8.1b). It is this circulation of representations that allows them to develop.

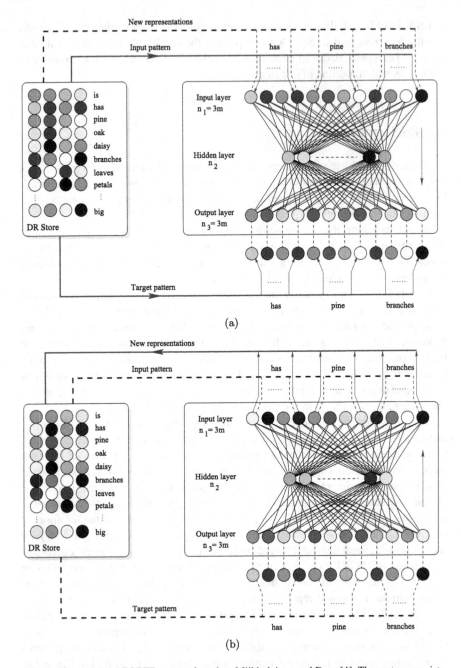

Fig. 8.1 A generalised FGREP system, based on Miikkulainen and Dyer [4]. The system consists of a symmetric, three-layer perceptron, a routing network, and a distributed representation store. In this example, the network is being taught to remember a three-term input sequence; at the same time, the representations used to construct the sequences are being updated by propagating the errors back to the input layer

While FGREP is, to the best of our knowledge, the first neural system that changes its own inputs to better reflect regularities in the training data, it is not the only one to use extended backpropagation. In particular, this rule, which was called *backpropagation-to-representation* by Rogers and McClelland [7], formed the basis of their model of human concept formation. What distinguishes FGREP from this approach is its use of a distributed store and a routing network. Even though backpropagation-to-representation develops a set of distributed representations, it does so across a hidden layer—the input and target patterns are still expressed as feature vectors and, for this reason, the number of terms it can support is completely determined by the size of its input and output layers.

In contrast, FGREP's capacity is largely independent of its network's geometry. The structure of FGREP's networks also means that it is able to form its representations from a series of declarative sentences (rather than a vector of frequencies or hand-picked features), and that it can do so *incrementally* and using an arbitrary sentence structure.

8.3 Methodology

In this paper, we investigate the types of clusters that FGREP can produce given either perceptual data or information about an entity's usage and interactions, we compare them to the reference hierarchies and those found using LSA, and we examine how well the resulting clusters generalise. This is in contrast to the earlier studies which used FGREP or extended backpropagation as part of a larger natural language processing system, or to study the process of human concept formation.

Our system extends that of [4] in order to focus on developing the distributed representations and improving their stability. We made several changes to the original network's structure, input–target pairs and learning rule. Firstly, we used a *symmetric* three-layer perceptron, or auto-encoder, in place of the original model's asymmetric feed-forward network. Secondly, we taught our network to reproduce its input sentences where earlier systems was trained to identify the agents, patients, actions and modifiers given a sentence's syntactic representation. (See Fig. 8.1 for a diagram of our generalised FGREP system.)

To improve the stability of the system, we implemented a batch learning algorithm alongside the existing sequential, or iterative, rule. (In sequential mode, the order of the training data varies from epoch to epoch, and the weights are changed after the presentation of each example; in batch mode, the weight changes are accumulated over the entire training set, and the differences are only applied at the end of each epoch.) Normally, the latter tends to yield better results and a faster rate of convergence; in our case, however, it also means that otherwise identical terms—i.e. those with the same initial position and usage—will develop divergent representations.

Finally, our system supports both dynamic and static representations. While the vast majority of representations continue to develop over time, this addition allows some terms, and the relations in particular, to be given a fixed representation. This use of static relations follows from [7], and initial tests have shown that the judicious use of fixed representations can help to 'ground' the system and improve both the quality of the other representations and their rates of convergence.

For evaluation, we used two distinct data sets. The first is based on the training corpus of [7, App. B2], but incorporates two changes: (1) the data was converted from a set of binary feature vectors to a sequence of `relation object attribute` sentences; and (2) information about the entities' class hierarchies was removed to produce a purely descriptive data set. The second set consists of the sentence templates and noun categories of [6, Tables 5.1 and 5.2]; to generate the training set proper, we went through the list of templates and enumerated every possible instance according to the list of nouns. These sets are shown in Tables 8.1a and 8.1b, respectively.

The data sets were chosen for two main reasons: (1) they allow us to verify the behaviour of our system using existing corpora, and (2) collectively, they allow us to see how FGREP performs under a variety of conditions. The first set is relatively small, contains only perceptual information, and represents knowledge using `relation object attribute` triples. In contrast, the second set is larger (containing nearly 700 sentences versus 60), consists of relational information, and uses five-part tuples to describe its entities and the roles they play.

The experiments consisted of 25 trials that ran for 100,000 epochs. We used default values for the distributed representation and hidden-layer sizes (12 and 18 neurons, respectively[1]), and sampled the representations periodically to track their development. Each trial within an experiment used the same declarations but seeded the random number generator with a different value; this meant that each trial had a unique set of distributed representations—both initial and learned. On a 2.8 GHz Pentium IV system, each trial took approximately six minutes to execute for the first set of data, and 190 minutes for the second.

8.4 Experiments and Results

The representations resulting from training were analysed using the open-source statistical package, R [8]. Firstly, dendrograms depicting the relationships between each of the terms were generated using hierarchical cluster analysis and the Euclidean measure of distance. The distances between each of the terms' distributed representations were averaged over all 25 trials to produce the 'typical' cluster plots shown in this paper; term–passage frequency vectors were also calculated using LSA in R, with stemming disabled, and an intermediate dimensionality that was half the number of terms in the corpus.

---

[1]The effects of the distributed representation and hidden layer sizes are the focus of ongoing experiments; for these trials, however, we used the default values, as recommended by Miikkulainen [6, p. 54].

Table 8.1 (a) The entity definitions, after Rogers and McClelland [7, App. B2]; and (b) the sentence templates and noun categories of Miikkulainen [6, Tables 5.1 and 5.2]

	pine	oak	rose	daisy	robin	canary	sunfish	salmon
is-pretty	0	0	1	1	0	0	0	0
is-big	1	1	0	0	0	0	0	0
is-living	1	1	1	1	1	1	1	1
is-green	1	0	0	0	0	0	0	0
is-red	0	0	1	0	1	0	0	1
is-yellow	0	0	0	1	0	1	1	0
can-grow	1	1	1	1	1	1	1	1
can-move	0	0	0	0	1	1	1	1
can-swim	0	0	0	0	0	0	1	1
can-fly	0	0	0	0	1	1	0	0
can-sing	0	0	0	0	0	1	0	0
has-skin	0	0	0	0	1	1	1	1
has-roots	1	1	1	1	0	0	0	0
has-leaves	0	1	1	1	0	0	0	0
has-bark	1	1	0	0	0	0	0	0
has-branch	1	1	0	0	0	0	0	0
has-petals	0	0	1	1	0	0	0	0
has-wings	0	0	0	0	1	1	0	0
has-feathers	0	0	0	0	1	1	0	0
has-gills	0	0	0	0	0	0	1	1
has-scales	0	0	0	0	0	0	1	1

(a)

human ate	human	boy girl man woman
human ate food	animal	bat chicken dog lion sheep wolf
human ate food with food	predator	lion wolf
human ate food with utensil	prey	chicken sheep
animal ate	food	carrot cheese chicken pasta
predator ate prey	utensil	fork spoon
human broke fragileobj	fragileobj	plate vase window
human broke fragileobj with breaker	breaker	ball bat hammer hatchet paperwt rock
breaker broke fragileobj	hitter	breaker vase
animal broke fragileobj	possession	ball bat dog doll hammer hatchet vase
fragileobj broke	object	food fragileobj possession utensil curtain desk paperwt
human hit thing	thing	animal human object
human hit human with possession	action	ate break hit move
human hit thing with hitter		
hitter hit thing		
human moved		
human moved object		
animal moved		
object moved		

(b)

Secondly, the average per-term reconstruction errors squared were calculated and plotted against time to visualise the relative rates of convergence. That is, the squared difference between each distributed representation and the corresponding signal at the network's output layer was calculated, and averaged over the 25 trials; these values were further averaged to determine the per-category and overall errors.

Finally, the classification accuracies of FGREP and LSA were compared. The statistic we chose for this analysis was *category intrusion*, defined as the number of *unrelated* terms whose representations fall within the hypersphere defined by the category's centroid and its furthest member (i.e. its radius). Not only does this metric provide us with an indication of the techniques' relative accuracy, but by calculating the intrusions for the super-classes as well as the categories, we can see the extent to which they overlap, and if the clusters, themselves, form meaningful class hierarchies.

8.4.1 Experiment 1: Identifying Concepts by Their Properties

The first experiment focuses on the nature and quality of the clusters that FGREP can form given only information about an entity's *perceivable* attributes, and its results are summarised in Figs. 8.2 and 8.3. From Fig. 8.2a, we can see that LSA tends to emphasise co-occurrences over semantic similarities; for example, it contains groups such as (robin (sing (fly feathers wings))), (swim gills scales), (petals pretty) and (move skin). It is, perhaps, for this reason that there are some inconsistencies within the entities' hierarchy: for example, there is a significant distance between the canary and the robin, and the trees are closer to the fish than the flowers.

In contrast, the clusters formed using FGREP capture both the terms' associations *and* type (Fig. 8.2b). Continuing with the above examples, we can see that FGREP was able to separate the entities, parts, attributes and actions into their own clusters, each of which could be further divided into plant and animal varieties:

```
(((salmon sunfish) (canary robin)) ((oak pine) (daisy rose)))
(((feathers wings) (skin (gills scales)))
 ((bark branch) (petals (leaves roots))))
((big green) (yellow (red (living pretty))))
((move (fly sing)) (grow swim))
```

Furthermore, in all of the 25 trials, this process of differentiation continued over time—i.e. the longer the training continued, the more distinct these clusters became.

Figure 8.3a shows the per-term and per-category errors-squared, averaged over the 25 trials. From this plot we can see not only how the reconstruction errors change over time, but how well the distributed representations capture the terms' nature. On the first point, we can see that, as a rule, the errors decrease for the first 40,000–50,000 epochs; after this point, however, they seem to plateau, or even oscillate. While determining the reason for this behaviour is beyond the scope of this paper, it

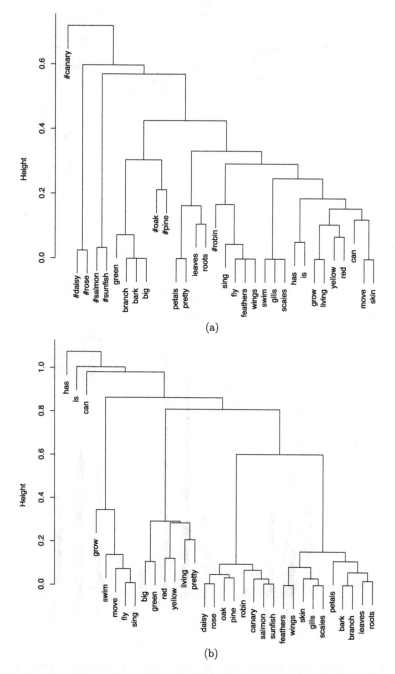

Fig. 8.2 Hierarchical cluster plots based upon: (**a**) the distances as estimated by LSA; and (**b**) the average distances, as estimated by FGREP, after 100,000 epochs. The average distances were calculated over all of the 25 trials, while the join heights indicate the Euclidean distance between term representations and/or clusters

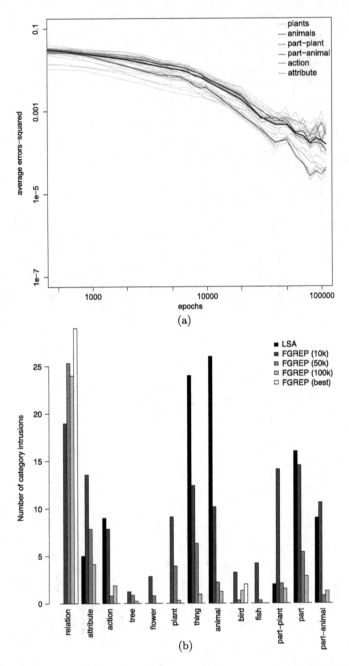

Fig. 8.3 (a) The average per-term and per-category reconstruction errors-squared, calculated over the 25 trials; and (b) The category intrusions when the terms were grouped according to their LSA and FGREP measures of similarity. The first three FGREP results are the average number of category intrusions after 10,000, 50,000 and 100,000 epochs, calculated over the 25 trials; the final column shows the number of intrusions, after 100,000 epochs, for the best individual trial

might be a sign that the network is trying to over-fit the representations, in turn, to each of the sentences in which they appear.

On the second point, we can see that the representations corresponding to the *entities*—that is, the plants and animals—had the lowest reconstruction errors while the actions and attributes (which spanned these categories) had the highest errors. In particular, there appears to be an inverse relationship between a term's reconstruction error and the information gain it offers. Take, for example, the animals and the actions (represented by the blue and red lines, respectively): given an animal, we can completely predict which other terms are about to appear; most actions, however, are associated with several plants and/or animals and, as such, they have less predictive power.

Looking at the category intrusions of Fig. 8.3b, and the errors associated with LSA, we can make several observations: (1) while LSA is able to pair the trees, flowers, birds and fish off into their own distinct clusters, it is unable to do the same for the parts, attributes and actions; (2) even though it is able to correctly group the entities, it has difficulty *generalising* these lower-level clusters into higher-level concepts such as animals, things and parts; and (3) it groups all of the relations together, even though they are used in completely different senses.

The subsequent four sets of errors in Fig. 8.3b were calculated from the results of the FGREP trials. The first three were based on the average inter-term distances after 10,000, 50,000 and 100,000 epochs; the last shows the results of the 'best' trial (i.e. the one with the fewest classification errors). In contrast to the LSA results, we can see that: (1) while FGREP is, on average, able to correctly group the trees, flowers, birds and fish, it does occasionally misclassify some of the entities; (2) unlike LSA, however, it *is* able to separate the parts, attributes and actions into their own clusters; (3) while the number of intrusions increases as we move up the class hierarchy, it is actually able to generalise the lower-level clusters into higher-level concepts; and (4) the relations remain relatively distinct.

Even with a relatively small data set, this experiment demonstrated that FGREP, and thus extended backpropagation, are able to form novel conceptual clusters from perceptual data alone. Like the network of Rogers and McClelland [7], our implementation was able to correctly identify the relationships between the birds, fish, trees and flowers. Unlike their system, our adaptation of Miikkulainen's feed-forward network was also able to develop representations for the other terms—representations that captured not only the terms' basic *types*, but their finer structure as well.

When compared to latent semantic analysis, FGREP seems to form more intuitive clusters, and is better able to arrange these groupings into meaningful hierarchies; in other words, while the average number of category intrusions increases as we move up the classification tree, it does so slowly enough for the resulting generalisations to still be useful. Of course, FGREP is a stochastic process and, thus, the quality of the representations it produces are affected by their initial values; this is in contrast to LSA, which is fundamentally deterministic.

8.4.2 Experiment 2: Identifying Concepts by Their Usage

In the first experiment, we focused on FGREP's ability to form clusters from perceivable attributes; the aim of the second experiment is to see how it performs given only information about an entity's usage and interactions, and its results are shown in Figs. 8.4 and 8.5. From Fig. 8.4a, we can see that LSA tends to favour co-occurrences and shared 'neighbourhoods' over similarities in the terms' usage. Consider, for example, the groups (ate (pasta carrot cheese)), (lion sheep wolf) and (paperwt vase). The first group contains both the unambiguous foods and the action ate. The second has placed the predators in with a prey, even though they play quite different roles in the training data. The final group consists of paperwt and vase; while both of these terms belong to the thing, object and hitter categories, there are some *semantically* significant differences too—vase is also a member of the fragileobj and possession groups while paperwt belongs to the breaker category.

In contrast, from Fig. 8.4b, we can see that FGREP appears to do a better job of capturing the terms' types and usage—unlike LSA, the foods have been separated from the term ate; the predators are now distinct from their prey; and the paperwt is now more closely associated with the hitters and breakers than it is with the vase. A related observation is that, while the ambiguous terms' relationships are somewhat unclear, FGREP was still able to separate them from amongst the other animals, hitters, fragileobjs and possessions. Unlike the previous experiment, the terms in this data set cannot be arranged into any meaningful hierarchies; rather the intrusions were calculated for each of the ten basic noun categories, the actions, and the catch-all category, 'thing'.

As with the first experiment, we can make two observations from the average reconstruction errors shown in Fig. 8.5a: (1) the reductions were only consistent for the first 40,000–50,000 epochs—after that they became much more erratic; and (2) a *category's* average error seems to reflect the diversity of its members' behaviour. Compare, for example, the categories human and thing: the former's members, which have amongst the lowest reconstruction errors, all exhibit the same behaviour, while the latter category consists of terms that assume ten distinct roles. The effects of these fluctuations can also be seen in Fig. 8.5b; while the number of intrusions decreases as we go from 10,000 to 50,000 epochs, the numbers after 100,000 epochs are actually higher for nine of the twelve categories.

Looking at the category intrusions of Fig. 8.5b and the errors associated with LSA, we can see that: (1) LSA was able to form distinct and disjoint clusters for each of the homogeneous categories—that is, the human, utensil and hitter categories whose members' behaviours are essentially identical; and (2) the categories with five or more intrusions each contained terms that played several different roles—for example, the concept animal includes terms that also belong to the hitter, breaker, possession, predator, prey and food categories.

FGREP's accuracy was less impressive in this experiment than it was for the first set of data. After 50,000 epochs, FGREP produced, on average, fewer intrusions than LSA for only three of the 12 categories, it was comparable for three, and

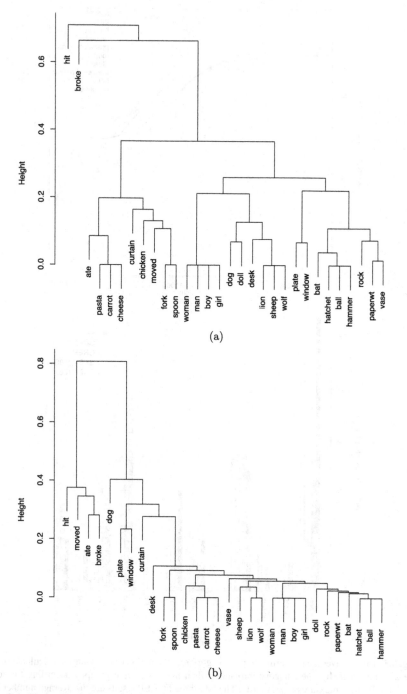

Fig. 8.4 Hierarchical cluster plots based upon: (**a**) the distances as estimated by LSA; and (**b**) the average distances, as estimated by FGREP, after 100,000 epochs, calculated over the 25 trials

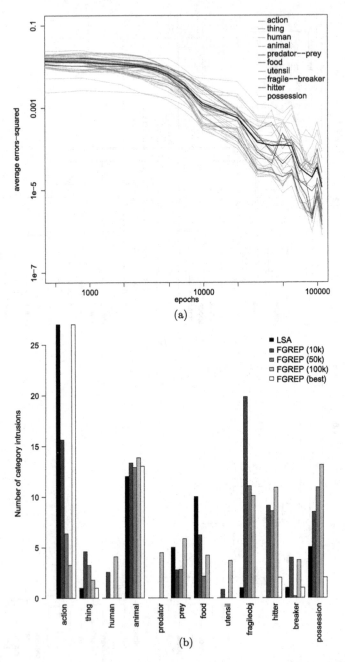

Fig. 8.5 (a) The average per-term and per-category reconstruction errors-squared, calculated over the 25 trials; and (b) The category intrusions when the terms were grouped according to their LSA and FGREP measures of similarity. The first three FGREP results are the average number of category intrusions after 10,000, 50,000 and 100,000 epochs, calculated over the 25 trials; the final column shows the number of intrusions, after 100,000 epochs, for the best individual trial

actually yielded *more* intrusions for the six other categories; after 100,000 epochs, FGREP's average accuracy was worse than LSA's for 11 of the 12 categories. However, its *best* case performance after 100,000 epochs was more encouraging, yielding fewer intrusions for four categories, and producing comparable results for five other categories.

While the results of this experiment were less conclusive than those of the first, they demonstrate that FGREP is able to identify conceptual clusters from information about the entities' usage alone; furthermore, they show that, in the best case, FGREP is able to capture most of an entity's behaviour, even when it assumes several distinct roles. This experiment, and the differences between the average and best-case results, also highlighted two of FGREP's less desirable qualities: (1) that the quality of its clusters can be quite sensitive to the distributed representations' initial positions; and (2) that this dependence is exacerbated for terms that exhibit multiple types of behaviour.

Where FGREP *did* perform better than LSA, it seemed to do so by capturing terms' roles as well as their co-occurrences. For example, LSA was unable to differentiate between the `predators` and the `sheep` because it cannot take term order or syntax into account; FGREP, on the other hand, was able to distinguish between these two categories because they assume different roles, and thus occupy different slots, in Miikkulainen's data (the `predators` are agents while the `prey` are patients). Similar results were also observed for the `food` and `breaker` categories.

8.5 Conclusions and Further Work

In this paper, we saw that FGREP is able to place terms into meaningful and intuitive clusters given either perceptual data or information about an entity's usage and interactions. While the quality of these clusters varied across trials and experiments, it is consistently able to: (1) arrange them into distinct and intuitive class hierarchies; (2) capture relationships and hierarchies that LSA cannot; and (3) distinguish between terms based upon the roles they assume.

These abilities, and at least some of FGREP's novelty as a technique, come from the fact that its fundamental unit of knowledge is the sentence, and that the order of terms *within* these sentences is both meaningful and well defined. Interestingly, the quality of its clusters seems to depend upon the number of distinct roles its members assume—i.e. the more complicated or ambiguous a term's usage, the harder it is to represent as a single point.

In the future, we will focus on characterising FGREP's typical behaviour, studying the stability of its learning rule, providing better support for terms that assume multiple roles, and extending FGREP to support discrete episodes and causal relationships; we will also use our analyses, both statistical and dynamical, to try to improve the average quality of its representations.

References

1. Cao, L.: In-depth behavior understanding and use: the behavior informatics approach. Inf. Sci. **180**, 3067–3085 (2010)
2. Han, J., Kamber, M.: Data Mining: Concepts and Techniques, 2nd edn. Elsevier, Amsterdam (2006)
3. Landauer, T.K., Foltz, P.W., Laham, D.: An introduction to latent semantic analysis. Discourse Process. **25**, 259–284 (1998)
4. Miikkulainen, R., Dyer, M.G.: Forming global representations with extended backpropagation. In: IEEE International Conference on Neural Networks, vol. 1, pp. 285–292 (1988)
5. Landauer, T.K., Dumais, S.: Latent Semantic Analysis. Scholarpedia **3**(11), 4356 (2008)
6. Miikkulainen, R.: Subsymbolic Natural Language Processing: An Integrated Model of Scripts, Lexicon, and Memory. MIT Press, Cambridge (1993)
7. Rogers, T.T., McClelland, J.L.: Semantic Cognition: A Parallel Distributed Processing Approach. MIT Press, Cambridge (2004)
8. Venables, W.N., Smith, D.M., The R Development Core Team: An Introduction to R (Version 2.9.0). Available from http://cran.r-project.org/doc/manuals/R-intro.pdf (2009)

Chapter 9
Scoring and Predicting Risk Preferences

Gürdal Ertek, Murat Kaya, Cemre Kefeli, Özge Onur, and Kerem Uzer

Abstract This study presents a methodology to determine risk scores of individuals, for a given financial risk preference survey. To this end, we use a regression-based iterative algorithm to determine the weights for survey questions in the scoring process. Next, we generate classification models to classify individuals into risk-averse and risk-seeking categories, using a subset of survey questions. We illustrate the methodology through a sample survey with 656 respondents. We find that the demographic (indirect) questions can be almost as successful as risk-related (direct) questions in predicting risk preference classes of respondents. Using a decision-tree based classification model, we discuss how one can generate actionable business rules based on the findings.

9.1 Introduction

Financial institutions such as banks, investment funds and insurance companies have been using surveys to elicit risk preferences of their customers.[1] They analyze the collected data to categorize their customer pool and to offer customized financial services. For instance, the institution can emphasize safety and predictability of investments for customers who are categorized as risk-averse, whereas it can emphasize potential gains to customers who are categorized as risk-seeking. Determining customers' risk preferences is a prerequisite for developing healthy financial plans. For this purpose, leading financial institutions often integrate the survey results into their Customer Relations Management (CRM) systems.

While the use of financial risk preference surveys is popular in practice, the survey questions are rarely determined using scientific reasoning. In addition, when

---

[1] See, for example http://www.paragonwealth.com/risk_tolerance.php.

G. Ertek (✉) · M. Kaya · C. Kefeli · Ö. Onur
Faculty of Engineering and Natural Sciences, Sabancı University, Orhanli, Tuzla, 34956, Istanbul, Turkey
e-mail: ertekg@sabanciuniv.edu

K. Uzer
School of Management, Sabancı University, Orhanli, Tuzla, 34956, Istanbul, Turkey

L. Cao, P.S. Yu (eds.), *Behavior Computing*,
DOI 10.1007/978-1-4471-2969-1_9, © Springer-Verlag London 2012

risk scores are calculated for survey respondents, questions are often given identical weights. Evaluating 14 risk surveys in France, [27] determines that *"Only a minority of the questionnaires in our sample rely on scoring techniques that attribute points for each answer. Furthermore, the questionnaires under review that do rely on scoring techniques generally fail to use sufficiently sophisticated econometric methods when setting their scoring rules. ... Consequently, the classification of investors is still based on subjective judgments, rather than on data and quantified findings."* Reference [27] also finds weak correlation between the risk scores of different surveys. That is, different financial institutions might be providing different financial advice to the same individual.

These observations indicate the need for scientific quantitative approaches for calculating risk scores using survey data. In this research, we offer a methodology to determine weights for the questions of a given risk survey, applying a regression-based iterative algorithm. Using these weights, we calculate a risk score for each survey respondent, which can be used for classification purposes.

Risk preference surveys include questions on two sets of respondent attributes: (1) *Direct attributes*, such as a choice between different hypothetical investment options, that are directly related to risk preferences; (2) *Indirect attributes*, such as demographic information, that are not directly related to risk preferences. The questions on direct attributes presumably provide more valuable information on respondents' risk preferences. However, since these questions aim at sensitive information and involve hypothetical scenarios, it may be difficult to elicit truthful information from respondents. This is particularly the case when the questions are numerous and framed too broadly. In contrast, indirect data is often readily available or can be collected easily. Our research offers a method to classify individuals based on their answers to indirect questions. One can use this classification to ask more tailored direct questions, if necessary.

The definition of risk, and risk preferences is context-dependent. Risk can be defined in many ways, including expected loss, expected disutility, probability of an adverse outcome, combination of events/consequences and associated uncertainties, uncertainty of outcome of actions and events, or a situation or event where something of a human value is at stake and the outcome is uncertain [3]. In this study, we focus on risk preferences of individuals in the context of financial investments.

The contributions of this work can be summarized as follows:

- We develop a novel behavior computing [5, 6] methodology for scoring and prediction of risk preferences. The two main components of the methodology are:
  - *Risk scoring algorithm*: Given a risk survey, this iterative algorithm determines which questions (attributes) to use and the weights for each direct question, and calculates risk scores for all respondents based on these weights.
  - *Classification model*: This model classifies respondents based on a set of (direct, indirect or both sets) attributes.
- We illustrate the use of methodology on a sample survey with 23 direct and 9 indirect questions applied to 656 respondents.

- We derive actionable business rules using a decision-tree-based classification model. These results can be conveniently integrated into the decision support systems of financial institutions.

In this section of the chapter, the study was introduced and motivated. In Sect. 9.2, an overview of the basic concepts in related studies is presented through a concise literature review. In Sect. 9.3, the proposed five-step methodology is presented and the methodology steps are illustrated through a sample survey study. In Sect. 9.4, the study is concluded with a thorough discussion of future research directions.

9.2 Literature

A number of researchers have evaluated the use of risk surveys by financial institutions to score the risk preferences of individuals and to classify them into categories. Reference's [27] evaluation of 14 risk surveys (questionnaires) used in France finds that only one third of the surveys try to quantify risk aversion, and those who quantify risk aversion fail to use sufficiently sophisticated econometric methods. Less than half of the institutions have developed scoring rules for the purpose of classification, and for most cases, classification is conducted based on subjective judgment rather than proper analytical methods. In addition, computed classes are only weakly correlated between different surveys. Our study addresses some of these issues.

Researchers have long discussed whether indirect attributes can be used effectively in classifying individuals into risk preference categories. See, for example, [14]. In particular, being male, being single, being a professional employee, younger age, higher income, higher education, higher knowledge in financial matters and having positive economic expectations are shown to be positively related to higher risk tolerance. However, blindly adopting such heuristics in classifying customers has its drawbacks. There is no consensus among researchers about the validity of these heuristics, which indicate the need for additional research (see [15] and the references therein). For example, using a survey with 20 questions, [14] finds older individuals to be more risk tolerant than younger ones, and married individuals to be more risk tolerant than single ones, which contradicts the common expectations. In a similar study, [15] uses the 1992 Survey of Consumer Finances (SCF) dataset, which contains the answers of 2626 respondents. Seven of the eight indirect attributes are found to be effective in classifying respondents into three risk tolerance categories. *The level of attained education* and *gender* are found to be the most effective attributes; whereas the effect of *age* attribute is found to be insignificant. Other related studies include [32] and [17].

A different but related problem is the *credit scoring* problem. Credit scoring can be defined as the application of quantitative methods to "predict the probability that a loan applicant or existing borrower will default or become delinquent" [23]. Credit scoring models are popular in finance, due to increasing competition in the industry and the high cost of bad debt. Reference [12] presents a review of credit scoring models based on statistical techniques and learning techniques, and their applications. Reference [34] provides a review of credit scoring and behavior scoring models, where the latter type of models use data on the repayment and ordering history of a given customer. Numerous novel credit scoring models have been published after the reviews of [12, 34], and are based on a variety of techniques; including neural networks [35], self-organizing maps [19], feature selection, case based reasoning, support vector machines (SVM) [20], discriminant analysis, multivariate adaptive regression splines (MARS), clustering, and combinations of these techniques [7]. Reference [31] develops a credit scoring framework and an expert system based on neuro-fuzzy logic to assess creditworthiness of an entrepreneur.

Different from our study, the mentioned studies do not focus on the individual's attitude towards risk, namely, his/her risk preference. Risk preference and being risky from a lender's perspective are different issues. For example, an individual who is very much risk-seeking may or may not have a high credit score (low credit risk). Also, these studies do not provide an algorithm for determining scores in the absence of a learning set.

Another research stream consists of the literature on customer segmentation as a part of Customer Relationship Management (CRM). Reference [25] presents a summary of the research on supervised classification for CRM. Reference [21] employs decision tree models for not only generating business rules regarding behavior patterns of customers, but also for dynamically tracking the changes in these rules.

We develop a numerical score for representing risk preferences with regards to financial decision making, using data from a field survey. However, risk preferences can also be estimated through controlled field experiments [18]. These experiments often identify deviations in human behavior from theoretical predictions, which is studied in the *behavioral finance* literature [4].

9.3 Methodology and Results

Our methodology is outlined below. In the following subsections, each step of the methodology is presented alongside the results we obtain based on our sample survey data.

1. Survey design
2. Survey conduct
3. Risk scoring
4. Classification
5. Insight generation

9.3.1 Survey Design

We investigated the risk scoring surveys of a number of financial institutions available on the Web, and developed our survey by choosing 23 direct (risk-related) and

9 indirect (demographic) questions among the popular ones. Appropriate selection of the direct attributes for a survey directly affects the risk scores and the subsequent data mining study, and hence is very important.

The questions in the survey were designed such that the *choices* given to respondents are sorted according to (hypothesized) risk preferences. For example, in the survey questions with three choices, selecting choice (a) is assumed to reflect risk-averse behavior, whereas selecting choice (c) is assumed to reflect risk-seeking behavior.

Examples of the questions on direct attributes include the number of times a person plays in the stock market, the investment types that a person would feel more comfortable with, and the most important investment goal of that person. A number of sample direct questions is provided in Appendix A. The complete survey (English version) is provided in Appendix A of the supplementary document for this chapter [11].

We used the following nine indirect attributes in our study:

– *Gender*: male or female
– *IsStudent*: whether the person an employee or a student
– *StudentLevel*: undergrad, masters
– *IncomeType*: fixed salary, incentive based, or both
– *SoccerTeam*: the soccer team that the person supports
– *HighschoolType*: public, private, public science, private science, other
– *EnglishLevel*: the level of English language skill
– *GermanLevel*: the level of German language skill
– *FrenchLevel*: the level of French language skill

The other indirect questions in the survey, such as the department that a student studies in, were not included in the scoring and prediction phases of the study because they were open-ended.

9.3.2 Survey Conduct

The survey was conducted in Turkish language on 656 respondents, with balanced distribution of working people (346) vs. students (250 undergraduates and 60 graduates), and gender (283 females vs. 373 males), from a multitude of universities and work environments. Among the working participants, 71 work only for commission, 204 work for fixed income and 71 work for both commission and fixed income. The distribution of values for the attributes are given in Appendix B of the supplementary document [11].

One challenge faced while conducting the survey was the communication of finance and insurance concepts, and the choices available to respondents. This is important for ensuring valid answers to the survey questions and hence improving the reliability of the sample study. To this end, all surveys were conducted through one-to-one interaction with individuals by our research team. One drawback of this approach is that communication may influence respondents' risk preferences. For example, [28] observes that farmers in Netherlands exhibit more risk-seeking behavior when they understand and trust the insurance tools through one-to-one interaction. The results of [28] confirm earlier findings in India, Africa, and South America. We do not analyze the effects of such a bias.

Once the survey was conducted, the data was assembled in a spreadsheet software and cleaned following the guidelines in the taxonomy of dirty data in [22]. Also at the data cleaning stage, data was anonymized, so that it can be shared with colleagues and students in future projects.

9.3.3 Risk Scoring

The survey data is fed into the *risk scoring* algorithm in the form of an $I \times J$ sized matrix, representing I respondents and J attributes. This algorithm determines which direct attributes are to be used in scoring, the weights for each attribute, and based on these, the risk scores for each respondent. The mathematical notation and the pseudo-code of the scoring algorithm are given in Appendix B.

The initialization step in the algorithm linearly transforms ordinal choice data into nominal values between 0 and 3. For example, if a question has five choices (a, b, c, d, e), the corresponding numerical values would be (0.00, 0.75, 1.50, 2.25, 3.00). This linear transformation is used for simplicity; however, there is no guarantee it is the most accurate representation.

Following the initialization phase, quantitative attribute values are fed into a regression-based iterative algorithm. The algorithm operates as a multi-pass self-organizing heuristic, which aims at obtaining converged risk scores. The stopping criterion is satisfied when the average absolute percentage difference in risk scores is less than the threshold provided by the analyst. At each iteration of the algorithm, the value vector for each of the selected attributes is entered into a linear regression model as factor, where the response is the incumbent risk score vector. Weights for the attributes are updated at the beginning of each iteration, such that the sum of the weights is equal to the number of included attributes. The algorithm allows for change in the direction of signs when the choices for an attribute should take decreasing—rather than increasing—values from choice (a) to the final choice. Hence, the algorithm not only eliminates irrelevant attributes, but also suggests the direction of risk preferences for the choices of a given attribute. The algorithm is an unsupervised algorithm, as it does not require any class labels or scores from the user. It is also a self-organizing algorithm [2], as it automatically converges to a solution at the desired error threshold.

After the risk scores are calculated for all respondents, a certain top percentage of them are labeled as *risk-seeking* and the rest as *risk-averse*. This is used in the subsequent *classification* step of the methodology.

The algorithm was coded in Matlab computational environment [26]. The mapping of the ordinal values $\mathbf{O} = [o_{ij}]_{656 \times 23}$ to nominal values in the initialization

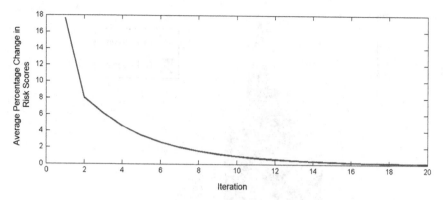

Fig. 9.1 The convergence of the algorithm, based on the average percentage change in risk scores

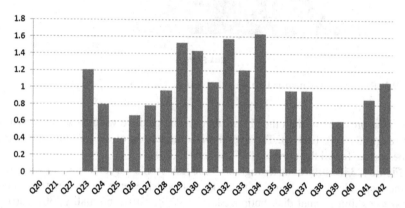

Fig. 9.2 Calculated weights for the direct attributes in the case study

step was performed in the spreadsheet software, and the Matlab code was run with the obtained matrix of nominal values $\mathbf{A} = [a_{ij}]_{656 \times 23}$. The parameters for the algorithm were selected as $E = 0.1$ and $\alpha = 0.05$. Running time for the algorithm was negligibly small (less than one second) for this sample.

9.3.3.1 Results on Scoring Algorithm

The average absolute percentage change $\bar{e}_k$ in risk scores is shown in Fig. 9.1. We observe $\bar{e}_k$ to halve in only two iterations, and to get very close to zero after the first 10 iterations. The algorithm converges to the given threshold E rapidly, in only 19 iterations.

Figure 9.2 shows the weights obtained for each of the 23 direct attributes. Five of the 23 direct attributes (Q20, Q21, Q22, Q38, Q40) are assigned a weight of 0 by the algorithm. That is, the algorithm removes these five questions from the risk score computations, because they fail to impact the scores in a statistically significant way,

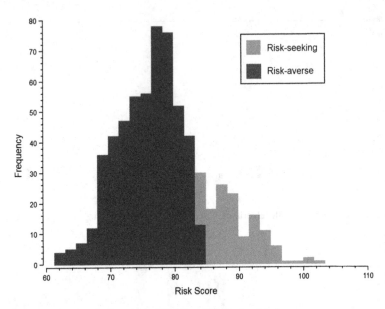

Fig. 9.3 Risk score histogram and the definition of (risk) class labels

given the presence of the other 18 attributes. The positive weights are observed in the range (0.2792, 1.6320). The hypothesized directions of choice ranks are found to be correct for all the attributes ($\Gamma_j = 1, \forall j \in \mathcal{J}$).

Figure 9.3 illustrates the histogram of the risk scores we calculate, labeling 20% of the respondents as risk-seeking and the rest as risk-averse. While the risk scores seem to exhibit normal distribution, Shapiro-Wilk test for normality [30], carried out in R statistical package [33], resulted in $p = 3.2\text{E}{-7} \ll 0.05$, very strongly suggesting a non-normal fit.

9.3.4 Classification

The next step in the methodology investigates whether risk preferences can be predicted through only direct, only indirect or both sets of attributes. To this end, we use five *classification algorithms* from the field of machine learning for predicting whether a person is risk-seeking or risk-averse, as labeled by the scoring algorithm of step 3. These algorithms, also referred to as *learners*, are Naive Bayes, k-Nearest Neighbor (kNN), C4.5, Support Vector Machines (SVM), and Decision Trees (DT)[2] [9].

In classification models, a *learning dataset* is used by the learner for supervised learning to later on predict the class label for new respondents. The predictors can

---

[2]Also referred to as *classification trees*.

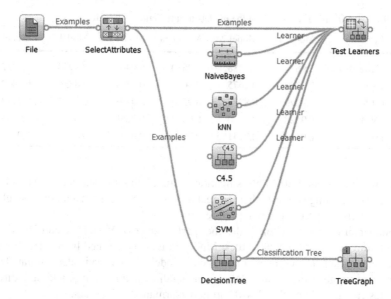

Fig. 9.4 Classification model for predicting risk preference behavior, together with the decision tree analysis widgets

have nominal or categorical values, whereas the predicted class attribute should have categorical (class label) values. The success of a learner is measured primarily through *classification accuracy* on a provided *test dataset*, besides a number of other metrics. Classification accuracy is defined as the percentage of correct predictions made by the classification algorithm on the test dataset.

Figure 9.4 illustrates the generic classification model we construct for risk preference prediction, as well as the widgets for decision tree analysis in the Orange data mining software [36]. In the classification model, some of the attributes in the full dataset are selected as the predictors and the risk-preference attribute (taking class label values of *risk-seeking* or *risk-averse*) is selected as the class attribute.

Classification accuracy is computed through 70% sampling with ten repeats. In other words, for each learner, ten experiments are carried out, with a random 70% of the sample being used as the training dataset each time, and the remaining used as the test dataset.

9.3.4.1 Results on Classification Models

Table 9.1 presents the classification accuracy results of the six models. We observe that in Model 1a, Naive Bayes learner successfully classifies (on the average) 96.35% of the respondents in the test dataset. This is not surprising, since the direct attributes that Model 1a uses were used in the computation of risk scores in the first place, which are eventually transformed into the risk preference class labels. Therefore, high classification accuracy for Model 1a is expected. What is surprising is the

Table 9.1 Classification accuracies of the models for predicting risk preferences

	Learner	Model 1a	Model 2a	Model 3a	Model 1b	Model 2b	Model 3b
1	Naive Bayes	0.9635	0.7888	0.9650	0.9467	0.8954	0.9452
2	kNN	0.9279	0.7675	0.9091	0.9391	0.8756	0.9452
3	C4.5	0.9279	0.8020	0.9269	0.9198	0.8985	0.9173
4	SVM	0.9528	0.8020	0.9452	0.9650	0.8985	0.9584
5	DT	0.9142	0.7741	0.9030	0.9239	0.8919	0.9137

relatively high (around 80%) classification accuracy that the learners in Model 2b achieve. This finding suggests that indirect attributes can be almost as successful as direct attributes in predicting risk preference.

Another surprising outcome is the poor performance of Models 3a and 3b, which use both direct and indirect attributes. Model 3a is outperformed by Model 1a with all but one learner. The comparison between Models 3b and 1b is also similar. This observation suggests that if one is already using the direct attributes, adding indirect attributes can deteriorate the classification performance of learners.

While not yielding the highest classification accuracy in any of the models, decision tree (DT) may be preferred over other (black-box) learners due to its strong explanatory capacity, in the form of explicit rules it generates. We discuss one decision tree application in the following step.

9.3.5 Insight Generation

In this step of the methodology, we aim to determine whether the answers to direct or indirect questions convey information about the risk preferences of respondents. To this end, a decision tree is constructed in the Orange model.

Decision trees summarize rule-based information regarding classes using trees. As opposed to the black-box operation of machine learning algorithms, decision trees return explicit rules, in the form *"IF Antecedent THEN Consequent"*, that can easily be understood and adopted for real world applications. For example, in the context of risk, [24] gives an example rule which states that credit card holders who withdrew money at casinos had higher rates of delinquency and bankruptcy. Such rules can also encapsulate the domain knowledge in expert systems development, in the form of *rule bases* [13]. Wagner et al. [37] state that knowledge acquisition is the greatest bottleneck in the expert system development process, due to unavailability of experts and knowledge engineers and difficulties with the rule extraction process. Our methodology offers a recipe for this important bottleneck of expert systems development.

In decision trees, branching is carried out at each node according to a *split criterion* and a tree with a desired depth is constructed. At each deeper level, the split that yields the most increase in the split criterion is selected. Reference [8] gives a

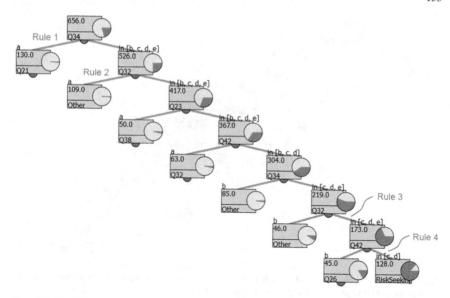

Fig. 9.5 Decision tree for Model 1a, where only direct attributes are used and the respondents with the top 20% highest scores are labeled as risk-seeking

concise review of algorithms for decision tree analysis, explaining the characteristics of each algorithm. In our decision tree analysis, we use the ID3 algorithm [29] in Orange software [36] that creates branches based on the *information gain* criterion. Each level in the decision tree is based on the value of a particular variable. For instance, in Fig. 9.5, the *root node* of the decision tree contains 656 respondents and the branching is based on question 34 ($Q34$). In each node of this decision tree, the dark slice of the pie chart shows the proportion of risk-seeking participants, and the remaining portion of the pie shows risk-averse respondents in that sub-sample. In decision trees, we are especially interested in identifying the nodes that differ significantly from the root node with respect to the shares of the slices, and the splits that result in significant changes in the slices of the pie chart compared to the *parent node* (the node above the split).

Figure 9.5 shows the decision tree for Model 1a, where only the direct attributes are used. We observe a significant branching based on the answer given to $Q34$ (question 34). When $Q34$ takes the value a, the percentage of risk-seeking respondents drops significantly from 20.00% (Definition "a") to just 1.53% (2 out of 130 respondents). Similarly, even if $Q34$ takes a value in $\{b, c, d, e\}$, if $Q32$ has the value of a, then again, the chances of that person being risk-seeking in this sample is much lower (actually 1, out of 109 respondents) than the root node (that represents the complete sample). Rule 1 and Rule 2, labeled on Fig. 9.5, reflect the aforementioned findings as below:

Rule 1: "**IF** $Q34 = a$ **THEN** *Proportion(RiskSeeking)* $= 1.53\%$."
Rule 2: "**IF** $Q34 \in \{b, c, d, e\} \wedge Q32 = a$ **THEN** *Proportion(RiskSeeking)* $= 0.92\%$."

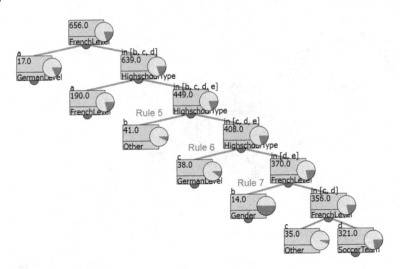

Fig. 9.6 Decision tree for Model 2a, where only indirect attributes are used and the respondents with the top 20% highest scores are labeled as risk-seeking

As $Q34$, $Q32$, and $Q23$ (questions with five choices) take values in $\{b, c, d, e\}$ (any value but a), and $Q42$ (a question with four choices) takes a value in $\{b, c, d\}$, the proportion of the risk-seeking respondents continues to increase compared to the root node. The next three splits are again related with these questions, and hence these four questions are the most important risk-related questions when deriving rules for Model 1a. $Q34$ and $Q32$ also had the largest weights in the scoring algorithm (as seen in Fig. 9.2), but $Q23$ and $Q42$ did not have the next two largest weights. This tells us that the weights obtained by the scoring algorithm are related, but not perfectly aligned with the results of the decision tree analysis. These questions ask about the volatility level that the person would be willing to accept ($Q34$), top investment priority ($Q32$), a self-assessment of risk preference compared to others ($Q23$), and the most preferred investment strategy ($Q42$).

The rules that corresponds to the splits marked with Rule 3 and Rule 4 in Fig. 9.5 are as follows:

Rule 3: "**IF** $Q34 \in \{c, d, e\} \wedge Q32 \in \{c, d, e\} \wedge Q23 \in \{b, c, d, e\} \wedge Q42 \in \{b, c, d\}$
        THEN *Proportion(RiskSeeking)* $= 68.21\%$."
Rule 4: "**IF** $Q34 \in \{c, d, e\} \wedge Q32 \in \{c, d, e\} \wedge Q23 \in \{b, c, d, e\} \wedge Q42 \in \{c, d\}$
        THEN *Proportion(RiskSeeking)* $= 86.72\%$."

The only difference between Rules 3 and 4 is that in Rule 4, $Q42$ takes values of c or d, rather than a value in $\{b, c, d\}$.

Next, we discuss the decision tree for Model 2a, where only the indirect attributes are used. As the decision tree in Fig. 9.6 suggests, *FrenchLevel* and *HighschoolType* are the attributes that result in the most fundamental splits. A significant change takes place in the pie structure when *HighschoolType* $= b$ (public science high school[3]), given that *FrenchLevel* ∈ {*b*, *c*, *d*}. Specifically, in the mentioned split, the pie slice that corresponds to risk-seeking respondents becomes much smaller (3 respondents out of 41) compared to its parent node. This is Rule 5, labeled in Fig. 9.6 and stated below.

Rule 5: "**IF** *FrenchLevel* ∈ {*b*, *c*, *d*} ∧ *HighschoolType* = *b* **THEN**
        Proportion(RiskSeeking) = 7.31%."

A similar split takes place when *HighschoolType* = *c* (private science high school), again resulting in a low proportion (4 out of 38) of risk-seeking respondents:

Rule 6: "**IF** *FrenchLevel* ∈ {*b*, *c*, *d*} ∧ *HighschoolType* = *c* **THEN**
        Proportion(RiskSeeking) = 10.53%."

It is striking that respondents who graduated from public and private science high schools are much more risk-averse, compared to other sub-groups. This has important implications for the business world: Our finding suggests that it is unlikely for respondents with a strong science background in high-school to establish risky businesses, such as high-technology startups. However, such startups are highly critical in the development of an economy, and are dependent on the know-how of technically competent people, such as professionals with a strong science background beginning in high school. Therefore, there should be mechanisms to encourage risk-taking behavior among science high school alumni, and to establish connections between graduates of science high schools and those with an entrepreneurial mindset.

In Fig. 9.6, another major split takes place in the split labeled as Rule 7. Here, the question that creates the split is *FrenchLevel* = *b* (French level is intermediate), given that *HighschoolType* ∈ {*d*, *e*} (private high school, and state high schools with foreign language). *FrenchLevel* = *b* results in a very high proportion (7 out of 14) of risk-seeking respondents. Yet, the number of respondents in the mentioned subsample is very small, only 14, and this rule has to be handled with caution.

Rule 7: "**IF** *HighschoolType* ∈ {*d*, *e*} ∧ *FrenchLevel* = *b* **THEN**
        Proportion(RiskSeeking) = 50.00%."

Upon further querying, we find that only 3 out of 17 respondents (17.65%) that have *FrenchLevel* = *a* are risk-seeking, whereas 8 out of 18 respondents (44.44%) that have *FrenchLevel* = *b* are risk-seeking. Risk-seeking behavior is minimal (9.33%) among the 75 respondents that have *FrenchLevel* = *c*. What could be the explanation for such a pattern? One possible explanation might be the following: In Turkey, individuals that have *FrenchLevel* = *a* typically come from wealthy families and learn French in expensive private high schools. Individuals that have *FrenchLevel* = *b* typically have strived to learn French by themselves, without going to such schools. They have aspirations to rise socio-economically, and are ready

---

[3]*Science high school*: specially designated high schools that heavily implement a math- and science-oriented curriculum.

to take the risks needed to achieve their aspirations. Definitely, the true explanation for this pattern is a research question for the field of sociology.

9.4 Conclusions

In this study, we develop a scoring algorithm, implement it with real world survey data, and obtain significant insights through mining risk scores for the sample. In particular, we find that demographic attributes of individuals can be used to predict their risk preference categories. This result has important practical implications: Without asking any risk-related questions, but by only obtaining demographic information, one can estimate with reasonable accuracy whether a particular respondent is risk-seeking or risk-averse. The data for those indirect attributes is often routinely collected on the Internet when registering for web sites. This would eliminate the need to collect extensive finance-related information or sensitive personal information [38] from customers. Another advantage is that, respondents would typically not distort their answers to indirect questions, whereas they could do so with direct ones. Hence, our methodology can feasibly be implemented in practice, and has the potential to bring significant predictive power to the institution at minimal effort.

Classification of customers into risk preference categories is an important problem for financial institutions. As argued in [15], incorrectly classifying a risk-averse customer as risk-seeking may later cause the customer to sell investments at a loss; whereas the opposite mistake may cause the customer to miss his investment objectives. Using our methodology, the institution can make a pre-classification of customers into risk-averse and risk-seeking categories. If necessary, these customers can then be given surveys with more tailored direct (risk-related) questions. The computational nature of our methodology makes it easy to be integrated into existing CRM systems in terms of data use and result feed.

The methodology and the scoring algorithm proposed in this work are actually *platforms* on which better methodologies and algorithms can be designed. There exists a rich possibility of future research on this area, mostly regarding the algorithm:

- The algorithm assumes that the risk score of each respondent can be computed with the same set of attribute weights. However, different weights may apply to different subgroups within the population. This can be analyzed by incorporating cluster analysis [39, 40] into the current study.
- The numeric values assigned to the ordinal values of the attributes were assumed to be linear and equally spaced; whereas the real relation may be highly nonlinear. Linearizable functions [10] or higher order polynomials can be assumed for attributes as a whole, or each attribute may be modeled flexibly to follow any of these functional forms. As an even more general model, weights can be computed not only for attributes, but for each choice of each attribute.
- In scoring, statistical techniques for feature selection and dimensionality reduction that exist in literature [16] may be adopted to obtain approximately the same results with fewer direct questions. This problem can be solved together with the

outlier detection problem, as in [1], where the authors present a hybrid approach combining case-based reasoning (CBR) with genetic algorithms (GAs) to optimize attribute weights and select relevant respondents simultaneously.

- The scoring algorithm can be developed such that consistent results are obtained for different samples. For example, in the ideal case, a respondent who answered the same question in a particular sample should have same score if he was a part of another sample. In our presented algorithm, each respondent's risk score is dependent on the answers of the whole sample. This will not pose a problem when the methodology is applied to large data sets, such as all customers of a financial institution.
- The proposed methodology eliminates irrelevant *direct* attributes in computing the risk scores, but it does not eliminate *indirect* attributes that are irrelevant or do not provide significant information. All the potential indirect attributes are considered in the classification models. Dimensionality reduction techniques can be used in this step of the methodology. This would allow asking as few indirect questions as possible, but still being able to predict risk preference with a high accuracy.

Acknowledgements The authors thank Sabancı University (SU) alumni Levent Bora, Kıvanc Kılınç, Onur Özcan, Feyyaz Etiz for their work on earlier phases of the study, and students Serpil Çetin and Nazlı Ceylan Ersöz for collecting the data for the case study. The authors also thank SU students Gizem Gürdeniz, Havva Gözde Ekşioğlu and Dicle Ceylan for their assistance. This chapter is dedicated to the memory of Mr. Turgut Uzer, a leading industrial engineer in Turkey, who passed away in February 2011. Mr. Turgut Uzer inspired the authors greatly with his vision, unmatched know-how, and dedication to the advancement of decision sciences.

Appendix A: Selected Survey Questions

Following are selected direct (risk-related) questions from the survey of the case study, which constitute the corresponding direct (risk-related) attributes.

Q34 Over the long term, typically, investments which are more volatile (i.e., that tend to fluctuate more in value) have greater potential for return (Stocks, for example, have high volatility; whereas government bonds have low volatility). Given this trade-off, what would be the level of volatility you would prefer for your investment?

a Less than 3%
b 3% to 5%
c 5% to 7%
d 7% to 13%
e More than 13%

Q32 What is your most important investment priority?

a I aim to protect my capital; I cannot stand losing money.

b I am OK with small growth; I cannot take much risk.
c I aim for an investment that delivers the market return rate.
d I want higher than market return; I am OK with volatility.
e Return is the most important for me. I am ready to take high risk for high return.

Q23 Compared to others, how do you rate your willingness to take risk?

a Very low
b Low
c Average
d High
e Very high

Q42 What is your most preferred investment strategy?

a I want my investments to be secure. I also need my investments to provide me
   with modest income now, or to fund a large expense within the next few years.
b I want my investments to grow and I am less concerned about income. I am comfortable with moderate market fluctuations.
c I am more interested in having my investments grow over the long-term. I am
   comfortable with short-term return volatility.
d I want long-term aggressive growth and I am willing to accept significant short-
   term market fluctuations.

Appendix B: ScoringAlgorithm

Following is the mathematical presentation of the developed scoring algorithm:

Sets

$\mathcal{I}$: set of respondents (observations, rows) in the sample; $i = 1, \ldots, I$
$\mathcal{J}$: set of attributes (questions, columns); $j = 1, \ldots, J$
$\mathcal{V}$: set of ordinal values for each attribute; $v = 1, \ldots, V$. For the presented case
    study, $\mathcal{V} = (a, b, c, d, e)$, where $a \leq b \leq c \leq d \leq e$

Inputs

$\mathbf{O} = [o_{ij}]_{I \times J}$: matrix of ordinal values of all attributes for all respondents
m_j: number of possible ordinal values for attribute j; $m_j \leq 5$ in this study

Internal Variables

$\mathbf{A} = [a_{ij}]_{I \times J}$: matrix of numerical (nominal) values of all attributes for all respon-
    dents
y_i: temporary adjusted risk score for respondent i, to be used in regression

Parameters

E: threshold on absolute percentage error (falling below this value will terminate the algorithm)

α: threshold for type-1 error (probability of rejecting a hypothesis when the hypothesis is in fact true)

M: a very large number

$\mathbf{B}$: transformation matrix for converting the ordinal input value matrix $\mathbf{O}$ into the numerical (nominal) value matrix $\mathbf{A}$

Outputs

z_j: whether attribute j is to be included in computing the risk score; $z_j \in \{0, 1\}$

w_j: weight for attribute j; $w_j \geq 0$

β_{0j}: intercept value for attribute j

β_{1j}: slope value for attribute j

Γ_j: sign multiplier for attribute j; $\Gamma_j \in \{-1, 1\}$

x_i: risk score for respondent i

Functions

$f(v, n) : (\mathcal{V}, \{2, \ldots, V\}) \to [0, 3]$: mapping function for an attribute with n possible values, that transforms the ordinal value v collected for that attribute to a nominal value $b_{v,n-1}$.

$$f(v, n) = b_{v,n-1}$$

where, for $V = 5$,

$$\mathbf{B} = [b_{vn}]_{V \times (V-1)} = \begin{bmatrix} 0.00 & 0.00 & 0.00 & 0.00 \\ 3.00 & 1.50 & 1.00 & 0.75 \\ \cdot & 3.00 & 2.00 & 1.50 \\ \cdot & \cdot & 3.00 & 2.25 \\ \cdot & \cdot & \cdot & 3.00 \end{bmatrix}$$

regression$(\mathbf{y}, \mathbf{a}')$

solve regression model $\mathbf{y} = \beta_0 + \beta_1 \mathbf{a}' + \varepsilon$ for vectors $\mathbf{y}$ and $\mathbf{a}'$

return (p, β_0, β_1), where p is the p-value for the regression model

preprocess()

// transform ordinal attribute values to nominal values

$a_{ij} = f(o_{ij}, m_j); \forall (i, j) \in \mathcal{I} \times \mathcal{J}$

Iteration-Related Notation

k: iteration count

N: number of attributes included in risk score computations at a given iteration

W: sum of weights for attributes

ε_k: absolute error at a given iteration k

e_k: absolute percentage error at a given iteration k

$\bar{e}_k$: average absolute percentage error at a given iteration k

ScoringAlgorithm(**O**, m_j)

BEGIN
// perform pre-processing to transform ordinal data to nominal data
preprocess()
// initialization:
// initially, all attributes are included in scoring,
// with unit weight of 1 and sign multiplier of 1.
// all of the regression intercepts are 0.
$z_j = 1, w_j = 1, \Gamma_j = 1, \beta_{0j} = 0; \forall j \in \mathcal{J}$
$N = \sum_j z_j$
// begin with iteration count of 1
$k = 1$
Begin_Iteration
// standardize the weights, so that their sum W will equal to N
$W = \sum_j w_j z_j$
$w_j \leftarrow (N w_j)/W; \forall j \in \mathcal{J}$
// compute the average of the intercepts
$\bar{\beta}_{0.} = (\sum_j \beta_{0j} z_j)/N$
// compute/update the risk scores at iteration k,
// which is composed of the average intercept value
// and the sum of weighted values for attributes
$x_{ik} = \bar{\beta}_{0.} + \sum_j \Gamma_j w_j a_{ij}; \forall i \in \mathcal{I}$
// compute total absolute error
$\varepsilon_k = \sum_i |x_{ik} - x_{i,k-1}|$
// correction for the initial error values
if $k = 1$ **then**
    $\varepsilon_0 = \varepsilon_1$
// termination condition
if $\varepsilon_k = 0$ **then**
    go to *Iterations_Completed*
// compute absolute percentage error,
// and then its average over the last two iterations
$\bar{x}_{.k} = \sum_i x_{ik}/I$
$e_k = 100\varepsilon_k/\bar{x}_{.k}$
$\bar{e}_k = (e_k + e_{k-1})/2$
// if the stopping criterion is satisfied, terminate the algorithm
if $\bar{e}_k < E$ **then**
    go to *Iterations_Completed*
// otherwise, continue with the regression modeling for each attribute j,
// and then go to next iteration
$\forall j \in \mathcal{J}$
    // if the attribute is included in the risk score calculation

if $z_j = 1$ **then**
   // first remove the attribute value from the incumbent score
   // to eliminate its effect
   $y_i = x_{ik} - a_{ij}$; $\forall i \in \mathcal{I}$
   // then define the vectors for the regression model of that attribute
   $\mathbf{y} = (y_i)$; $\mathbf{a}' = (\Gamma_j a_{.j})$
   $(p, \beta_0, \beta_1) = regression(\mathbf{y}, \mathbf{a}')$
   // if the regression yields a high p value
   // that is greater than the type-1 error,
   // this means that attribute j does not contribute significantly
   // to the risk scores
   if $p > \alpha$ **then**
      // and the attribute should not be included in risk calculations
      $z_j = 0$
   else
      // else it will be included (will just keep its default value)
      $z_j = 1$
      // and weight for the attribute will be the slope value
      // obtained from the regression
      $w_j = \beta_1$
      // the sign of the slope is important;
      // if it is negative, this should be noted
      if $\beta_1 < 0$ **then**
         // record the sign change in the sign multiplier
         $\Gamma_j = -1$
      else
         $\Gamma_j = 1$
// advance the iteration count and begin the next iteration
$k++$
go to *Begin_Iteration*
Iterations_Completed
$x_i = x_{ik}$
return x_i, z_j, w_j, Γ_j, β_{0j}
END

References

1. Ahn, H., Kim, K., Han, I.: Hybrid genetic algorithms and case-based reasoning systems for customer classification. Expert Syst. **23**(3), 127–144 (2006)
2. Ashby, W.R.: Principles of the self-organizing system. In: Principles of Self-organization, pp. 255–278 (1962)
3. Aven, T., Renn, O.: Risk Management and Governance: Concepts, Guidelines and Applications. Springer, Berlin (2010)
4. Barberis, N., Thaler, R.H.: A survey of behavioral finance. In: Constantinides, G.M., Harris, M., Stulz, R.M. (eds.) Handbook of the Economics of Finance, vol. 1, Part 1, pp. 1053–1128. Amsterdam, Elsevier (2003)

5. Cao, L.: Behavior informatics and analytics: Let behavior talk. In: ICDMW'08. IEEE International Conference on Data Mining Workshops, pp. 87–96 (2008)
6. Cao, L.: In-depth behavior understanding and use: the behavior informatics approach. Inf. Sci. **180**, 3067–3085 (2010)
7. Chen, F.L., Li, F.C.: Combination of feature selection approaches with SVM in credit scoring. Expert Syst. Appl. **37**(7), 4902–4909 (2010)
8. Chien, C.F., Chen, L.F.: Data mining to improve personnel selection and enhance human capital: A case study in high-technology industry. Expert Syst. Appl. **34**(1), 280–290 (2008)
9. Clarke, B., Fokoué, E., Zhang, H.H.: Principles and Theory for Data Mining and Machine Learning. Springer, Berlin (2009)
10. Daniel, C., Wood, F.S.: Fitting Functions to Data. Wiley, New York (1980)
11. Ertek, G., Kaya, M., Kefeli, C., Onur, C., Uzer, K.: Supplementary document for "Scoring and Predicting Risk Preferences". Available online under http://people.sabanciuniv.edu/ertekg/papers/supp/03.pdf (2011)
12. Galindo, J., Tamayo, P.: Credit risk assessment using statistical and machine learning: Basic methodology and risk modeling applications. Comput. Econ. **15**(1), 107–143 (2000)
13. Giarratano, J.C., Riley, G.: Expert Systems: Principles and Programming. Brooks/Cole, Pacific Grove (1989)
14. Grable, J.E.: Financial risk tolerance and additional factors that affect risk taking in everyday. J. Bus. Psychol. **14**(4), 625–630 (2000)
15. Grable, J.E., Lytton, R.H.: Investor risk tolerance: Testing the efficacy of demographics as differentiating and classifying factors. Financ. Couns. Plan. **9**(1), 61–74 (1998)
16. Guyon, I., Elisseeff, A.: An introduction to variable and feature selection. J. Mach. Learn. Res. **3**, 1157–1182 (2003)
17. Hallahan, T.A., Faff, R.W., Mckenzie, M.D.: An empirical investigation of personal financial risk tolerance. Financial Services Review **13**(1), 57–78 (2004)
18. Harrison, G.W., Lau, M.I., Rutstrom, E.E.: Estimating risk attitudes in Denmark: A field experiment. Scand. J. Econ. **109**(2), 341–368 (2007)
19. Hsieh, N.C.: An integrated data mining and behavioral scoring model for analyzing bank customers. Expert Syst. Appl. **27**(4), 623–633 (2004)
20. Huang, C.L., Chen, M.C., Wang, C.J.: Credit scoring with a data mining approach based on support vector machines. Expert Syst. Appl. **33**(4), 847–856 (2007)
21. Kim, J.K., Song, H.S., Kim, T.S., Kim, H.K.: Detecting the change of customer behavior based on decision tree analysis. Expert Syst. **22**(4), 193–205 (2005)
22. Kim, W., Choi, B.J., Hong, E.K., Kim, S.K., Lee, D.: A taxonomy of dirty data. Data Min. Knowl. Discov. **7**(1), 81–99 (2003)
23. Koh, H.C., Wei, C.T., Chwee, P.G.: A two-step method to construct credit scoring models with data mining techniques. Int. J. Bus. Inf. **1**(1), 96–118 (2006)
24. Kuykendall, L.: The data-mining toolbox. Credit Card Manag. **12**(7) (1999)
25. Lessmann, S., Voß, S.: Supervised classification for decision support in customer relationship management. In: Bortfeldt, A. (ed.) Intelligent Decision Support, p. 231 (2008)
26. MathWorks: Matlab, http://www.mathworks.com (2011)
27. Palma, A., Picard, N.: Evaluation of MiFID questionnaires in France. Technical report, AMF (2010)
28. Patt, A., Peterson, N., Carter, M., Velez, M., Hess, U., Suarez, P.: Making index insurance attractive to farmers. Mitig. Adapt. Strategies Glob. Chang. **14**(8), 737–753 (2009)
29. Quinlan, J.R.: Induction of decision trees. Mach. Learn. **1**(1), 81–106 (1986)
30. Shapiro, S.S., Wilk, M.B.: An analysis of variance test for normality (complete samples). Biometrika **52**(3/4), 591–611 (1965)
31. Sreekantha, D.K., Kulkarni, R.V.: Expert system design for credit risk evaluation using neurofuzzy logic. Expert Syst., (2010). doi:10.1111/j.1468-0394.2010.00562.x
32. Sung, J., Hanna, S.: Factors related to risk tolerance. Financ. Couns. Plan. **7**, 11–20 (1996)
33. The R Foundation for Statistical Computing: R Project, http://www.r-project.org (2011)

34. Thomas, L.C.: A survey of credit and behavioural scoring: forecasting financial risk of lending to consumers. Int. J. Forecast. **16**(2), 149–172 (2000)
35. Tsai, C.F., Wu, J.W.: Using neural network ensembles for bankruptcy prediction and credit scoring. Expert Syst. Appl. **34**(4), 2639–2649 (2008)
36. University of Ljubljana, Bioinformatics Laboratory: Orange, http://orange.biolab.si/ (2011)
37. Wagner, W.P., Najdawi, M.K., Chung, Q.B.: Selection of knowledge acquisition techniques based upon the problem domain characteristics of production and operations management expert systems. Expert Syst. **18**(2), 76–87 (2001)
38. Wang, X.T., Kruger, D.J., Wilke, A.: Life history variables and risk-taking propensity. Evol. Hum. Behav. **30**(2), 77–84 (2009)
39. Xu, R., Wunsch, D.: Survey of clustering algorithms. IEEE Trans. Neural Netw. **16**(3), 645–678 (2005)
40. Zakrzewska, D., Murlewski, J.: Clustering algorithms for bank customer segmentation (2005)

Chapter 10
An Introduction to Prognostic Search

Nithin Kumar M and Vasudeva Varma

Abstract Implicit relevance feedback has received wide attention recently, as a means to capture the search context in improving search accuracy. However, implicit feedback is usually not available to public or even research communities at large for reasons like being a potential threat to privacy of web users. This makes it difficult to experiment and evaluate web search related research and especially web search personalization algorithms. Given these problems, we are motivated towards an artificial way of creating user relevance feedback, based on insights from query log analysis. We call this simulated feedback. We believe that simulated feedback can be immensely beneficial to web search engine and personalization research communities by greatly reducing efforts involved in collecting user feedback. The benefits from "Simulated feedback" are—It is easy to obtain and also the process of obtaining the feedback data is repeatable and customizable. In this chapter, we describe a simple yet effective approach for creating simulated feedback. We evaluated our system using clickthrough data of a set of real world users and achieved 65% accuracy in generating click-through data of those users.

10.1 Motivation

Personalization of search has been a growing topic of interest for a while, but has stayed out of the radar for most people until recently. There are two ways to personalize search: (i) Personalization through implicit relevance feedback and (ii) Personalization through explicit relevance feedback. Explicit relevance feedback is collected by asking the users explicitly to mark their preferences. But this method is very expensive and it is highly impractical to collect large amounts of data [1]. In general, users do not prefer to put the extra amount of effort needed to give their preferences to the search engine [13].

N. Kumar M (✉) · V. Varma
Search and Information Extraction lab, International Institute of Information Technology,
Hyderabad, India
e-mail: nithin_m@research.iiit.ac.in

V. Varma
e-mail: vv@iiit.ac.in

L. Cao, P.S. Yu (eds.), *Behavior Computing*,
DOI 10.1007/978-1-4471-2969-1_10, © Springer-Verlag London 2012

Methods involving implicit relevance feedback consisting of user actions [3] like click-through data, dwelling time, text selection on a page etc have been proposed of late [2]. It is very easy and feasible to collect large amount of implicit feedback. Hence, using implicit relevance feedback has become very popular and widely accepted for personalization of search.

A simple implicit relevance feedback would be the search engine query logs. Query logs can be very useful in personalizing search results and improving ranking algorithms [8, 16]. A user's intentions and preferences can be mined from his query logs. However, such data is not available for public use for reasons like the privacy concerns of the users involved. Thus unavailability of relevance feedback has stood as a major roadblock for research in personalization.

In addition, personalization through implicit feedback has been plagued by other problems viz., in web-search domain, ranking algorithms change rapidly and hence the results keep changing constantly for a given query. Consequently, click-through data collected may become stale very quickly and hence unfit for evaluation purposes. And also due to the dynamic nature of the web, the contents of a particular URL tend to change rapidly. This might bring up anomalies in the click-through data.

In this chapter, we address the above problems by generating artificial relevance feedback using prognostic search techniques. *Prognostic search* is a process of simulating user's search process and emulating their actions, through the preferences captured in their profile. Such generated feedback can be used for research in personalization techniques and analyzing personalization algorithms and search ranking functions [7]. The main advantage with this system is that we can create data on the fly and hence not fear of it becoming stale. Since it does not involve user's actions, it is feasible to generate large amounts of data in this way.

10.2 A Brief Overview of Existing Techniques

Implicit feedback methods for personalization of search and modeling the user have received wide attention in the recent past. In [12], they explore the effectiveness of implicit feedback in improving the web search. They performed experiments which mined out the association between explicit and implicit relevance feedback. They found out that the best models for individual pages combined click-through and time spent on the search results page which can be categorised as implicit feedback.

Ample work has been done in building user profiles of a search user to personalize the search results. References [14, 15] propose novel methods of user modeling for automatic identification of user interests and preferences. Reference [4] proposes topic sensitive page ranking methods to re-rank the search results based on the user's click history. Thus, implicit feedback techniques have always stood effective in user modeling and personalizing search.

Implicit feedback techniques have also been used for query expansion in order to personalize search [7]. But to our knowledge, not much work has been done in generating artificial relevance feedback through web search simulations. In [11], they

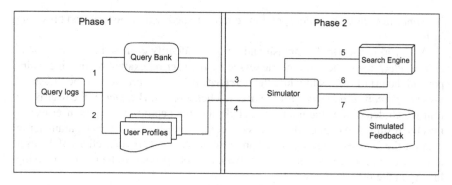

Fig. 10.1 System architecture

show that clickthrough data and other implicit data of a user can be used to build user models to effectively personalize the search results. White et al. [12] created searcher simulations for evaluating implicit feedback models. Assuming that the relevant and irrelevant documents for a query are given, it generates the searcher simulations. The system creates simulation of searches following the relevance paths and different strategies in generating the clicks were experimented. Their research concentrates mainly on the result selection part and does not discuss about the role of the user. But search process is incomplete without discussing or characterizing the user participating in the search. In this paper, we propose a method to create simulated feedback which would be helpful in developing and testing personalization algorithms.

10.3 Our Approach to Generate Simulated Feedback

Simulated feedback is created by observing and analyzing real world search log data. We propose a two phase process to create artificial relevance feedback as follows: In phase 1, we process real world click-through data of search users and build their profiles. In phase 2, we simulate the user's search process and emulate their actions based on their preference stored in their profile. This process is called "*Prognostic Search*". We explain each of the above steps in the later part of this chapter.

10.3.1 Creating Profiles

After closely examining and analyzing the semantics of the query log, we have chosen the following parameters to characterize a user: an anonymous user-id, *perceived relevance threshold, patience,* previous queries issued, search history, and browsing history.

A user-id is used to distinguish and uniquely identify each and every user. *Perceived relevance* is the relevance assessed by the user on examining the title, snippet and the url of the result while the *perceived relevance threshold* is the threshold value of *perceived relevance* for the user to click a result. *Patience* of the user is the trait which determines the number of clicks and the depth to which the user examines the results. We explain the process of computing a user's patience parameter in detail in the later sections. We stored previous queries and previous clicks of the user to capture the preferences of the user. We have not introduced the use of browsing history in the present paper.

We used the search history of the user like the previous queries issued by the user and the previous results clicked by the user. We store the titles and snippets of those results to capture the interests of the user. Here, our aim is to generate artificial implicit relevance feedback which is very close to the real world data. To generate artificial relevance feedback, we instantiate these parameters with appropriate values using real world data. However, in an experimental setup, these parameters can be customized and accordingly data can be generated based on the needs of the personalization algorithm that is being tested.

10.3.2 Prognostic Search

Prognostic search is simulation of a user's search process and emulating their actions based on their interests and preferences captured in their user profile. Simulating search process involves four steps viz., (a) Query formulation, (b) Searching (c) Browsing the results and (d) Generating Clicks. Each of these processes are explained below.

10.3.2.1 Query Formulation

Query formulation involves cognitive process of the user and requires background knowledge about the user like their interests, preferences and the knowledge they have in different areas. It is highly impossible to capture the cognitive thought process of a user and emulate their method of generating a query. To solve this problem, we randomly select a search session from the user's history and send the queries in it sequentially to the search engine. This helps us to preserve the inter query relations that naturally exist between the subsequent queries in a session.

10.3.2.2 Searching

This step involves retrieving documents relevant to the query generated in the previous step. We used yahoo search engine which is very much similar to the search engine from which the test data is collected.

Algorithm 10.1 User browsing model in real world

Step1: Start browsing with the top-most result.
Step2: Examine title, snippet and URL of the result.
Step3: Click if the result looks promising.
Step4: If(user has patience) go to step 5, else go to step 6.
Step5: Select next result and go to step 2.
Step6: Start examining the clicked results.
Step7: If(information need satisfied) end the process, else go to step 8.
Step8: Reformulate the query and go to step 1.

Algorithm 10.2 Simulated User browsing model

Step 1: Determine the number of results to be browsed.
Step 2: Browse the results in increasing order of their ranks and examine them.
Step 3: Compute the perceived relevance score of the results.
Step 4: In the same order, generate clicks based on the perceived relevance scores of the results.
Step 5: If(session has more queries) go to step 6, else end the process.
Step 6: Select next query in the session and go to step 1.

10.3.2.3 Browsing Results

In this step, we simulate the manner in which a user browses the results in the real world. Based on the observations in [6, 9], we assume that the user in the real world follows the browsing model explained in Algorithm 10.1. In real world, a user may follow any other browsing model, but presently we have considered this browsing model to simplify things.

Accordingly, to simulate the browsing process of the user explained in 10.1, we followed the Algorithm 10.2.

Thus based on the *patience* parameter, we determine the number of results that the user browses. After a deep statistical analysis of the query log parameters, we learned the *patience* value of a user can be characterized by the following parameters: number of clicks per session, maximum page rank clicked in a session, time spent in a session and the number of queries issued per second. We found out that the patience of the user is directly proportional to the maximum page rank of the result he has clicked in a session. We also found out that the number of clicks a user generates is inversely proportional to the number of queries issued per minute and directly proportional to the amount of time he spends per session. Thus, a user with more *patience* tends to examine more search results and thus generate more clicks based on their relevance. We explain these dependencies in detail in the experiments section. So in order to learn the Patience parameter of the user, we devised the following formula:

$$Patience = \alpha \times \frac{(MPR \times T \times C)}{Q} \tag{10.1}$$

Here MPR denotes the average of maximum page rank of the results clicked by the user per session, T denotes the average time spent per session, C is the average number of clicks per session and Q denotes the average number of queries issued per session. "α" is an equalization constant.

10.3.2.4 Generating Clicks

This is the most important step in our simulation process. Typically, a user observes the visual information viz., title, snippet and the URL of a result [10]. Then based on their interests, they choose the results relevant to them. Similarly, we closely examine the results selected in the previous step and then score them according to their relevance to the user. We consider the title, snippet and the page-rank of the result and determine its relevance to the user known as *perceived relevance*.

We first compute the semantic distance between the title and snippet of the present result from the titles and snippets of previously clicked results of the user. Here, our idea is to calculate the nearness of the present result to the previous results that the user clicked. So, we calculate the semantic distance between them. But, sometimes it may so happen that the user has not searched for similar things previously but the present document will fully satisfy his need. In such case, if we compare it with the search history of the user, it may lead to lowering the perceived relevance score of that result which is not desired. So, it is highly important to know when to personalize the results. To avoid such undesired situations, we compute the semantic distance of the present query to the queries he issued in past. We use this distance (β) to gauge the influence of the user's previous clicks in the present context.

We used latent semantic analysis (LSA) to compute the semantic distance between the results. LSA does not take the dictionary meaning of the words as input; it rather extracts the contextual meaning of the word with respect to all other words in semantic space [5]. This property of LSA is very much useful in the present context. A particular word may have a lot of meanings but we are concerned about only those meanings of the word which the user interprets. And those meanings of the word are captured in the sentences present in the user's click history. Hence, we used LSA to compute the semantic distance between the results.

We also consider the page-rank of the result which has proven to be an important factor in taking the decision of a click. In our study, we found that for about 89% of the queries with clicks, the top ranked document has been clicked and for 56% of the queries second ranked document has been clicked. In Fig. 10.2, we show the click ratio for each of the top ten ranked documents.[1] Thereby, we derive that the

---

[1]In Fig. 10.2, we have normalized the clicks statistics with the number of clicks for top ranked document. So, the click-ratio for the top ranked document will be 1.

Fig. 10.2 Ranks vs clicks-ratio

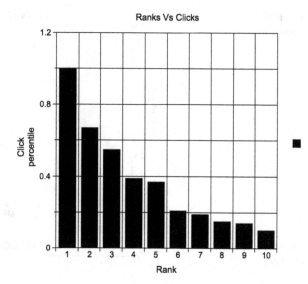

page rank of the result is also a very important factor in deciding whether a result has to be clicked or not. We also consider the distance of the present result from the previous click of the user. In [10], it is shown that the user is more biased to click the result that immediately follows the result he previously clicked. In our simulation process, if this distance for any result exceeds 10, then we terminate the browsing process and reformulate the query.

Considering the above observations, we formulated *Perceived relevance* score of the result as follows:

> Perceived relevance of the result
>
> $= W_1 \times \beta \times$ page-rank of the result
>
> $+ (1 - \beta) \times W_2 \times$ *relevance score* of the result
>
> $+ W_3 \times$ distance from previous click $\qquad$ (10.2)

Here $'\beta'$ is determined by the relevance of the present query to the previous queries issued by the user. W_1, W_2, W_3 are normalization constants.

The factor 'β' leverages the impact of the relevance score of the result in the final perceived relevance score. Thus *Perceived relevance* of a result is calculated. Then, we compare this score with the *Perceived relevance threshold* of the user and generate the clicks accordingly. *Perceived relevance threshold* of a user is calculated by averaging the relevance score of the results previously clicked by the user. We used 30% of training data to determine the *Perceived relevance threshold* of a user while the rest of the data is used to build profile.

$$\text{Perceived relevance threshold} = \frac{\sum_{i=0}^{n} PR_i}{n} \qquad (10.3)$$

Fig. 10.3 MPR vs
queries/min

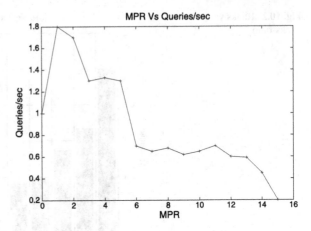

Fig. 10.4 Clicks vs
queries/min

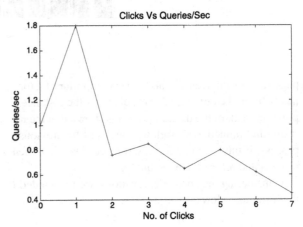

Here PR_i denotes the *Perceived relevance* score of ith clicked result and "n" is
the total number of clicked results.

10.4 Experiments

Query logs are a valuable source for user information. We have done a deep statistical analysis on query logs to determine the influence of different parameters
of query logs on our user model variables. In our analysis, we have found that the
page-rank of a result can highly influence the user to make a click which can be
seen in Fig. 10.2. Previously, we mentioned that the patience of a user depends on
the number of queries he issues per second in a session. In Figs. 10.3 and 10.4, it can
be clearly seen that the *Patience* of the user is inversely proportional to the user's
number of Queries/min. These graphs show the influence of the factor Queries/min
on the number of clicks the user generates for a query and the maximum page rank
clicked by the user in a session. Since maximum page rank clicked and the num-

Table 10.1 System configurations

System	Patience	Clicks
System1	Random	Random
System2	Random	Proposed method
System3	Proposed method	Proposed method

ber of clicks per session directly affect the *Patience* parameter, we can say that Queries/min is inversely proportional to the *Patience* of the user.

Both the graphs show occasional phases of increasing behaviour which can be attributed to a variety of reasons. While plotting the graphs, for a given value of MPR/number of clicks, we take observations from numerous sessions of the user and average the queries/min value. Thus, presence of some outlier values may affect the overall output of the graph. It can also be attributed to the low quality of results that the search engine might have returned due to various reasons.

10.5 Evaluation

In this section, we present the evaluation procedure of our approach. We first collected query log data of 60 users using a browser plug-in for around two months. We split the data into two parts: first part consisting of the latest query log data (data of duration one week) and the second part containing the rest. We built profiles of the searchers using the second part of the data. Then, we simulated the prognostic search process of these users and generated the clicks using the queries in the first part of the data. Based on our empirical analysis, we set the values of W_1, W_2 and W_3 as -0.1, 10, -0.1 respectively, to normalize the influence of each of those factors. We used the clicks of the corresponding queries and measured the accuracy of our system as follows:

$$\text{Accuracy} = \frac{\text{No. of simulated clicks in the user clicked pool of results}}{\text{Total no. of clicks generated by the system}} \quad (10.4)$$

We found our system to be 65% accurate. We also built two more systems which we considered as the baseline systems. The first system gives a random value for the patience value of the user—random value is used to determine the number of documents to be browsed during the prognostic search process—and random value is given for the user's *Perceived relevance threshold* parameter. The second system generates the patience value of the user according to the process described by us in Sect. 10.3.2.3 and gives a random value for the *Perceived relevance threshold* value of the user. And the third system is the one described in this paper. Systems built by us can be summarized as shown in Table 10.1.

Figure 10.5 shows a comparison of the accuracies of the three systems. Here, we can see that the baseline 1 which uses random values for *patience* and *generating clicks* is only 10% accurate in generating clickthrough data. However, with the addition of our *generating clicks* approach to the baseline 1, the performance increased

Fig. 10.5 Results
comparison

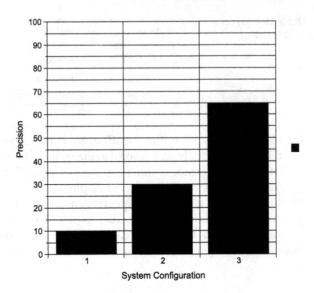

by 200%. And the system 3 which uses our proposed models for both *patience* and *generating clicks* generates 65% accurate data which is a 550% improvement over the baseline 1.

10.6 Conclusion and Future Work

In this chapter, we proposed Simulated Feedback based on insights from query logs and using artificial methods to generate feedback. There is a lot of scope for interesting future directions to the current work. It would be an interesting experiment to see the use of the simulated feedback in evaluation of personalized search algorithms. Consider a personalized search algorithm, and use it to learn a user model from existing explicit/implicit feedback data. Learn a user model using the same algorithm from simulated feedback and compare the results. We plan to pursue the same in future.

As an extension to the current work, we aim to improve the web search process especially the query formulation step with insights from a user study. We are working towards incorporating much richer and complex models for query formulation like HMMs etc. Ability of the system to automatically create query reformulations of the original when no clicks are found is another interesting future work. We also plan to dig more information about the user by analysing the query log data. For example, the difference in the time between the clicks and the distance between the clicks can be used to analyze the browsing behaviour of the user. These observations can inturn be used in generation of simulated feedback thus reducing its gap with real world implicit feedback.

References

1. Belkin, N.J.: Reading time, scrolling and interaction: Exploring implicit sources of user preferences for relevance feedback during interactive information retrieval. In: Proceedings of the 24th Annual International Conference on Research and Development in Information Retrieval, SIGIR, pp. 408–409 (2001)
2. Brown, D.: Implicit interest indicators. In Intelligent User Interfaces, pp. 33–40 (2001)
3. Cao, L.: In-depth behavior understanding and use: the behavior informatics approach. Inf. Sci. **180**, 3067–3085 (2010)
4. Cho, J.: Automatic identification of user interest for personalized search. In: Proceedings of WWW (2006)
5. Dennis, S.: Handbook of Latent Semantic Analysis. Lawrence Erlbaum Associates, Mahwah (2007)
6. Gay, G.: Eyetracking analysis of user behavior in www search. In: Conference on Research and Development in Information Retrieval, SIGIR (2004)
7. Harman, D.: Towards interactive query expansion. In: The 11th Annual ACM SIGIR Conference on Research and Development in Information Retrieval, pp. 321–331 (1988)
8. Joachims, T.: Optimizing search engines using clickthrough data. In: Proceedings of the Eighth ACM SIGKDD International Conference on Knowledge Discovery and Data Mining, pp. 133–142 (2002)
9. Joachims, T.: Evaluating the robustness of learning from implicit feedback. In: ICML Workshop on Learning in Web Search (2005)
10. Pan, B.: Accurately interpreting clickthrough data as implicit feedback. In: Proceedings of 28th Conference on Research and Development in Information Retrieval, SIGIR (2005)
11. Ragno, R.: Learning user interaction models for predicting web search result preferences. In: Proceedings of 29th Conference on Research and Development in Information Retrieval, SIGIR, pp. 3–10 (2006)
12. van Rijsbergen, C.J.: Evaluating implicit feedback models using searcher simulations. In: ACM Transactions on Information Systems, ACM TOIS, pp. 325–361 (2005)
13. Rocchio, J.J.: The smart retrieval system experiments in automatic document processing (1999)
14. Tan, B.: Implicit user modeling for personalized search. ACM Trans. Inform. Sys. (2005)
15. White, T.: Evaluating implicit measures to improve web search. ACM Trans. Inf. Sys. **23**, 147–168 (2005)
16. Yoshikawa, M.: Adaptive web search based on user profile constructed without any effort from users. In: Proceedings of WWW, pp. 675–684 (2004)

Part III
Behavior Mining

Chapter 11
Clustering Clues of Trajectories for Discovering Frequent Movement Behaviors

Chih-Chieh Hung, Ling-Yin Wei, and Wen-Chih Peng

Abstract In this chapter, we present a new trajectory pattern mining framework, namely, *Clustering Clues of Trajectories (CCT)*, for discovering *trajectory routes* that represent frequent movement behaviors of a user. In addition to spatial and temporal biases, we observe that trajectories contain *silent durations*, i.e., the time durations when no data points are available to describe movements of users, which bring many challenge issues in clustering trajectories. We claim that a movement behavior would leave some *clues* in its various sampled/observed trajectories. These clues may be extracted from spatially and temporally co-located data points from the observed trajectories. Based on this observation, we propose *clue-aware trajectory similarity* to measure the clues between two trajectories. Accordingly, we further propose the *clue-aware trajectory clustering* algorithm to cluster similar trajectories into groups to capture the movement behaviors of the user. We validate our ideas and evaluate the proposed CCT framework by experiments using both synthetic and real datasets. Experimental results show that CCT is more effective in discovering trajectory patterns than the state-of-the-art techniques in trajectory clustering.

11.1 Introduction

Owing to the pervasiveness of GPS-equipped mobile devices today, the locations of users can be easily determined and shared with friends via social networking websites, such as Google Latitude and Foursquare. Notice that movements of mobile users can be captured as trajectories. For example, time-stamped and geotagged photos shared by a user on Flickr may provide the trajectory of her photo-shooting trip. Moreover, some time-sorted check-in records of a user on Foursquare also form a

C.-C. Hung (✉) · L.-Y. Wei · W.-C. Peng
Department of Computer Science, National Chiao Tung University, Hsinchu, Taiwan
e-mail: hungcc@nctu.edu.tw

L.-Y. Wei
e-mail: lywei.cs95g@nctu.edu.tw

W.-C. Peng
e-mail: wcpeng@nctu.edu.tw

L. Cao, P.S. Yu (eds.), *Behavior Computing*,
DOI 10.1007/978-1-4471-2969-1_11, © Springer-Verlag London 2012

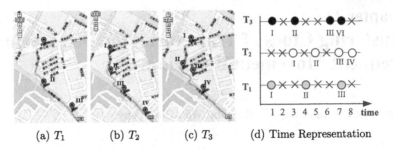

(a) T_1 (b) T_2 (c) T_3 (d) Time Representation

Fig. 11.1 An example of trajectories from CarWeb dataset

trajectory. Furthermore, a number of websites have been hosting the sharing of user-generated trajectory data [1–3]. Obviously, user trajectories provide very valuable information that can be useful for various location-based services such as trip planning, personalized navigation routing service, mobile commerce, and location-based recommendation services. In these application domains, techniques for mining trajectory patterns and frequent trajectory routes are very important. In this chapter, we study data mining techniques for discovering frequent trajectory patterns and routes of individual users.

Mining trajectory patterns and routes is an important task in behavior computing [4] and is very challenging due to inherent noises and limitation of trajectory acquisition technology. Generally speaking, a trajectory consists of sequential data points recording the locations and associated occurrence time sampled from the movement of a user. Given the logged historic trajectories of the user, we not only aim to identify the sequential relationships, also termed as the *movement behavior*, among regions where the user frequently passing by but also to construct detailed trajectory routes that represent these movement behaviors. Nevertheless, as reported in many prior works, trajectories exhibit certain spatial/temporal biases and temporal shifts. In other words, the locations and occurrence time of data points in two trajectories are usually not the same even if these two trajectories capture the same movement behavior of the user. We use Fig. 11.1 as an example to illustrate the spatial/temporal biases, temporal shifts, and a phenomenon called *silent duration* that frequently appears in trajectories. Figures 11.1(a)–(c) show three trajectories T_1, T_2 and T_3, respectively, collected from our CarWeb service. These three trajectories actually logged the same movement behavior of a user. Notice that the roman numbers associated with data points represents their sampling order in the corresponding trajectories. Figure 11.1(d) shows the timestamps of all data points (and their corresponding trajectories) in the time line. It can be seen in Fig. 11.1 that data points of T_1 and T_2 are not exactly aligned in either spatial and temporal dimensions even though these two trajectories capture the same movement behavior of the user. Compared to T_1 and T_3, T_2 is delayed for approximately 2 time slots. This is referred to as the temporal shift of trajectories in this chapter. Consider the data points collected at time slot 7 in T_1 and T_3. While these two trajectories do not have time shifts, there is a spatial bias since the location of data point III in Fig. 11.1(a) is clearly different from that of data point VI in Fig. 11.1(b). Additionally, trajectory

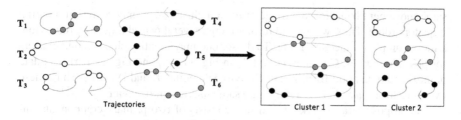

Fig. 11.2 Clustering and aggregating for deriving frequent trajectory routes

data may also exhibit the silent duration phenomenon that, to our best knowledge, has not been explicitly addressed previously in the literature. The silent duration denotes a time duration when there is no data point in presence due to data loss or sampling strategies employed in forming a trajectory. As shown in Fig. 11.1(d), data points are not available at all the time slots. Thus, detailed information between two data points of trajectories are missing. For example, the route information between data point I and data point II of trajectory T_1 shown in Fig. 11.1(a) is not available. Consequently, it is a challenge task to infer trajectory routes from trajectories with spatial and temporal bias, temporal shifts and silent durations.

As mentioned, given the trajectories of a user, we aim to determine frequent trajectory routes of this user in addition to his/her trajectory patterns. Consider again the three trajectories in Fig. 11.1 that capture the same movement behavior of a user. It's intuitively challenging to identify the actual routing path behind these three trajectories. Furthermore, a user usually has more than one movement behaviors hidden in the logged trajectories. An idea is to cluster the trajectories of a user into several groups. Each group of similar trajectories is supposed to represent one movement behavior of the user.

To illustrate the research tasks and issues faced to achieve our goal, we consider Fig. 11.2, where six trajectories representing two movement behaviors (i.e., circle and S-shape movements) are given. Our goal is to derive trajectory patterns and routes as shown in the right-hand side of Fig. 11.2. With the presence of silence durations in these trajectories, it is very difficult to determine whether two given trajectories are similar, i.e., capturing the same movement behavior, or not. Therefore, there is a need to define a new similarity measure for comparing a pair of trajectories with silent durations. Based on the similarity among pairs of trajectories, one could cluster similar trajectories into groups (as depicted in Fig. 11.2).

In this chapter, a new trajectory pattern mining framework, called *Clustering Clues of Trajectories (CCT)*, is presented for discovering frequent movement behaviors. As suggested by its name, the framework is built upon an important notion of *clues*. We observe that trajectories obtained by sampling the same movement behavior are likely to consist of some spatially and temporally close data points which capture certain common partial movement behavior of the user. These spatially and temporally close data points provides important clues hinting the movement behavior behind. Thus, trajectories with the close clues should be clustered together. However, even if we are able to cluster together the trajectories sampled from the same movement behavior, the trajectories in a cluster may only represent a partial trajectory route. Thus, we need to further merge several partial trajectory routes into a more complete trajectory route. To do so, one may identify clues among clusters for possible merges. Consider T_1 and T_2 in Fig. 11.1(a) and Fig. 11.1(b). The data point II and data point III in T_2 are close to data points II and IV in T_3. Thus, these two trajectories are likely to refer to the same movement behavior.

The proposed framework CCT which consists of two primary components, including (i) *clue-aware trajectory similarity (CATS)*, and (ii) *clue-aware trajectory clustering (CATC)*. Notice that CATS identifies clues among trajectories first and use them to measure the pair-wise similarity among trajectories. In light of CATS, CATC exploits clues between trajectories to group similar trajectories into clusters such that each cluster represents one movement behavior of the user. Specifically, given a set of trajectories with pair-wise similarity measures, a *clue-graph* is constructed to capture the similarity relationship among trajectory pairs. By deriving cliques in the clue-graph, CATC is able to obtain clusters of trajectories that have high similarity and further merge similar clusters into a larger cluster based on the notion of clues. Notice that the proposed trajectory pattern mining framework discovers trajectory patterns to capture the frequent movement behaviors of a user. Extensive experiments using both real and synthetic datasets are conducted for performance evaluation. We compare the proposed algorithms with existing clustering algorithms and trajectory pattern mining algorithms. The results show that CCT can discover trajectory patterns effectively, even if trajectories with silent durations do not fully capture the complete movement route of a user.

11.2 Related Works

The existing work on trajectory clustering could be classified in the following two categories: (i) clustering on the entire trajectories; and (ii) clustering on sub-trajectories. For the first category, a model-based approach for trajectory clustering is proposed in [5], which assumes that each trajectory is smooth such that each data point could be estimated by a probability density function. In this work, each cluster is assumed to have a density probability function to "generate" trajectories. Therefore, the probability that a trajectory belongs to some clusters can be modeled as the mixture of these density probability functions. Therefore, each trajectory is first represented by a regression mixture model, which is used to determine its cluster memberships based on maximum-likelihood principle. However, a major deficiency of this technique is that the probability density function of each cluster and the number of clusters need to be given a priori. In [9], a density-based clustering algorithm is adapt to trajectory data based on a simple notion of distance between trajectories. In this work, the Euclidean distance between location points at each time slot is calculated. Then, the average Euclidean distance over all time slots obtained. Accordingly, a density-based clustering algorithm, OPTICS, is used to cluster trajectories. Moreover, an empirical comparison with several traditional k-means and hierarchical algorithms showed that OPTICS is the most suitable clustering algorithm for clustering trajectories. An advantage of this work is that the number of clusters does not need to be specified in advance. However, data points at every time slot need to be available (or be well-approximated) to compute the proposed distance function. Thus, this approach may not work very well when silent durations appear in trajectories. For the second category, the state-of-art approach is TraClus [7]. The primary goal of this work is to discover common sub-trajectories from a set of trajectories. A key observation in this work is that clustering based on the whole trajectory could miss some common sub-trajectories, which are very useful in many applications, especially when regions of special interest are considered for analysis. Therefore, a partition-and-group framework for clustering sub-trajectories is proposed. This framework first decomposes a trajectory into a set of line segments, and then groups similar line segments together into a cluster. In our work, we cluster the whole trajectories instead of sub-trajectories. For comparison, we adapt TraClus to find common sub-trajectories among trajectories and then formulate the similarity of trajectories using these common sub-trajectories.

11.3 Clue-Aware Trajectory Similarity

In this section, we first analyze the unique characteristics of trajectories and then present our design of similarity measure. Finally, we discuss some interesting properties of the proposed clue-based similarity measure.

11.3.1 Characteristics of Trajectories

To cluster "similar" trajectories together, it is essential to formulate a similarity measure between two trajectories. Before we proceed to present the clue-based similarity measure, we summarize some characteristics of trajectories as follows:

- *Spatial and temporal bias*: In practice, trajectories are obtained by trajectory acquisition devices/schemes, which unfortunately may introduce spatial and temporal bias to data points of trajectories. For example, the position accuracy of GPS (Global-Position System) has inherent spatial bias. Moreover, the occurrence times of data points sampled from exactly the same movement behavior are not always the same. Consider a worker goes to his office from his home at 8:00 a.m. every day. Data points of trajectories recording this movement behavior usually do not have the same occurrence time. One reason is that the positioning device takes some time to determine the position. Thus, even if this user leaves his home at 8:00 a.m. every day, data points of trajectories have some temporal bias.
- *Temporal shifts*: Due to varied speeds and delays of users movements, trajectories may have temporal shifts. For example, although a user follows the same movement path to his office every day, some sub-trajectories have shifted occurrence time.

- *Noise*: Positioning devices can be easily affected by environment factors, such as buildings, shelters, and weathers. Hence, data points of trajectories usually have some *noises*.
- *Silent duration*: The length of a trajectory is mainly decided by the time and the sampling rate. For the same movement path, even if the same sampling rate is used, trajectories collected may still have different lengths due to environmental factors (e.g., the weather) and the limitation of position devices (i.e., the capability of positioning devices in computing and networking). In this chapter, a silent duration refers a time duration when there are no logged data points about user movements.

11.3.2 Design of the Clue-Aware Trajectory Similarity

Trajectories that capture the same movement behavior of a user are likely to have some "clues" referring those spatially and temporally co-located data points among trajectories. Due to that trajectories have silent durations, these spatially and temporally co-located data points should be carefully identified and utilized to infer trajectory routes. Thus, given two trajectories, the Clue-Aware Trajectory Similarity (CATS) aims to couple as many spatially and temporally co-located data points between two trajectories as possible. CATS overcomes the impacts resulted from the aforementioned characteristics of trajectories via a spatial decaying function, clue scores of data points, and a new mapping scheme.

To identify whether two trajectories comes from the same movement behavior, we could observe whether there are many spatially and temporally co-located data points between two trajectories. The concept of clues is to evaluate how many such co-located points exist between two trajectories. To achieve this goal, some technical issues should be dealt with. First, for any two points in the different trajectories, the closer two points are, the stronger clues they reveal. That is, the movement routes of two trajectories tend to pass the similar area. Moreover, two co-located points reveal some clues only if their occurrence times are close enough. Due to the nature of temporal shifting, two points may appear in the similar area in a certain time delay. Thus, we should tolerate such a temporal shifting when identifying the clues between two trajectories. To conclude, by tolerating certain temporal shifting, the strength of clues between two trajectories is decided by not only the number of spatial co-located points but also how close they are. Thus, each component of CATS in the following is designed to capture the strength of clues between two trajectories.

A trajectory T_i of a user is a time-ordered sequence of data points, expressed as $T_i = \langle p_{i,1}, p_{i,2}, \ldots, p_{i,n} \rangle$, where $p_{i,j} = (\ell_{i,j}, t_{i,j})$ represents the location of the user (i.e., $\ell_{i,j}$) at the time $t_{i,j}$, $t_{i,j} < t_{i,j+1}$ and n is the length of trajectory T_i. The location $\ell_{i,j}$ is usually a two-dimensional or three-dimensional data point. Since trajectories usually contain spatial bias, we first use a spatial decaying function to measure the degree of bias as follows.

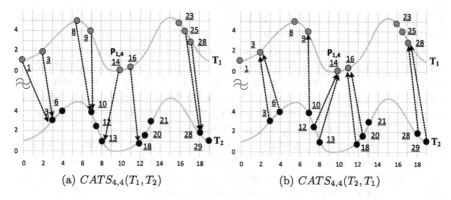

(a) $CATS_{4,4}(T_1, T_2)$ (b) $CATS_{4,4}(T_2, T_1)$

Fig. 11.3 An illustrative example for clue-aware similarity of T_1 and T_2

Definition 11.1 (Spatial Decaying Function) Given a spatial threshold ϵ, and two data points $p_{i,\ell} = (l_{i,\ell}, t_{i,\ell})$ and $p_{j,k} = (l_{j,k}, t_{j,k})$ from two trajectories (i.e., T_i and T_j), a spatial decaying function for two points $p_{i,\ell}$ and $p_{j,k}$ is defined as

$$f_\epsilon(p_{i,\ell}, p_{j,k}) = \begin{cases} 0, & \text{if } dist(p_{i,\ell}, p_{j,k}) > \epsilon \\ 1 - \frac{dist(p_{i,\ell}, p_{j,k})}{\epsilon}, & \text{otherwise} \end{cases}$$

where $dist(\cdot, \cdot)$ denotes Euclidean distance between two data points.

The value of spatial decaying function is ranged from 0 to 1. Obviously, the closer the two data points, the larger the value is. If the locations of two data points are exactly the same, the value is 1. On the other hand, if the distance between two points is greater than ϵ, the value is 0. For example, Fig. 11.3 shows two trajectories T_1 and T_2, where the underlying grey lines are actual movements and the circles represent the data points of T_1 and T_2. The underlined number of each data point is the occurrence time of this data point. Consider $\epsilon = 4$ and Euclidean distance as the distance function. Given two points $p_{1,2} = (2, 2, 3)$ and $p_{2,1} = (3, 3, 3)$, it can be derived that $f_4(p_{1,1}, p_{2,4}) = 1 - \frac{\sqrt{2}}{4} = 0.65$. On the other hand, given $p_{1,4} = (7, 4, 9)$ and $p_{2,1} = (3, 3, 3)$, we can derive that $f_4(p_{1,4}, p_{2,1}) = 0$ since $dist(p_{1,4}, p_{2,1}) > \epsilon = 4$.

In the spatial decaying function, a parameter ϵ is given to tolerate the spatial bias and shifting of data points.[1] Basically, the spatial decaying function performs a continuous space quantization (i.e., from 0 to 1), which reflects the closeness between two data points, in contrast to the discrete space quantization employed in LCSS and EDR (i.e., 0 or 1). For example, if the ϵ is set as 10 meters, consider two cases: (i) two points with a distance of 1 meter, and (ii) two points with a distance of 9 meters. As aforementioned, the closer two points are, the stronger clues they have since they are highly likely to co-locate at the nearby area. Therefore, in this case,

---

[1]Since the parameter ϵ will decide the size of hot regions in trajectory patterns, this parameter should be set according to application requirements (the desirable size of hot regions).

Case (i) reveals stronger clue than case (ii). LCSS and EDR do not distinguish these two cases since the distances in both cases are smaller than 10 meters.

According to the spatial and temporal information of data points, we give a score for data points with respect to a *reference trajectory*. For a data point, there are many possible way to evaluate the clues between this point and the reference trajectory. An obvious way is to map this data point to the closet point on the reference trajectory by tolerate some temporal shifting. That is, a data point is aligned to the point that can reveal the strongest clues in the reference trajectory. To realize this idea, we define the clue score of data points with respect to the reference trajectory. Specifically, given a data point $p_{i,\ell} \in T_i$ and a reference trajectory T_j, the clue score of data point $p_{i,\ell}$ with respect to trajectory T_j is used to identify the best mapping point on T_j for $p_{i,\ell}$ in spatial and temporal dimensionality.

Definition 11.2 (Clue Score of Data Points) Given a point $p_{i,\ell}$, a reference trajectory T_j, a spatial threshold ϵ, and a temporal threshold τ, the clue score of data point $p_{i,\ell}$ to trajectory T_j is defined as $score_{\epsilon,\tau}(p_{i,\ell}, T_j) = \max\{f_\epsilon(p_{i,\ell}, p_{j,k}) | p_{j,k} \in T_j$ and $t_{j,k} \in [t_{i,\ell} - \tau, t_{i,\ell} + \tau]\}$.

Figure 11.3 illustrates the computation of clue scores. Assume that $\epsilon = 4$, $\tau = 4$. Figure 11.3(a) shows two trajectories, T_1 and T_2 where the underlying grey line is the real movement. Consider a data point $p_{1,5}$ of T_1 as an example. The clue score of data point $p_{1,5}$ with respect to trajectory T_2 finds the best mapping data points of T_2 within a time interval between $14 - 4$ and $14 + 4$. As shown in Fig. 11.3(a), four data points $p_{2,3}$, $p_{2,4}$, $p_{2,5}$ and $p_{2,6}$ of T_2 are possible mapping data points for $p_{1,5}$ since their time are within $14 - 4$ to $14 + 4$. Since $f_4(p_{1,5}, p_{2,5}) = 1 - \frac{\sqrt{5}}{4} = 0.44$ is the largest value among that of all other points, the clue score of $p_{1,5}$ with respect to trajectory T_2 is thus $score_{\epsilon=4,\tau=4}(p_{1,5}, T_2) = 0.44$.

In the definition of clue scores, a temporal parameter τ is used to retrieve data points of trajectories whose occurrence times are within a particular time interval. Using this parameter, our proposed CATS can deal with temporal shifting. Since the time of data points to be mapped are constrained by the temporal parameter, the clue score is sensitive to time. Likewise, the clue score is also sensitive to locations. However, it still tolerates spatial and temporal biases within the degree specified by spatial and temporal thresholds.

In light of the clue score defined for data points, we define the clue-aware similarity between two trajectories as follows.

Definition 11.3 (Clue-Aware Trajectory Similarity) Given a spatial threshold ϵ and a temporal threshold τ, the clue-aware trajectory similarity from T_i to T_j is defined as $CATS_{\epsilon,\tau}(T_i, T_j) = \frac{1}{|T_i|} \times \sum_{p_{i,\ell} \in T_i} score_{\epsilon,\tau}(p_{i,\ell}, T_j)$.

For example, let $\epsilon = 4$ and $\tau = 4$. The arrows in Fig. 11.3(a) show the mapping relationships from each data point of T_1 to data points of T_2. Consequently, the clue-based similarity measurement from T_1 to T_2 is derived as $CATS_{4,4}(T_1, T_2) = \frac{1}{9} \times (score(p_{1,1}, T_2) + score(p_{1,2}, T_2) + \cdots + score(p_{1,9}, T_2)) = 0.58$.

11.3.3 Properties of Clue-Based Similarity Measurements

From the definition of clue scores, both ϵ and τ thresholds are used to overcome spatial and temporal biases. Moreover, these two thresholds could deal with both the spatial and temporal shiftings. Since noisy data results in a larger distance value, our spatial decay function can easily filter out noises. For the mapping scheme, our clue-based similarity measurement allows many data points to map to the same data point on the reference trajectory. Consider the clue-based similarity from T_i to T_j as an example. It is possible that some data points of trajectory T_i map to the same data point on T_j if the mapped data point of T_j are not filtered by both the spatial and temporal thresholds ϵ and τ. This mapping scheme is referred as n-to-1 mapping (abbreviated as $n - 1$). If data points on T_i have higher clue scores to T_j than the vice versa, the data points of T_i in fact provide more detailed information about the movement behavior than T_j does. Thus, these data points are very helpful for dealing with the silent duration in T_j. Furthermore, via the spatial and temporal thresholds, data points of T_i may not get mapped to any data point on T_j. In that case, the clue score is 0. Given two trajectories of different lengths which have some clues, our CATS may still derive a high clue score for these two trajectories, showing that CATS is able to overcome silent durations of trajectories.

Note that CATS has an asymmetry property. For example, $CATS_{4,4}(T_1, T_2) = 0.57$ is not equal to $CATS_{4,4}(T_2, T_1) = 0.65$ in Fig. 11.3. From the aforementioned example, it can be seen that T_3 provides only a limited number of clues for the movement behaviors. On the other hand, with a large CATS values, trajectory T_4 provides more detailed movement information for T_3. Hence, this asymmetry property of CATS is helpful for identifying relationships between two trajectories. Based on our observations, there are two relationships between two trajectories: the first one is that both of two trajectories can provide sufficient clues to each other. The second case is that from perspective of T_i, trajectory T_i is likely to provide some detailed sub-trajectories to trajectory T_j, but from the perspective of T_j, T_j does not have a similar movement behavior with T_i. The asymmetric property reflects two kinds of relationships between two trajectories. The first relationship usually happens when silent durations of two trajectories are distributed in the similar way and the two trajectories have only some spatial and temporal shifting. For the first case, these two trajectories can be recognized as to have the same movement behavior. Figure 11.3 illustrates such a case. Figure 11.3(a) shows that $CATS_{4,4}(T_1, T_2) = 0.58$ and Fig. 11.3(b) shows that $CATS_{4,4}(T_2, T_1) = 0.65$. These two values shows that both T_1 and T_2 have almost the same amount of clues to each other. Thus, the two trajectories are very likely to follow the same movement behavior. For the second relationship, one trajectory may be part of the other trajectory. For example, in Fig. 11.4, the actual movement of T_3 and T_4 are the same (i.e., the underlying grey lines). Figure 11.4(a) shows that mapping data points of T_4 have clues to T_3. However, in Fig. 11.4(b), most data points of T_3 have no clues to T_4. In this case, we can see that by compensating T_3 with the data points of T_4, the movement behavior can be revealed in more detail.

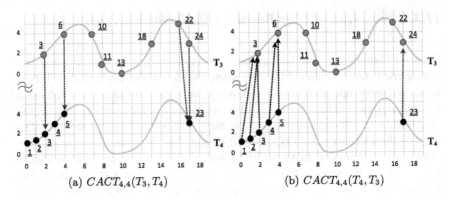

(a) $CACT_{4,4}(T_3, T_4)$ (b) $CACT_{4,4}(T_4, T_3)$

Fig. 11.4 An example to show the asymmetric property of CATS

With the design of clue-aware trajectory similarity, we are able to compute the pair-wise clue-aware similarity values for a set of trajectories. Due to silent durations of trajectories, though trajectories have strong clues, trajectories are likely to represent some partial movement behavior. One should fully utilize clues among trajectories to infer the complete trajectory route. Thus, in the next section, we propose a clue-aware trajectory clustering algorithm to cluster similar trajectories into groups, where each group represents one frequent movement behavior.

11.4 Clue-Aware Trajectory Clustering Algorithm

In this section, we describe the Clue-Aware Trajectory Clustering (CATC) algorithm for clustering trajectories based on CATS (clue-aware trajectory similarity) values.

11.4.1 Design of Clue-Aware Trajectory Clustering Algorithm

Algorithm CATC consists of three phases: (1) Clue-graph generation phase, (2) Core set identification phase, and (3) Cluster discovery phase. Explicitly, in the clue-graph generation phase, a graph structure is used to represent CATS values among every pair of trajectories. In the core set identification phase, we cluster trajectories which have sufficient large CATS values to each other as core sets. According to core sets derived, in the cluster discovery phase, core sets are merged as clusters. As such, each cluster represents one movement behavior. Clusters that have a sufficient number of supporting trajectories are identified as frequent movement behaviors. The details of the algorithm are described as follows:

Phase 1: Clue-Graph Generation Phase Given a set of trajectories, the CATS values between any two trajectories can be computed pairwisely. Then, a graph structure can be used to represent their clue-aware trajectory similarities. The definition of a clue-graph is given below:

Definition 11.4 (Clue-Graph) Given a set of trajectories $T = \{T_1, T_2, \ldots, T_n\}$ and a threshold λ, a clue-graph is a weighted directed graph $G = (V, E)$. In the clue-graph G, a set of vertices $V = \{v_1, v_2, \ldots, v_n\}$ represents the set of all trajectories and a set of edges is defined as $E = \{(v_i, v_j) | CATS_{\epsilon, \tau}(v_i, v_j) \geq \lambda\}$ where an edge (v_i, v_j) is weighted as $CATS_{\epsilon, \tau}(v_i, v_j)$.

Figure 11.5 shows a clue-graph with $\lambda = 0.4$. A clue-graph is used to represent the strength of clues between trajectories. There is no edge between two vertices if the CATS values between them is smaller than a threshold λ. Once two vertices have no edge in the clue graph, these two trajectories do not have enough clues (i.e., the spatially and temporally co-located points) to show they follow the same movement behavior. On the other hand, once the CATS values of two vertices exceed the threshold λ, the corresponding trajectories can be viewed to have enough clues to show that they follow the same movement behavior. In this case, two trajectories are called to have strong clues between them. In the clue-graph, we could further define a directly clue-reachable relationship among trajectories as follows.

Definition 11.5 (Directly Clue-Reachable) A vertex v_i is directly clue-reachable to a vertex v_j, denoted as $v_i \rightsquigarrow v_j$, if $(v_i, v_j) \in E$.

Phase 2: Core Set Identification Phase In light of the clue-graph and Definition 11.5, in this phase, we aim at deriving core sets in which trajectories in the core set have strong clues to each other. Note that with our clue-aware similarity measurement, if the CATS values of two trajectories exceed a threshold in both directions, these two trajectories likely infer the same movement behavior. In other words, these two trajectories are directly clue-reachable to each other in the clue-graph, i.e., they are one-hop neighbors. To capture trajectories with strong clue correlations, we define the notion of *core set* as follows.

Definition 11.6 (Core Set) Given a clue-graph $G = (V, E)$, a core set is a directed complete subgraph of G, where any two vertices v_i and v_j in a core set are directly clue-reachable to each other.

Each vertices in the core set has high clue values with respect to other vertices, indicating that these vertices in this core set capture the same movement behavior. Clearly, these core sets are viewed as seeds and these seeds could further merge with nearby seeds in the clue-graph. The detailed merging procedure among core sets is presented later. According to Definition 11.6, core sets could be as cliques in the clue-graph. Thus, we may adopt a clique-covering algorithm to derive cliques. Since most clique-covering algorithms are executed in an undirected graph [6], to facilitate the generation of core sets by using existing clique-covering algorithms, a Strong Clue-Graph (abbreviated as SC-G) from the clue-graph G is defined as follows.

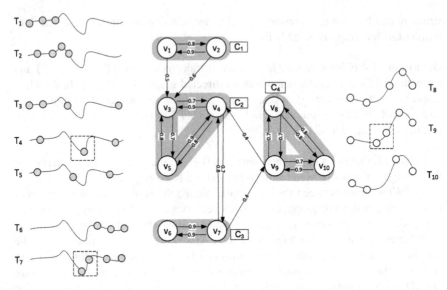

Fig. 11.5 An illustrative example for CATC

Definition 11.7 (Strong Clue-Graph) Given a clue-graph $G = (V, E)$, a Strong Clue-Graph, denoted as SC-G, is a undirected graph, represented as SC-G$= (V, E')$, where $(v_i, v_j) \in E'$ if both $(v_i, v_j) \in E$ and $(v_j, v_i) \in E$.

By performing an existing clique-covering algorithm in SC-G, a set of core sets is derived. Then, vertices in the same core set are labeled in the clue-graph as well. Figure 11.5 shows an example of trajectories with their CATS values and a set of cliques $\{C_1, C_2, C_3, C_4\}$, where vertices in the same shaded region belong to a clique. As such, these cliques are viewed as core sets. The trajectories of each core set intend to represent the similar movement behaviors because they have many co-located points to each other. For example, in the core set C_1, all data points of T_1 and T_2 are in the beginning of the movement route (in grey underline). Therefore, in the same core set, these two trajectories are likely to represent the same movement behavior.

Phase 3: Cluster Discovery Phase   Although each core set refers to one movement behavior, the movement behaviors of some core sets may be merged into a more complete movement behavior. For example, consider the core set C_1, and C_3 in Fig. 11.5. The core set C_1 indicates that this user moves in the beginning of the route (i.e., most data points of these trajectories in C_1 are in the beginning of the route), whereas the core set C_3 represents the movement behavior that this user tends to move in the end of the route. In this case, once we could merge the movement behaviors of C_1 and C_3, we can derive one movement behavior that is much closer than the actual route. Therefore, in this phase, the main goal is to use clues between core sets to infer which core sets can be merged into a cluster that represent the same movement behavior.

Core sets may be merged with other core sets as candidate clusters if two core sets have some reachable relationships, where candidate clusters are likely to infer one complete moving behavior. Since a core set may have reachable relationships with more than one core set, one should judiciously decide which core sets are selected for merging. To infer whether two core sets capture the same movement behavior or not, the number and the weights of edges between two core sets should be considered. Intuitively, if both the number of edges among vertexes of two core sets and edge weights are large, these two core sets are likely to reflect the same movement behavior. Furthermore, two core sets may still have clue reachable relationships via other core sets between them. To define the clue reachable relationship among two core sets (referring to as *clue-connected*), we should define a clue-reachable relationship between two vertex as follows.

Definition 11.8 (Clue-Reachable) A vertex u is clue-reachable to a vertex v, denoted as $u \leadsto^* v$, if there exists a chain of vertices $v = v_1, v_2, \ldots, v_n = u$ such that $v_i \leadsto v_{i+1}$ for all $i = 1, 2, \ldots, n - 1$.

With the definition of clue-reachable, we could define the clue-connected relationship between two core sets as follows.

Definition 11.9 (Clue-Connected) Given two core sets C_u and C_v, C_u can clue-connect to C_v, denoted as $C_u \Rightarrow C_v$, if there exists a core set C_w such that $x \leadsto^* y$ for all $x \in C_u$ and for some $y \in C_w$, and $y' \leadsto^* z$ for all $y' \in C_w$ and for some $z \in C_v$.

For example, in Fig. 11.5, C_1 can clue-connect to C_2. Let $C_v = C_1$, $C_w = C_1$, and $C_u = C_2$. We have all vertices in C_1 clue-reachable to some vertices in C_1 since C_1 is a core set, and $v_1 \leadsto^* v_3$ and $v_2 \leadsto^* v_3$. Two clue-connected core sets demonstrate the same movement behavior if these two core sets have connected core sets between these two core sets. Clearly, if two core sets have a clue-connected relationship, these two core sets should be grouped in the same cluster.

Thus, the result of algorithm CATC is a set of clusters consisting of core sets and the number of vertices in each cluster is larger than *min_sup*. To facilitate our presentation, a candidate cluster is used to represent our merging results of core sets. Initially, each core set is viewed as one candidate cluster. We iteratively merge candidate clusters until no further merge operation is needed. One criterion for stopping this merge operation is to measure the quality of cluster results. Traditional clustering algorithms use *cohesion* and *separation* to evaluate the quality of clusters. However, since our clustering algorithm is performed on the clue-graph, we develop clue-cohesion and clue-separation as follows.

Definition 11.10 (Clue-Cohesion) Given a candidate cluster K, the clue-cohesion of K, denoted by $CCOH(K)$, is defined as the minimum weight that for every core set $C_i \in K$, there exists a core set $C_j \in K$ such that $C_i \Rightarrow C_j$.

Definition 11.11 (Clue-Separation) Given two candidate clusters K_m and K_n, the clue-separation from K_m to K_n, denoted $CSEP(K_m, K_n)$, is defined as the total weight of all edges from K_m to K_n.

For example, consider C_2 and C_4 in Fig. 11.5. If we want to make $C_4 \Rightarrow C_2$, two extra edges (e.g., (v_8, v_3) and (v_{10}, v_5)) with the total weight $2 \times \lambda = 0.8$ should be added since there already exists an edge (v_9, v_4) from C_4 to C_2. Thus, if a cluster K contains C_2 and C_4, the clue-cohesion of this cluster can be derived as $CCOH(K) = 0.8$. The clue-separation $CSEP(C_4, C_2) = 0.4$ since there is one edge (v_9, v_4) from C_4 to C_2 with the total weight $\lambda = 0.4$. Note that the clue-cohesion $CCOH(K_m, K_n)$ is zero if K_m can clue-connect to K_n. As such, if two core sets from different candidate clusters have the clue-connected relationship, these two candidate clusters are likely to be merged. By merging these two candidate clusters, a larger candidate cluster has more trajectories for inferring the whole frequent trajectory route.

With the definitions of clue-cohesion and clue-separation, we intend to derive a set of clusters such that the cluster result should have smaller clue-cohesions and clue-separations among clusters. In other words, trajectories within the same cluster have as many clues as possible and trajectories from different clusters have as few clues as possible. Therefore, the desired cluster result is defined as follows:

Definition 11.12 (Objective Function for Clusters) Given a clue-graph $G = (V, E)$, the clustering algorithm aims to derive a set of clusters $\mathcal{K} = \{K_1, K_2, \ldots, K_m\}$ such that (1) K_i contains a set of core sets, (2) minimize $(\sum_{K_i \in \mathcal{K}} CCOH(K_i) + \sum_{K_i, K_j \in \mathcal{K}} CSEP(K_i, K_j))$, and (3) $|K_i| \geq min\_sup$ for all $K_i \in K$.

Based on the above objective function for clustering, we design a benefit function to evaluate whether merging two candidate clusters is able to reduce the value of the objective function or not. The benefit function is formulated as follows.

Definition 11.13 (Benefit Function) Given two candidate clusters K_m and K_n, the benefit function is defined as $Benefit(K_m, K_n) = DesCSEP(K_m, K_n) - IncCCOH(K_m, K_n)$. $DesCSEP(K_m, K_n) = (CSEP(K_m, K_n) + CSEP(K_n, K_m))/2$ and $IncCCOH(K_m, K_n) = CCOH(K_m) + CCOH(K_n) + \sum_{C_i \in K_m} \min_{C_j \in K_n} \{I(C_i, C_j) \times \lambda\}$, where $I(C_i, C_j)$ denotes the number of vertices that have no edge from core set C_i to C_j and λ is the threshold used in the clue-graph.

Generally speaking, merging two candidate clusters will increase the clue-cohesion while decreasing the clue-separation. Intuitively, if merging two candidate clusters could lead to more decrease in clue-separation than increase in clue-cohesion, the merging operation is able to minimize the objective function, which brings a benefit to achieve a better clustering result. Consequently, the benefit function is to evaluate whether merging two candidate clusters is able to reduce the value of the objective function (i.e., the sum of the clue-cohesion and the clue-separation after merging two candidate clusters). Assume that we intend to derive the benefit of merging two candidate cluster K_m and K_n. The first term of the benefit function (i.e., DesCSEP) represents how much amount of clue-separation could be reduced by merging K_m and K_n. The average of the clue-separations between two candidate clusters aims to prevent the scenario in which only one cluster has a lot of edges to the other one but there's no edges in reverse. For example, in Fig. 11.5, assume that four core sets are candidate clusters and a candidate cluster set, denoted as $\mathcal{K}$, $\mathcal{K} = \{K_1 = C_1, K_2 = C_2, \ldots, K_4 = C_4\}$. Then, we illustrate how to derive $Benefit(K_2, K_3)$. It can be derived that $CSEP(K_2, K_3) = 0.7$ since the only edge crosses from K_2 to K_3 is (v_4, v_7) with weight 0.7. Similarly, we can obtain that $CSEP(K_3, K_2) = 0.8$. Thus, after merging K_2 and K_3, two edges (v_4, v_7) and (v_7, v_4) are in the same cluster such that the clue-separation is decreased by $0.7 + 0.8 = 1.5$. Thus, the value of $DesCSEP(K_2, K_3)$ can be derived as $1.5/2 = 0.75$. The second term of the benefit function (i.e., IncCCOH) is to evaluate the amount of increase in clue-cohesion by merging K_m and K_n. For two candidate clusters K_m and K_n, they need $CCOH(K_m)$ and $CCOH(K_n)$ to make their core sets clue-connected. The last term of IncCCOH refers to the minimum weight that every core sets in K_m can clue-connect to some core set in K_n. Specifically, if a core set C_i is required to clue-connect to the other core set C_j, every vertex in C_i should have an edge to some vertex in C_j. One could imagine that the cost for adding an edge is λ since there is an edge between two vertices if the CATS value between them is at least λ. Therefore, to make C_i clue-connect to C_j, the total cost can be derived by multiplying λ and the number of vertices that have no edge from core set C_i to C_j. Recall the example above where two candidate-clusters $K_2 = \{C_2\}$ and $K_3 = \{C_3\}$ are given. Since these two candidate-clusters contain only one core sets, we can obtain that $CCOH(K_2) = 0$ and $CCOH(K_3) = 0$. If we consider the case that K_2 needs to clue-connect to K_3, it needs to add extra two edges from C_2 to C_3 (e.g., adding $(v_5, v_6$ and $(v_3, v_6))$. Therefore, the total increase of the clue-cohesion value is $IncCCOH(K_2, K_3) = 0 + 0 + 2 \times \lambda = 0.8$. Finally, $Benefit(K_2, K_3) = 0.75 - 0.8 = -0.05$, which shows that by merging K_2 and K_3, there is no benefit to minimize the objective function.

By exploiting this benefit function, algorithm CATC iteratively selects two candidate clusters with the maximum benefit value until the value of the benefit function is smaller than zero. Once the merging operation is finished, candidate clusters that have more than *min_sup* vertices will become final clusters. The reason for having at least *min_sup* vertices is that each cluster represents one frequent movement behavior. Note that the computation of the benefit function could be implemented by dynamic programming. Since candidate clusters are expanded in a bottom-up fashion, the clue-cohesions and clue-separations of any two candidate clusters are computed at previous rounds. Therefore, we explore a dynamic programming strategy in algorithm CATC.

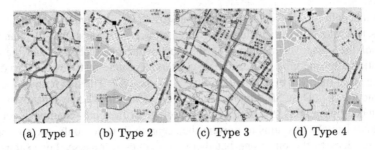

(a) Type 1   (b) Type 2   (c) Type 3   (d) Type 4

Fig. 11.6 Trajectories in *CarWeb* dataset

11.5 Performance Evaluation

A series of experiments has been conducted to evaluate the proposed algorithms using real datasets. The real dataset consists of trajectories obtained from the *CarWeb* platform [8]. Through the CarWeb dataset, we have access to the ground truth of user movement behaviors, which makes the quality evaluation of the discovered trajectory patterns feasible. We select trajectories capturing four types of frequent movement behaviors (as shown in Fig. 11.6). Note that both Type 2 and Type 4 have similar movement paths but their times are different. Trajectories of Type 2 present a movement behavior happening on an early morning, whereas trajectories of Type 4 capture another movement behavior in the late afternoon.

Here we compare our proposed clustering algorithm CATC with existing trajectory clustering algorithms (i.e., PISA [9] and TraClus [7]). PISA uses its proposed distance function and then applies OPTICS (a self-tuning DBSCAN) to cluster trajectories. TraClus aims to discover clusters of "sub-trajectories", which does not align to our goal. Thus, we slightly modify the TraClus algorithm as follows. Suppose that the set of all sub-trajectory clusters is $\{sc_1, sc_2, \ldots, sc_n\}$, which can be treated as features of trajectories. Thus, each trajectory T_i could be represented as a vector $\vec{v_i} = \langle x_1, \ldots, x_n \rangle$, where $x_j = 1$ if T_i passes sc_j. Accordingly, *cosine similarity* can be employed to measure the similarity of two vectors. Finally, DBSCAN is used to cluster trajectories. The following experimental results are the best results by fine tuning the parameters of the corresponding algorithms.

The clustering results of CATC, TraClus, and PISA are shown in Table 11.1. As can be seen in Table 11.1, each cluster generated by examined algorithm contains trajectories of certain trajectory types. For cluster #1, PISA contains all types of trajectories, whereas CATC and TraClus have only two types of trajectories. To visualize trajectories in cluster #1, Fig. 11.7 shows trajectories in cluster #1, where each color represents one type of trajectory. TraClus could derive several common sub-trajectory clusters since trajectories of all types have several common areas. By representing trajectories into vectors of sub-trajectory clusters, the cosine similarity values of the same type trajectories and different type trajectories could be similar. Table 11.1 shows that cluster #1 contains many trajectories of both Type 1 and Type 2. The reason is that trajectories of Type 1 and Type 2 have overlaps in the beginning of trajectories. As shown in Fig. 11.7(a), they will have many common

Table 11.1 Clustering results for different clustering algorithms; I (II, and III) represent trajectories of Type 1 (2, and 3)

Cluster	#1	#2	#3	#4	#5
CATC	I:{0–6} II:{14,22}	I:{10–13} II:{15,24}	I:{7–9}	II:{16, 17–21, 23, 25, 26}	III:{27–31}
TraClus	I:{0–13} II:{14–25}	II:{26} III:{29}	III:{27, 28, 30, 31}		
PISA	I:{1} II:{18, 19} III:{27–31}	I:{0, 3, 5–10} II:{14–17, 20–26}	I:{2, 11, 13}	I:{4, 12}	

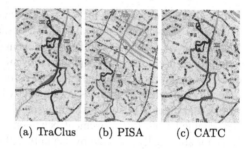

Fig. 11.7 Visualization of cluster #1 derived by different clustering algorithms

(a) TraClus   (b) PISA   (c) CATC

sub-trajectory clusters such that TraClus may put these two types into one. Figure 11.7(b) shows the clustering results of PISA. This cluster has trajectories from three types. In PISA, trajectories are viewed as piece-wise lines and the distance function aims to compare the average of the sum of Euclidean distance in each time slot. This distance function may not be appropriate for computing the distance value between trajectories with silent durations. Moreover, since PISA uses DBSCAN to cluster trajectories, the size of a cluster becomes large easily such that a cluster may contain trajectories of different types. Compared with the results above, Fig. 11.7(c) shows the clustering results of CATC, where there are only two Type 2 trajectories. The reason is that these two trajectories are highly overlap to some Type 1 trajectories such that CATC mistakes them as the Type 1 trajectories.

To investigate the scalability of three clustering algorithms to the size of trajectory datasets, Fig. 11.8 shows the results of the execution time (the unit is *second*) by increasing the number of trajectories in the datasets, which are generated by randomly duplicating existing trajectories. In all cases, CATC spends less execution time than TraClus and performs competitively in comparison with PISA. However, notice that CATC achieves the highest purity and the lowest entropy. Thus, we conclude that CATC is the best choice for clustering trajectories with silent duration in terms of effectiveness and efficiency.

Fig. 11.8 Scalability
comparison of trajectory
clustering algorithms

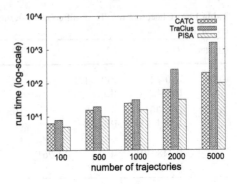

11.6 Conclusions

In this chapter, we propose the CCT framework to discover trajectory patterns. In addition to spatial and temporal bias, we observe that trajectories usually contain silent durations, during which detailed movement information are missing. Furthermore, Since users may have multiple movement behaviors, trajectories should be clustered before hot regions are identified for trajectory pattern mining. Existing trajectory clustering techniques do not deal with trajectories with silent durations. Notice that clues about the same movement behavior are usually left in trajectories sampled from this movement behavior. We argue that clues of a movement behavior are usually reflected by spatially and temporally co-located data points in trajectories. Thus, we formulate a clue-aware trajectory similarity and a clue-aware clustering algorithm to cluster similar trajectories into groups. We evaluate CCT using both real and synthetic datasets by experiments. Experimental results show that CCT is able to effectively discover trajectory patterns even if trajectories only capture fragments of movement behaviors.

References

1. EveryTrail—GPS Travel Community: http://www.everytrail.com/
2. MapMyRun Website: http://www.mapmyrun.com
3. Run GPS Community Server: http://www.gps-sport.net/
4. Cao, L.: In-depth behavior understanding and use: the behavior informatics approach. Inf. Sci. pp. 3067–3085 (2010)
5. Gaffney, S., Smyth, P.: Trajectory clustering with mixtures of regression models. In: Proc. of KDD, pp. 63–72 (1999)
6. Gramm, J., Guo, J., Huffner, F., Niedermeier, R.: Data reduction, exact, and heuristic algorithms for clique cover. In: Proc. of SIAM Workshop on Algorithm Engineering and Experiments (2006)
7. Lee, J.-G., Han, J., Whang, K.-Y.: Trajectory clustering: a partition-and-group framework. In: Proc. of SIGMOD (2007)
8. Lo, C.-H., Peng, W.-C., Chen, C.-W., Lin, T.-Y., Lin, C.-S.: CarWeb: a traffic data collection platform. In: Proc. of MDM (2008)
9. Nanni, M., Pedreschi, D.: Time-focused clustering of trajectories of moving objects. J. Intell. Inform. Syst. **27**(3), 267–289 (2006)

Chapter 12
Linking Behavioral Patterns to Personal Attributes Through Data Re-Mining

Gürdal Ertek, Ayhan Demiriz, and Fatih Cakmak

Abstract A fundamental challenge in behavioral informatics is the development of methodologies and systems that can achieve its goals and tasks, including behavior pattern analysis. This study presents such a methodology, that can be converted into a decision support system, by the appropriate integration of existing tools for association mining and graph visualization. The methodology enables the linking of behavioral patterns to personal attributes, through the re-mining of colored association graphs that represent item associations. The methodology is described and mathematically formalized, and is demonstrated in a case study related with retail industry.

12.1 Introduction

This study aims at understanding the behavioral patterns exhibited by people in relation to their personal attributes. The research is conducted in the context of retail industry, where consumers engage in purchase transactions at retail shops and stores. The traditional data mining technique for identifying the patterns in these transactions is association mining, which enables the discovery of interpretable and actionable results related with item associations. However, straightforward application of association mining returns only item purchase patterns. An important question, whose answer has been ignored in literature, is how these patterns are related to consumer attributes, such as demographic attributes and physical state of the consumer during the purchase. In other words, the link between the behavioral pattern

G. Ertek (✉)
Faculty of Engineering and Natural Sciences, Sabancı University, Orhanli, Tuzla, 34956, Istanbul, Turkey
e-mail: ertekg@sabanciuniv.edu

A. Demiriz
Department of Industrial Engineering, Sakarya University, 54187, Sakarya, Turkey
e-mail: ademiriz@gmail.com

F. Cakmak
Faculty of Arts and Social Sciences, Sabancı University, Orhanli, Tuzla, 34956, Istanbul, Turkey

L. Cao, P.S. Yu (eds.), *Behavior Computing*,
DOI 10.1007/978-1-4471-2969-1_12, © Springer-Verlag London 2012

(consumer purchase behavior) and the person (consumer) him/herself is missing. Establishing this link requires a methodology, as well as domain knowledge to enable domain-driven data mining [6].

A graph-based visualization methodology, namely AssocGraphRM, is proposed for presenting association mining results, together with summary statistics regarding the associations. The methodology suggests a visual data *re-mining* process, based on the results generated by association mining. In the graph-representation, items and itemsets are represented as vertices, set membership are represented through edges, and attribute statistics are linearly mapped to the colors of vertices. The applicability and usefulness of the methodology is demonstrated through a market basket analysis (MBA) case study where data from a consumer survey is analyzed. The survey contains a multitude of personal attributes, as well as preferences for items at Starbucks coffee stores. Several actionable insights are derived regarding the relationship between the behavioral patterns (item purchases) and the personal attributes, and their policy implications are discussed.

The remainder of the chapter is organized as follows: In Sect. 12.2, an overview of the basic concepts in related studies is presented through a concise literature review. In Sect. 12.3, the AssocGraphRM methodology for visual re-mining on colored association graphs is described, and framed as an algorithm using mathematical formalism. The methodology and its applicability is then demonstrated in Sect. 12.4, using survey data from food retail industry. The validity of the methodology is discussed in Sect. 12.5. Finally, in Sect. 12.6, the study is summarized and future directions are discussed.

12.2 Literature

12.2.1 Behavior Informatics

The field of *behavior informatics* is introduced by Cao [4, 5], and suggests the analysis of behavioral patterns and impacts following *behavior explicitation*, through the extraction of behavior elements masked in transactional data. The main goals and tasks of behavior informatics are listed in [4] as behavior modeling and representation, construction of behavioral data, behavior impact modeling, behavior pattern analysis, and behavior presentation. The main idea in behavioral informatics is to organize the transactional data into a new form that is constructed in terms of behavior, rather than entity relationships. With the explosion of data that is collected electronically in massive amounts, the main challenge in behavioral informatics is the development of methodologies and systems that can achieve its goals and tasks. This study presents such a methodology, that can be converted into a decision support system, by the appropriate integration of existing tools for association mining and graph visualization.

12.2.2 Association Mining

Association mining is an increasingly used data mining and business tool among practitioners and business analysts [11], due to the interpretable and actionable results it generates. Association mining results can be classified based on several criteria, as outlined in [19]. In this chapter, we focus on frequent itemset, which are the sets that define single-dimensional, single-level boolean association rules. Efficient algorithms such as Apriori [1] enable the analysis of very large transactional data, frequently from transactional sales data, resulting in a large collection of frequent itemsets.

Association mining is typically presented and discussed in the context of one of its most common applications, namely market basket analysis (MBA), which can be used in product recommender systems [11]. Let $I = \{i_1, i_2, \ldots, i_m\}$ be a set of items considered in MBA. Each transaction (basket) t will consist of a set of items where $t \subseteq I$. Let $D = \bigcup t$ be the database of all transactions. The *support sup(f)* of an itemset (and also of the rule that contains the items in that itemset) is defined as the percentage of the transactions in D that contain all the items of the itemset f:

$$sup(f) = \frac{\sum_{t \in D} 1_{\{f \subseteq t\}}}{|D|} \tag{12.1}$$

A *frequent itemset* is an itemset that has support value greater than or equal to a given minimum support threshold: $sup(f) \geq min\_sup$.

12.2.3 Visualizing Frequent Itemsets

Commonly, finding the frequent itemsets and association rules from very large data sets is heavily emphasized, since it is considered as the most challenging step in association mining [7, 18, 42, 43]. Results are typically presented in a text (or table) format with certain degree of querying and sorting functionalities. However, the real objective of the association mining analysis is to foster the discovery of insights, and there exists considerably less work that focuses on the interpretation of the association mining results [15].

Information visualization is the branch of computer science that investigates how data and information can be visualized to obtain significant, deep, actionable insights [10, 20, 24, 28]. Within information visualization, graph visualization can be a significant source of insights, as demonstrated by numerous case studies in a multitude of disciplines [9, 27, 31–34, 36, 37, 39]. There is a broad literature on graph visualization, including the literature on graph drawing, but the use of graphs for the visualization of association mining results is not well-formalized in academic literature. Still, data analysis systems such as MS SQL Server [29] and SAS [35] can generate association graphs.

This study is an extension of earlier work by Ertek and Demiriz [15], where a graph-based methodology is proposed to visualize and interpret the results of well-known association mining algorithms as directed graphs. According to the methodology in [15], the items (also referred to as 1-*itemsets*) and the itemsets are represented as vertices on an association graph. The vertices that represent the items are shown with no color, whereas the vertices that represent the itemsets are colored reflecting the cardinality of the itemsets. The sizes (the areas) of the vertices show the support levels. The directed edges symbolize which items constitute a given frequent itemset.

The main idea in the methodology is to exploit already existing graph drawing algorithms [38] and software in the information visualization literature [20] for visualizing association mining results which are generated by already existing algorithms and software in the data mining literature [19].

In the current study, the methodology in [15] is extended to incorporate additional attributes, and mathematical formalism is introduced for describing the methodology. Color is now used to represent the values of a selected additional attribute, instead of the cardinality of the itemset. The cardinality of the itemset is instead reflected by the thickness of the vertices. So, this extended study enables linking association mining results with attributes of the person that carried out the transaction. In the context of behavior informatics, the methodology establishes the critical link between behavioral patterns and personal attributes.

12.2.4 Re-Mining

The re-mining methodology was first introduced by Demiriz et al. [12]. *Re-mining* process is defined as "combining the results of an original data mining process with a new additional set of data and then mining the newly formed data again". Re-mining is fundamentally different from post-mining [8, 23, 26, 44, 45]: post-mining only summarizes the data mining results, such as visualizing the association mining results [15, 22]. The re-mining methodology extends and generalizes post-mining. Re-mining can be considered as an additional data mining step of Knowledge Discovery in Databases (KDD) process [25] and can be conducted in explanatory/exploratory, descriptive, and predictive manners.

In another study, Demiriz et al. [13] elaborate on the re-mining concept, introduce mathematical formalism and present the algorithm for the methodology. Reference [13] also extends the application of predictive re-mining in addition to exploratory and descriptive re-mining, and presents a complexity analysis.

Quantitative and multi-dimensional association mining (QAM&MAM) are well-known techniques within association mining [19] that can integrate additional attribute data into the association mining process. The associations among the additional attributes, and among them and itemsets are computed. However, both techniques introduce significant additional complexity, since association mining is carried out with the complete set of attributes rather than just the market basket data.

The techniques work directly towards the generation of *multi-dimensional* rules. They relate all the possible categorical values of all the attributes to each other, which is *NP-hard*.

Re-mining, on the other hand, conveniently expands *single dimensional* rules with additional attributes. In re-mining, attribute values are investigated and computed only for the associated item pairs, with much less computational complexity that can be solved in polynomial running time. Running time of QAM&MAM increase exponentially with the number of additional attributes and the number of transactions, and re-mining is even more preferable in such situations.

Demiriz et al. [12, 13] propose a practical and effective methodology that efficiently enables the incorporation of attribute data (e.g. price, category, sales timeline) in explaining positive and negative item associations, which respectively indicate the complementarity and substitution effects.

The work closest to re-mining is by Yao et al. [41], where a framework of a learning classifier is proposed to explain the mined results. Unlike in [41], the re-mining [12, 13] and visual re-mining approaches (this study) are applied to real world datasets as a proof of their applicability.

12.3 Methodology: Re-Mining on Association Graphs

The fundamental idea in re-mining is to exploit the domain specific knowledge in a new analysis step. Thus, re-mining is a recipe for domain-driven data mining [6]. The AssocGraphRM methodology proposed and described in this chapter is a special type of re-mining: Re-mining is conducted through visually mining colored association graphs. By introducing mathematical formalism the methodology is presented in the form of an algorithm.

In AssocGraphRM, following the execution of conventional association mining and the generation of the frequent itemsets, a new database R is formed from the frequent itemsets F and additional attributes A, and then exploratory visual analysis is performed. Visual re-mining consists of the following main steps:

Step 1: Carry out association mining
Step 2: Compute the statistics for the frequent itemsets
Step 3a: Represent the frequent itemsets as a directed graph
Step 3b: Map the computed statistics linearly to colors
Step 3c: Color the graph with respect to this coloring scheme
Step 4: Visually explore the colored graphs and discover actionable insights

The graph is constructed by following the design specifications below:

a. Each item(set) is represented as a vertex.
b. Area of each vertex is proportional to the support of the corresponding item(set).
c. For itemsets with more than single item, edges are drawn from the (vertices of the) items of that itemset to the (vertex of the) itemset.

d. Line thickness of each vertex is proportional to the number of items in the corresponding itemset.
e. Color of each vertex reflects the value of a selected attribute for the corresponding itemset.

The inputs for the algorithm, parameters to be decided before the algorithm run, and additional definitions involving sets and functions are presented below:

Inputs

I: set of items; $i \in I$

D: set of transactions, containing only the itemset information

A_n: additional numerical attribute n introduced for re-mining; $n = 1, \ldots, N$. Any non-numerical (categorical/ordinal) attribute can be converted into a numerical attribute by computing the percentage of transactions for a specific value of the categorical/ordinal attribute. For example, if the original attribute is the gender of the person involved in the transaction, then let A_n be the percentage of transactions containing the given itemset and involving a female.

A: set of all attributes introduced for re-mining; $A = \bigcup A_n$

Parameters

min_sup: minimum support required for frequent itemsets
min_items: minimum number of items in the itemsets
max_items: maximum number of items in the itemsets
min_diameter: diameter for the item with *min_sup*. Note that the diameter of an itemset may be smaller than this value, but that of an item can not.

Definitions

F_1: set of frequent items (1-itemsets); $f \in F_1$

$F_{>1}$: set of frequent k-itemsets $(k > 1)$ that have positive association; $f \in F_k$

F: set of all frequent itemsets (k-itemsets, with $k = 1, \ldots, max\_items$); $f \in F$

min_value$_n$: minimum value for attribute n, over all frequent itemsets F
max_value$_n$: maximum value for attribute n, over all frequent itemsets F

R: set of records for re-mining, that contain *positive* associations; $r \in R$

$G(V, E)$: association graph that represents the frequent itemsets with vertices V and edges E

$\mathcal{G}$: graph collection that will be used for visual exploration in the re-mining phase

Functions

apriori$(D, min\_items, max\_items, min\_sup)$: apriori algorithm that operates on D and generates frequent itemsets with minimum of *min_items* items, maximum of *max_items*, and a minimum support value of *min_sup*

sup(f): support of itemset f

$\psi(A_n, f)$: function that computes the value of attribute A_n for a given frequent itemset f, $f \in F_k$. Once the function runs for a particular itemset, it stores the information in its corresponding record **record_of**(f), and returns from the record next time it is run.

create_new_vertex(v): function that creates a new vertex v

vertex_of(f): function that returns the vertex that corresponds to itemset f

create_new_edge(e): function that creates a new edge e

record_of(f): function that returns the record that corresponds to itemset f

clone_graph(G): function that creates a clone of graph G

compute_color(δ): function that computes color based on darkness $\delta \in [0, 1]$. Besides the argument δ, the RGB value for the computed color depends on the base RGB values for $\delta = 0$ and $\delta = 1$.

The complete methodology is formalized as an algorithm below:

Algorithm: AssocGraphRM

1. *Perform association mining.*
   $$F = \mathbf{apriori}(D, 1, max\_items, min\_sup)$$
2. *Define frequent items and itemsets.*
   $$F_1 = \{f \in F : |f| = 1\}$$
   $$F_{>1} = F - F_1$$
3. *Label the item associations accordingly and append them as new records.*
   $$R = \{r : r = (f), \forall f \in F\}$$
4. *Expand the records with the cardinality value for the itemsets, and additional attributes for re-mining.*
   for all $r = (f) \in R$
      $r = (f, |f|)$
      for $n = 1 \ldots N$
         $r = (r, \psi(A_n, f))$
5. *Create the vertices of the graph, with area being linearly proportional to the support, and thickness being proportional to the cardinality of the itemset.*
   $$V = \{\}$$
   for all $f \in F$
   create_new_vertex(v)
      $v.itemset = f$
      $v.record = \mathbf{record\_of}(f)$
      $v.diameter = min\_diameter\sqrt{\dfrac{sup(f)}{min\_sup}}$
      $v.thickness = |f|$
      vertex_of(f) $= v$
      $V \bigsqcup v$
6. *Create the edges of the graph, emanating from the items in the itemsets and terminating at the itemsets.*

$E = \{\}$
for all $f \in F_{>1}$
   for all $i \in f$
      create_new_edge(e)
      $e.from = $ **vertex_of**(i)
      $e.to = $ **vertex_of**(f)
      $E \sqcup e$

7. *Apply organic layout on* G.
8. *Compute the minimum and maximum values for each of the attributes.*

$$min\_value_n = min_{n=1...N, f \in F} \psi(A_n, f)$$
$$max\_value_n = max_{n=1...N, f \in F} \psi(A_n, f)$$

9. *Color the vertices in* G *with respect to each additional attribute, and add to the graph collection* $\mathcal{G}$. *The closer the value for attribute n gets to the maximum value of that attribute* max_value_n, *the darker the vertex will be colored.*

   for $n = 1, \ldots, N$
      $G' = $ **clone_graph**(G)
      for all $v \in V$
         $f = v.itemset$
         $v.darkness = \frac{\psi(A_n, f) - min\_value_n}{max\_value_n - min\_value_n} \in [0, 1]$
         $v.color = $ **compute_color**$(v.darkness)$
      $\mathcal{G} \sqcup G'$

10. *Perform visual re-mining through human-involved exploratory examination of the graphs in* $\mathcal{G}$.

12.4 Case Study

The proposed methodology is demonstrated through a case study using a survey dataset collected from coffee retail industry. Several insights are discovered regarding the relationships among the frequent itemsets and personal attributes, and suggestions are made on how these actions might be used as operational policies.

12.4.1 Retail Industry

Recent research has positioned association mining as one of the most popular tools in retail analytics [3]. Market basket analysis is considered as a motivation, and is used as a test bed for these algorithms. Additional data are readily available either within the market basket data or as additional data, thanks to loyalty cards in retailing, which enable linking transactions to personal data, such as age, gender, county of residence, etc.

As of November 2010, the retail industry in US alone runs on a monthly sales of $377.5 billion, with food services and food retail industry constituting 10% share in it [40]. Due to its gigantic size, retail industry has been selected as the domain of the case study. Starbucks is one of the best-known brands in food retail, and the best-known brand in coffee retail / specialty eateries industry, with 137,000 employees and a global monthly revenue of nearly $1 billion [40]. Due to the company's visibility, the products of Starbucks have been considered for constructing the survey data.

12.4.2 The Data

A survey has been conducted with 644 respondents, that contain nearly equal distribution of working people vs. students (all students were assumed non-working), women vs. men, and a multitude of universities and working environments. Each respondent was questioned for 22 attributes that reflect their demographic characteristics and life style preferences.

The fields in the dataset include demographic attributes (YearOfBirth, Gender, EmploymentStatus, IncomeType, etc.), attributes related with life style (Favorite-Color, SoccerTeam, FavoriteMusicGenre, etc.), educational background (University, EnglishLevel, FrenchLevel, etc.), perceptional and intentional information (Reason-ForGoing, etc.), and physical status at the time the survey was conducted (Hunger-Level, ThirstLevel). The number of additional attributes to be used in Step 8 of the algorithm totalled to 21.

As the transaction data, each respondent was also asked which items they would prefer from the menu of Starbucks Turkey stores if they had 15 TL Turkish Liras (approximately $10). The menu considered was the menu as of October 2009, and the respondents were limited to select at most four items without exceeding the budget limit. While the original survey distinguished between the different sizes (tall, grande, venti), in the data cleaning and preparation phase, the sizes for each type of item (such as Cafe Latte or Capuchino) were aggregated and considered as a single item (group). The original survey also contained preferences under a budget of 20 TL, but those preferences were not analyzed in the case study.

Even though market basket analysis is carried out in retail industry with transactions data that is logged through sales, preference data was collected in the survey and used instead of the transactions data. This is due to the well-known difficulty of obtaining real world transactions data from companies, which they consider highly confidential, even when masked.

12.4.3 The Process

Data was assembled in MS Excel and cleaned following the guidelines in the taxonomy of dirty data by Kim et al. [21]. The transactions were given as input into Borgelt's apriori software [2], and the apriori algorithm was run with a support value of 2%. The sizes of the vertices (based on the support values), the statistics for the frequent itemsets, and the corresponding vertex colors were computed in MS Excel spreadsheet software through distributed manual processing by 30 Sabancı University students, and assembled through the EditGrid [14] online spreadsheet service. The association graph was manually drawn in yEd Graph Editor software and an organic layout was applied. yEd implements several types of graph drawing algorithms, including those that create hierarchical, organic, orthogonal, and circular layouts, and allows customization of the layouts through structure and parameter selections. Past experience with the software in applied research projects has shown that in Classic Organic Layout in yEd is especially suitable for displaying associations. Organic layout is generated based on force-directed placement algorithms [17] in graph drawing. This layout selection ends up placing items that belong to similar frequent itemsets and in close proximity of each other. After constructing the base graph in yEd, the yEd graphml file was cloned, and each copy was colored manually according to a different additional attribute. Then the resulting collection of graphs were analyzed through brain-storming sessions and actionable insights, together with their policy implications, were determined.

12.4.4 Analysis and Results

Visual re-mining was performed on the colored association graphs in the collection $\mathcal{G}$ through human-involved exploratory examination of the graphs. Figures 12.1– 12.4 are selected graphs from $\mathcal{G}$ that are constructed for this case study. In the figures, regions of the graphs are highlighted for illustrating the insights and policies.

Figure 12.1 shows only the frequent item preferences of the participants in the case study. In this figure, the large region shows that Mosaic Cake and White Chocolate Mocha are selected by a large percentage of the people, and they are purchased together frequently, as represented by the large size of the corresponding vertex F11. The small region suggests that Chai Tea Latte and Lemon Cake are purchased frequently with each other, but not with other items. In the context of retailing, these are referred to as *complementary items*, and the classic operational policy is to use each of these items for increasing the sale of the other(s):

Policy 1 "Bundle Mosaic Cake, White Chocolate Mocha and Water together, and/or target cross-selling by suggesting the complementary item when the other is ordered."

Policy 2 "Bundle Chai Tea Latte and Lemon Cake together, and/or target cross-selling by suggesting the complementary item when the other is ordered."

Besides identifying the most significant and interdependent items, one can also observe items that are independent from all items but one. The central item is an *attractor*, drawing attention to less popular items related to them. In retail industry,

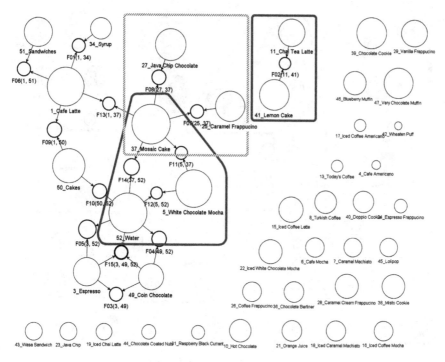

Fig. 12.1 Association graph that displays the frequent itemsets

attractor items are placed at visible locations (ex: the aisle ends) to attract customers to the items in nearby but less visible locations (ex: inside the aisles).

Another type of insight that can be derived from Fig. 12.1 is the identification of items which form frequent itemsets with the same item(s) but do not form any frequent itemsets with each other. One such pattern is indicated with the dashed borderline. Caramel Frappucino and Java Chip Chocolate each independently form frequent itemsets with Mosaic Cake, but do not form frequent itemsets with one another. These items may be *substitute items* and their relationship deserves further investigation.

While Fig. 12.1 illustrates behavioral patterns with regards to item purchases, it does not tell us how these patterns relate to attributes of people. The mapping of attributes to colors on the graph in the proceeding figures solves this problem, and enables the discovery of deeper additional insights in more dimensions.

Figure 12.2 displays the items and itemsets together with gender attribute (percentage of females selecting that itemset, computed from the data field Gender) mapped to color. Darker vertices (itemsets) indicate that among the people that selected that itemset, the percentage of females is higher compared to males. Even though the vertex sizes and locations are exactly the same as the earlier figure, new insights are derived due to the coloring. The selected region shows that Mosaic Cake is purchased with either Java Chip Chocolate, as represented by the itemset F08, or with Caramel Frappucino, as represented by the itemset F07. However, F08 and F07

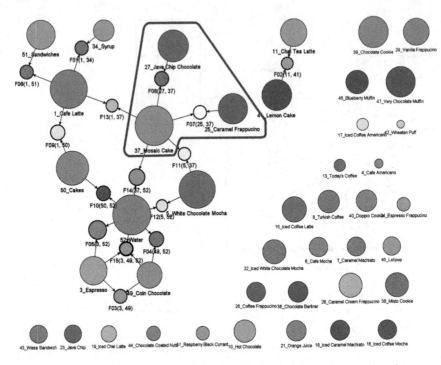

Fig. 12.2 Gender attribute (percentage of females selecting each itemset) mapped to color

are colored clearly differently, with F08 being darker. This means that the percentage of women among those that prefer the itemset F08 (Mosaic Cake and Java Chip Chocolate) is higher. So, if only the content of the selected region is considered, the following policy can be applied:

Policy 3 "If a male customer orders Mosaic Cake, try to cross-sell to him Caramel Frappucino by suggesting that item; otherwise, if a female customers orders it, try to cross-sell to her Java Chip Chocolate."

Figure 12.3 displays the knowledge of French language (FrenchLevel) mapped to color. Darker vertices indicate that among the people that selected that itemset, the level of knowledge for the French language is higher on the average. The k-itemset ($k > 1$) with the darkest color is F07, which is the preference for the items Mosaic Cake and Caramel Frappucino together. Items that have the darkest colors, in order of decreasing support, are Very Chocolate Muffin, Doppio Cookie, Orange Juice, Caramel Machiato, and Java Chip. This discovery can be used in conjunction with geographical location of the stores, as the next policy suggests:

Policy 4 "If the store is located near a university or high-school where the language of instruction is French, then emphasize Very Chocolate Muffin, Doppio Cookie, Or-

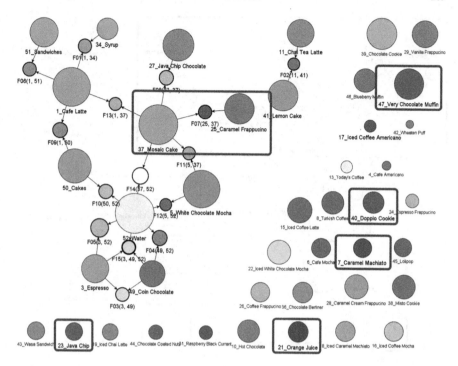

Fig. 12.3 Knowledge of the French language (average French knowledge of those selecting each itemset) mapped to color

ange Juice, Caramel Machiato and Java Chip as stand-alone products, and the item pair Mosaic Cake and Caramel Frappucino as a bundle."

Figure 12.4 displays the hunger level of the customer (HungerLevel) mapped to color, where darker colors denote higher hunger level on the average. It is observed that none of the k-itemsets with $k > 1$ have white color. This is consistent with what would be expected, since a person who is not hungry is unlikely to order many items. The items that can be offered to a person, even if he is not hungry at all, are suggested in the next policy:

Policy 5 "If, at the point of sale (POS), a customer does not seem to be hungry, suggest Iced White Chocolate Mocha, Java Chip or Orange Juice."

The coloring of the vertices revealed insights on the behavioral patterns, explaining them through personal attributes. The fact that the association graphs can be interpreted even by the least technical analysts is a big advantage and a great motivation for using the methodology in the real world.

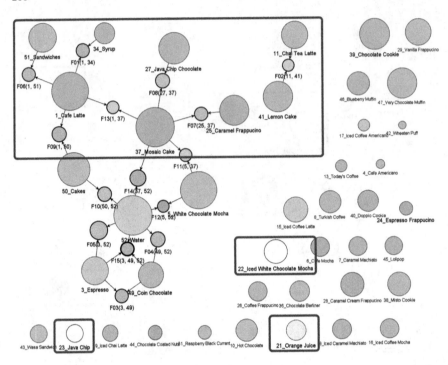

Fig. 12.4 Hunger level attribute (average hunger level for those selecting each itemset) mapped to color

12.5 Validity

A fundamental question, regarding the validity of the proposed AssocGraphRM methodology, can be posed in the line of the classical dilemma of statistical data analysis [30]: "Is the discovered relation a result of causality, or is it just correlation?" For example, considering Policy 4 in Sect. 12.4, does a person who is fluent in French order Orange Juice due to his knowledge of French, or due to some other reason, which would explain both his language proficiency and preference for Orange Juice? For practical purposes, this is not a problem. Even if the underlying attribute for the behavior is hidden, the visible attribute FrenchLevel signals the presence of the detected specific behavior. The person should still be offered Orange Juice in the store (Policy 4), especially if he does not seem to be hungry (Policy 5).

A notable shortcoming of the methodology, related with the above issue, is that it enables re-mining only on a single additional attribute. However, there may be conflicting outcomes, or deeper interactions at deeper levels of analysis that would make the policies suggested at single-level depth invalid. For example, Policy 4 in the case study suggests Orange Juice to a person who is fluent in French. However, Policy 5 suggest the same item to customers who are not hungry at all. So what should be done when a *hungry* French-speaking person arrives? Due to FrenchLevel attribute, he should be offered Orange Juice, but due to HungerLevel attribute, he should definitely not be offered that item. One way to resolve this conflict would be to offer only "safe" items, which do not create a conflict. For the described example, Doppio Cookie and Caramel Machiato are two items that cater to both French-speaking people and to hungry people. So these items can be suggested to the mentioned customer.

A threat to validity of the analysis in the case study is the validity of the collected data. Preferences for food items are heavily influenced by the time of the day, as well as temperature and other conditions under which the data was collected. A French-speaking person would most probably prefer to drink Orange Juice in a warmer day, and a hot Caramel Machiato on a colder day, but Policy 4 does not currently differentiate between the two situations. The survey did not record all such conditions and thus the threat to the validity of the listed policies is indeed pertinent. However, the main contribution of this study is the AssocGraphRM methodology, rather than the specific policies, and this threat to validity does not affect the main contribution.

It is crucial that the policies obtained through AssocGraphRM be handled through a scientific analysis-based approach: They should be put into computational models for justifying their feasibility quantitatively. For example, Policy 3 in the case study of Sect. 12.4 suggests that Caramel Frappucino should be offered to a male customer, rather than Java Chip Chocolate, since he has a higher chance of accepting the former offer. But what if the profit margin of the latter was much higher? Then it might be feasible to offer to the customer the same item that is offered to the female customer. So the superiority of this policy can not be guaranteed by our methodology alone, without a formal fact-based numerical analysis. Thus, the policies should not be applied in isolation, but taking into consideration other critical information, and their interactions.

12.6 Conclusion and Future Work

A novel methodology was introduced for knowledge discovery from association mining results. The applicability of the methodology was illustrated through a market basket analysis case study, where frequent itemsets derived from transactional preference data were analyzed with respect to the personal attributes of survey participants. The theoretical contribution of the study is the methodology, which is formally described as an algorithm. The practical contribution of the study is the proof-of-concept demonstration of the methodology through a case study.

In every industry, especially food retail industry, new products emerge and consumer preferences change at a fast pace. Thus one would be interested in laying the foundation of an analysis framework that can fit to the dynamic nature of retailing data. The presented methodology can be adapted for analysis of frequent itemsets and association rules over time by incorporating latest research on evolving graphs [16] and statistical tests for measuring the significance of changes over time.

The main motivation of the chapter is the discovery of behavioral patterns in relation to the human that exhibited the behavior. The methodology can be applied to similar data from different fields that study the behavior of agents individually and in relation to each other, including psychology, sociology, behavioral economics, behavior-based robotics, and ethology.

Acknowledgements The authors thank İlhan Karabulut for her work that inspired this research, Ahmet Şahinöz for creating colored graphs with the earlier datasets, that inspired the final form of the graphs. The authors thank Samet Bilgen, Dilara Naibi, Ahmet Memişoğlu, and Namık Kerenciler for collecting the data used in the study, and to Didem Cansu Kurada for her insightful suggestions regarding the study.

References

1. Agrawal, R., Srikant, R.: Fast algorithms for mining association rules in large databases. In: Proceedings of the 20th International Conference on Very Large Data Bases, pp. 487–499. Morgan Kaufmann, San Mateo (1994)
2. Borgelt, C.: http://fuzzy.cs.uni-magdeburg.de/~borgelt/apriori.html (2011)
3. Brijs, T., Swinnen, G., Vanhoof, K., Wets, G.: Building an association rules framework to improve product assortment decisions. Data Min. Knowl. Discov. **8**(1), 7–23 (2004)
4. Cao, L.: Behavior informatics and analytics: Let behavior talk. In: IEEE International Conference on Data Mining Workshops, 2008, ICDMW'08, pp. 87–96 (2008)
5. Cao, L.: In-depth behavior understanding and use: the behavior informatics approach. Inf. Sci. **180**, 3067–3085 (2010)
6. Cao, L., Zhang, C.: The evolution of KDD: towards domain-driven data mining. Int. J. Pattern Recognit. Artif. Intell. **21**(4), 677–692 (2007)
7. Ceglar, A., Roddick, J.F.: Association mining. ACM Comput. Surv. **38**(2), 5 (2006)
8. Changchien, S.W., Lu, T.-C.: Mining association rules procedure to support on-line recommendation by customers and products fragmentation. Expert Syst. Appl. **20**(4), 325–335 (2001)
9. Chatti, M., Jarke, M., Indriasari, T., Specht, M.: NetLearn: social network analysis and visualizations for learning. In: Learning in the Synergy of Multiple Disciplines, pp. 310–324 (2009)
10. Chen, C.: Information visualization. Wiley Interdiscip. Rev.: Comput. Stat. **2**(4), 387–403 (2010)
11. Demiriz, A.: Enhancing product recommender systems on sparse binary data. Data Min. Knowl. Discov. **9**(2), 147–170 (2004)
12. Demiriz, A., Ertek, G., Atan, T., Kula, U.: Re-mining positive and negative association mining results. Lect. Notes Comput. Sci. **6171**, 101–114 (2010)
13. Demiriz, A., Ertek, G., Atan, T., Kula, U.: Re-mining item associations: Methodology and a case study in apparel retailing. Decis. Support Syst., (2011). doi:10.1016/j.dss.2011.08.004
14. EditGrid: http://www.editgrid.com (2011)
15. Ertek, G., Demiriz, A.: A framework for visualizing association mining results. Lect. Notes Comput. Sci. **4263**, 593–602 (2006)
16. Erten, C., Harding, P.J., Kobourov, S.G., Wampler, K., Yee, G.: GraphAEL: Graph animations with evolving layouts. In: Lecture Notes in Computer Science, vol. 2912, pp. 98–110. Springer, Berlin (2004)
17. Fruchterman, T.M.J., Reingold, E.M.: Graph drawing by force-directed placement. Softw. Pract. Exp. **21**(11), 1129–1164 (1991)
18. Grahne, G., Zhu, J.: Fast algorithms for frequent itemset mining using fp-trees. In: IEEE Transactions on Knowledge and Data Engineering, pp. 1347–1362 (2005)
19. Han, J., Kamber, M.: Data Mining: Concepts and Techniques. Morgan Kaufmann, San Mateo (2006)

20. Herman, I., Melançon, G., Marshall, M.S.: Graph visualization and navigation in information visualization: A survey. IEEE Trans. Vis. Comput. Graph. **6**(1), 24–43 (2000)
21. Kim, W., Choi, B.J., Hong, E.K., Kim, S.K., Lee, D.: A taxonomy of dirty data. Data Min. Knowl. Discov. **7**(1), 81–99 (2003)
22. Kimani, S., Lodi, S., Catarci, T., Santucci, G., Sartori, C.: VidaMine: a visual data mining environment. J. Vis. Lang. Comput. **15**(1), 37–67 (2004)
23. Liu, B., Hsu, W., Ma, Y.: Pruning and summarizing the discovered associations. In: Proceedings of the Fifth ACM SIGKDD International Conference on Knowledge Discovery and Data Mining, pp. 125–134. ACM, New York (1999)
24. Ltifi, H., Ayed, B., Alimi, A.M., Lepreux, S.: Survey of information visualization techniques for exploitation in KDD. In: IEEE/ACS International Conference on Computer Systems and Applications, 2009, AICCSA 2009, pp. 218–225. IEEE Press, New York (2009)
25. Maimon, O.Z., Rokach, L.: Data Mining and Knowledge Discovery Handbook. Springer, New York (2005)
26. Mansingh, G., Osei-Bryson, K.M., Reichgelt, H.: Using ontologies to facilitate postprocessing of association rules by domain experts. Inform. Sci. (2010)
27. Mansmann, F., Fischer, F., Keim, D.A., North, S.C.: Visual support for analyzing network traffic and intrusion detection events using TreeMap and graph representations. In: Proceedings of the Symposium on Computer Human Interaction for the Management of Information Technology, pp. 19–28. ACM, New York (2009)
28. Mazza, R.: Introduction to Information Visualization. Springer, New York (2009)
29. Microsoft: MS SQL Server, Analysis Services. http://tinyurl.com/6gudq23 (2011)
30. Nowak, S.: Some problems of causal interpretation of statistical relationships. Philos. Sci. **27**(1), 23–38 (1960)
31. O'Hare, S., Noel, S., Prole, K.: A graph-theoretic visualization approach to network risk analysis. Vis. Comput. Secur. **60**–67 (2008)
32. Pavlopoulos, G.A., Wegener, A.L., Schneider, R.: A survey of visualization tools for biological network analysis. Biodata Min. **1**, 12 (2008)
33. Perer, A., Shneiderman, B.: Integrating statistics and visualization: case studies of gaining clarity during exploratory data analysis. In: Proceeding of the Twenty-Sixth Annual SIGCHI Conference on Human Factors in Computing Systems, pp. 265–274. ACM, New York (2008)
34. Santamaría, R., Therón, R.: Overlapping clustered graphs: co-authorship networks visualization. In: Smart Graphics, pp. 190–199. Springer, Berlin (2008)
35. SAS: http://www.sas.com (2011)
36. Shen, Z., Mobivis, K.L.Ma.: A visualization system for exploring mobile data. In: IEEE Pacific Visualization Symposium, 2008, PacificVIS'08, pp. 175–182. IEEE Press, New York (2008)
37. Tamassia, R., Palazzi, B., Papamanthou, C.: Graph drawing for security visualization. In: Graph Drawing, pp. 2–13. Springer, Berlin/Heidelberg (2009)
38. Tollis, I., Eades, P., Di Battista, G., Tollis, L.: Graph Drawing: Algorithms for the Visualization of Graphs. Prentice Hall, New York (1998)
39. Wattenberg, M.: Visual exploration of multivariate graphs. In: Proceedings of the SIGCHI Conference on Human Factors in Computing Systems, pp. 811–819. ACM, New York (2006)
40. WolframAlpha: http://www.wolframalpha.com (2011)
41. Yao, Y., Zhao, Y., Maguire, R.: Explanation-oriented association mining using a combination of unsupervised and supervised learning algorithms. Lect. Notes Comput. Sci. **2671**, 527–532 (2003)
42. Zaki, M.J.: Scalable algorithms for association mining. IEEE Trans. Knowl. Data Eng. **12**(3), 372–390 (2002)
43. Zaki, M.J., Hsiao, C.J.: Efficient algorithms for mining closed itemsets and their lattice structure. IEEE Trans. Knowl. Data Eng. **17**(4), 462–478 (2005)
44. Zhao, Y., Zhang, H., Cao, L., Zhang, C., Bohlscheid, H.: Combined pattern mining: from learned rules to actionable knowledge. In: Proc. of the 21st Australasian Joint Conference on Artificial Intelligence (AI 08), pp. 393–403 (2008)

45. McNicholas, P.D., Zhao, Y.: Association rules: an overview. In: Zhao, Y., Zhang, C., Cao, L. (eds.) Post-Mining of Association Rules: Techniques for Effective Knowledge Extraction, May 2009, pp. 1–10. Springer, Berlin (2009). ISBN 978-1-60566-404-0. Information Science Reference

Chapter 13
Mining Causality from Non-categorical Numerical Data

Joaquim Silva, Gabriel Lopes, and António Falcão

Abstract Causality can be detectable from categorical data: hot weather causes dehydration, smoking causes cough, etc. However, in the context of numerical data, most of the times causality is difficult to detect and measure. In fact, considering two time series, although it is possible to measure the correlation between both associated variables, correlation metrics don't show the cause-effect direction and then, *cause* and *effect* variables are not identified by those metrics.

In order to detect possible cause-effect relationships as well as measuring the strength of causality from non-categorical numerical data, this paper presents an approach which is a simple and efficient alternative to other methods based on regression models.

13.1 Introduction

Astronomers work with data from several time series produced by satellite and ground based observations. Each of these data sets may be seen as a parameter/variable. Correlation metrics can be used to assess how strongly related two variables are, but they don't show the causality direction, i.e., which one is the cause, which one is the effect, a very important issue in behavior informatics [7].

We were invited to develop an efficient approach to detect all possible cause-effect relationships between variables/parameters, as it would be an useful tool for astronomers, taking into account the large number of available time series they work with. This motivated us to work on a more generic approach to deal with data sets containing numerical data, not only time series, in the sense that sets didn't have to

---

J. Silva (✉) · G. Lopes
FCT/Universidade Nova de Lisboa, 2829-516 Caparica, Portugal
e-mail: jfs@di.fct.unl.pt

G. Lopes
e-mail: gpl@di.fct.unl.pt

A. Falcão
Uninova, 2829-516 Caparica, Portugal
e-mail: ajf@uninova.pt

L. Cao, P.S. Yu (eds.), *Behavior Computing*,
DOI 10.1007/978-1-4471-2969-1_13, © Springer-Verlag London 2012

be sequences of data points measured at successive times spaced at uniform time intervals; and not only for the astronomy domain.

Considering that these sets contain non-categorical numerical data, Bayesian Networks or other methodologies using conditional probabilities would not be the most adequate approaches to use. In fact, taking variables such as *DeathReason* and *Smoker* assuming possible values like "lung cancer", "heart disease", "Parkinson's disease" or "other reason" for *DeathReason*, and "yes" or "no" for *Smoker*, it is possible to calculate, for example, $p(DeathReason = $ "lung cancer"$|Smoker = $ "yes"), meaning the conditional probability that someone has died with "lung cancer" given that he/she was a smoker; but it wouldn't make sense to calculate $p(Precipitation = 3.92|Temperature = 10.8)$, standing for the probability of the precipitation in a certain town being equal to exactly 3.92 millimeters given that the average temperature is exactly 10.8 degrees Celsius. In fact, though some numerical data can be discretized—for example, some medical tests have standard discretization: "below normal", "normal" or "above normal"—a great number of time series contain data that is not suitable to be put into ranges: Solar Radio Flux, Sunspot Number, Atmospheric Pressure, Precipitation, etc.

Some causality approaches use regression models. They assume that there is a functional dependency that rules the variables cause and effect, and a regression function is calculated to fit that functional dependency. However, sometimes the real quality of the regression functions are not good.

In this paper we present an approach which is not dependent on regression models. The core idea lays on the assumption that, if X causes Y then, for all particular values of X, say $X = x$, to small deviations from x there must correspond small dispersions in Y values. Then, a χ^2 test decides if those dispersions are small enough so that we may believe that there is a cause-effect relationship.

Section 13.2 presents the related work. Section 13.3 describes our approach. Results are presented in Sect. 13.4 and the last section presents conclusions.

13.2 Related Work

In [1], Bayesian Networks are used to infer cause-effect relationships from gene expression profiles of cancer versus normal cells. This approach needs to calculate conditional probabilities using categorical variables such as $Y = 0$ and $Y = 1$ meaning *normal cells* and *cancer cells*. However, considering that we are interested in detecting and measuring cause-effect relationships between, for example, time series Solar Radio Flux and Neutron Monitor, data is not categorical in these contexts: Solar Radio Flux values can be for example, 75.5, 76.1, 76.1; and 6405, 6468, 6453 are some values taken from the Neutron Monitor time series. So, unless adaptations can be introduced, approaches based on Bayesian Networks are not adequate to deal with this type of data we need to work with.

References [2] and [3] contain very important investigation in causality. The author defines the *total effect* and *direct effect* concepts: *The total effect of X on Y is given by* $P(y|do(x))$, *namely, the distribution of Y while X is held constant at x and all other variables are permitted to run their natural course.* The second concept is also defined with conditional probabilities; see [2] and [3] for details. These probabilities are calculated for categorical variables such as $X = men$ or $X = woman$, etc. Clearly, these concepts are not adequate to detect cause-effect relationships between variables containing non-categorical numerical data.

Authors in [4] and [5] research on association rules. In [4], association rules are mined in large relational tables containing both quantitative and categorical attributes. In [6], user-specified constraints on those association rules are included. An example of such an association might be $\langle Age : 30..39\rangle$ *and* $\langle Married : Yes\rangle \Rightarrow$ $\langle NumCars : 2\rangle$ *with 100% confidence*, meaning that 100% of married people between age 30 and 39 have at least 2 cars. These approaches deal with quantitative attributes such as *Age, NumCars, Income*, etc., by fine-partitioning the values of the attribute and then combining adjacent partitions as necessary. This way, original quantitative attributes can be dealt as they were categorical attributes. So, supposing that the rule of the example given above was mined from a large relational table, we would say that "if someone between 30 and 39 is married it *implies/causes* that he/she has at least two cars". So, for some applications, these approaches could be used for detection of cause-effect relationships. However, they could not be used when data range in variables is unknown or when the nature of that data doesn't allow a fine-partitioning of their values, as it happens for many time series.

In [8], authors present a model that predicts an agent's ascriptions of causality between two events in a chain, based on background knowledge about the normal course of the world. Unfortunately, this approach is not applicable to non-categorical numerical data. In [9, 10] and [11] we find causality approaches that lay on regression models for prediction. However, regression calculations may generate *unacceptable* regression functions. Let us take, for instance, two data sets composed of the following (x, y) pairs: $\{(10, 8.04), (8, 6.95), (13, 7.58), (9, 8.81), (11, 8.33), (14, 9.96), (6, 7.24), (4, 4.26), (12, 10.84), (7, 4.82), (5, 5.68)\}$ for data set (a); and $\{(8, 6.58), (8, 5.76), (8, 7.71), (8, 8.84), (8, 8.47), (8, 7.04), (8, 5.25), (19, 12.5), (8, 5.56), (8, 7.91), (8, 6.89)\}$ for data set (b).

Each data set yields the same standard output from a typical regression program, namely: Number of observations $(n) = 11$; Mean of the x's $(\bar{x}) = 9.0$; Mean of y's $(\bar{y}) = 5.5$; Regression coefficient $(b1)$ of y on $x = 0.5$; Equation of regression line: $y = 3 + 0.5x$; Sum of squares of $x - \bar{x} = 110.0$; Regression sum of squares = 27.5 (1 d.f.); Residual sum of squares of $y = 13.75$ (9 d.f.); Estimated standard error of $b1 = 0.118$; Multiple correlation coefficient $R^2 = 0.667$. By this summary, the regression function would be accepted for both data sets. The data sets are graphed in Fig. 13.1, together with the fitted regression line. Although, after looking at Fig. 13.1, while the regression line seems to *follow* the scatter of data set (a), the same is not true for data set (b). In fact, there is something unsatisfactory about data set (b): there are only 2 different x values and all the information about the slope of the regression line resides in one observation (for $x = 19$)—if that observation was deleted the slope could not be estimated. Besides, there are significant differences among the 10 values of y for the other x value ($x = 8$). Thus, the suggested

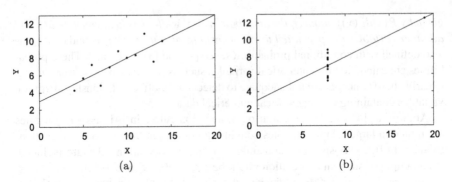

Fig. 13.1 The same calculated regression line for two very different data sets

regression line could be a good tool as a base to detect a cause-effect relationship between the corresponding variables X and Y for data set (a), but obviously not for data set (b). As we will see in Sect. 13.3, we propose a simple alternative approach which does not depend on the quality of the regression functions as there is no need of regression models.

13.3 Detecting Cause-Effect Relationships and Measuring Their Strengths

Considering two variables X and Y populated with non-categorical numerical data, in our approach we assume that X determines Y if, for all possible small ranges of values in X domain there correspond small dispersions in Y domain. In other words, if X *causes* Y then, if $Y = y$ when $X = x$, it is also the case that $Y = y1$ when $X = x1$ being $x1$ and $y1$ very close values to x and y respectively.

Thus, let variables X and Y represent two numerical data sets. So, in order to decide if X causes Y we follow 3 steps: Partitioning Y data according to small range sets in X; Measuring the dispersion of each partition in Y; Testing cause-effect relationship from dispersions.

13.3.1 Partitioning Y Data According to Small Range Sets in X

Let X and Y be data sets such that $X = \{x_1, x_2, \ldots, x_N\}$ and $Y = \{y_1, y_2, \ldots, y_N\}$. And be XY composed from X and Y such that, $XY = \{(x_i, y_i) : x_i = X(i), y_i = Y(i)\}$ where $X(i)$ and $Y(i)$ stand for the ith elements of X and Y respectively. Then, in order to partition Y data we follow Algorithm 13.1, shown ahead. Algorithm 13.1 lays on *partitioning*($XY, p, OutputList$), a procedure which receives: data set XY; the p value to control the range of values in X data; and the *OutputList* which is an empty list when procedure *partitioning* is called. This procedure returns *OutputList*

Algorithm 13.1 Partitioning Y data according with small range sets in X

{Parameters: XY (initially composed from X and Y data sets); p controls the range of values in X data subsets; *OutputList* is the list of partitions of Y data}

$partitioning(XY, p, OutputList)$
$X \leftarrow \{x | (x, y) \in XY\}$
$Y \leftarrow \{y | (x, y) \in XY\}$
if $max(X) > min(X).(1 + p) \wedge \|X\| > MinSize$ **then**
   $XY_{left} \leftarrow \{(x, y) | (x, y) \in XY \wedge x \leq median(X)\}$
   $XY_{right} \leftarrow \{(x, y) | (x, y) \in XY \wedge x > median(X)\}$
   $partitioning(XY_{left}, p, OutputList)$
   $partitioning(XY_{right}, p, OutputList)$
   return OutputList
else
   append Y to OutputList
   return OutputList
end if

---

containing a list of subsets of values from data set Y such that for each of these subsets, the *index-corresponding* subset in data set X contains a very small dispersion among its values. Ranges of values in subsets of X are controlled by parameter p which must be small enough to assure small ranges, but also big enough so each subset must have at least *MinSize* elements. *MinSize* has been set to 2 and p has been chosen by the user.

Thus, procedure *partitioning* starts by obtaining a copy of the X *part* and a copy of the Y *part* of data set XY. So, if the range of values in the X part exceeds the permitted value specified by p (the 4th line of the procedure), then XY is split into two new *data sets*: a left one and a right one, both having similar length. Thus, the left one, XY_{left}, is made from all elements in XY where the values in the X part are lower or equal than the middle value in X after sorting it. We used the *median* to obtain the middle value. Obviously, the right one, XY_{right}, is made from all elements in XY where the values in the X part are greater than the middle value in X (the 5th and 6th lines of the procedure). Then, taking XY_{left} and XY_{right}, procedure *partitioning* is called recursively (the 7th and 8th lines) until the range of values are small enough (the 4th line). If the range of values in the X part is small enough, the Y part of XY is appended to the *OutputList*. As an example, let us consider that XY is composed from data sets X and Y such that $XY =$ {(537.21, 7.70), (537.06, 7.73), (536.75, 7.75), (535.81, 7.78), (534.80, 7.80), (534.56, 7.81), (328.01, 21.03), (327.45, 21.94), (326.30, 22.03), (325.04, 22.15), (207.03, 29.03), (206.44, 29.47), (205.91, 29.61), (205.39, 30.51), (205.18, 30.54), (204.72, 30.60)}; now let us suppose we want to partition Y according to ranges in X, such that $p = 1\%$. By Algorithm 13.1, we would obtain the following list of subsets (partitions) from Y: {30.60, 30.54, 30.51, 29.61}, {29.47, 29.03}, {22.15, 22.03}, {21.94, 21.03}, {7.78, 7.75, 7.73, 7.70} and { 7.81, 7.80}. Taking the first partition as an example, { 30.60, 30.54, 30.51, 29.61}, the reader may notice that the corresponding values in the X part are $\{204.72, 205.18, 205.39, 205.91\}$, which respect the p range restriction, since $205.91 \neq 204.72.\,(1 + p)$, for $p = 1\%$ (see the 4th line in Algorithm 13.1).

13.3.2 Measuring the Dispersion of Each Partition in Y

Now we need to measure the dispersion of the values of each partition contained in the list returned by Algorithm 13.1. To do that, first we considered measuring the range of values in each partition in Y, that is, similarly to what is done for the X part, comparing the extreme values in the partition. However, though a simple tool, we concluded that comparison wouldn't be a good metric since it is too sensitive to outlier values in the partition and may hide the *average* dispersion among its values. In other words, a partition having 10 values may be made of 9 very close values and another one which deviates considerably from the rest. This outlier may happen because values in data set Y may be determined not only by data set X, but also by other parameters. However, in this work we just want to check the existence of the cause-effect relationship from X to Y. For example, precipitation is determined not only by minimum air temperature, but also by atmospheric pressure and other parameters. So, by comparing extreme values in a partition, it wouldn't be possible to smooth the outlier effect and assess the real low dispersion of the partition. Then, in order to measure the dispersion, we considered the well-known *sample standard deviation*:

$$\sigma(P) = \sqrt{\frac{1}{\|P\|}\sum_{p_i \in P}(p_i - \mu(P))^2}, \quad \mu(P) = \frac{1}{\|P\|}\sum_{p_i \in P}p_i \qquad (13.1)$$

being $p_1, p_2, \ldots, p_n$ the elements of the sample, that is, in our case, the elements of a partition P; and $\|P\|$ stands for the number of elements of P. However, since this metric is scale-sensitive it didn't show to be a good tool to assess the dispersion considering our purpose. For example, be P_1 and P_2 two partitions such that $P_1 = \{2, 2.01, 2.02, 2.03, 2.04\}$ and $P_2 = \{200, 201, 202, 203, 204\}$; their respective sample standard deviations are $\sigma(P_1) = 0.0141$ and $\sigma(P_2) = 1.41$, that is, according to this metric, the dispersion is 100 times greater in P_2 than in P_1; this difference does not capture the similar *relative dispersion* shown by both partitions. In fact, apart from the scale factor, both sets show equal relative deviations among their values. Then we tried to apply the *sample coefficient of variation* which is a normalized measure of dispersion, defined by the ratio of the sample standard deviation $\sigma(P)$ to the mean $\mu(P)$.

$$Cv(P) = \sigma(P).\mu(P)^{-1}. \qquad (13.2)$$

Now $Cv(P_1) = Cv(P_2) = 0.00700$. This showed to be a good metric for the most cases we worked with. However, it isn't defined when mean is zero. Moreover, mean may be very close to zero, not because the elements of the set may be very small positive values, but because the sum of the positive elements may be similar to the absolute value of the sum of the negative elements. In these cases, Cv doesn't work as we need. For example, let partitions $P_3 = \{-0.0100, -0.0110, 0.0100, 0.01101\}$ and $P_4 = \{-0.0100, -0.0110, 0.0100, 0.01095\}$; then $Cv(P_3) = 4205.805$ and $Cv(P_4) = -839.906$, which don't reflect the magnitude of the relative dispersion values we expected for these partitions. In order to overcome this problem, we propose in this paper a metric which is defined by the ratio of the sample standard deviation $\sigma(P)$ to the mean of the absolute values of all elements of the sample:

$$Rd(P) = \frac{\sigma(P)}{\frac{1}{\|P\|} \sum_{p_i \in P} |p_i|}. \tag{13.3}$$

Now, $Rd(P_3) = 1.00108$ and $Rd(P_4) = 1.00114$ which reflects the magnitude of the relative dispersion we expected for these partitions. Besides, Rd and Cv return the same values when partitions contain just positive elements: for example $Rd(P_1) = 0.007001 = Cv(P_1)$ and $Rd(P_2) = 0.007001 = Cv(P_2)$. Contrary to Cv, Rd is defined when mean is zero, as long as at least one of the elements of the partition is not zero.

13.3.3 Testing Cause-Effect Relationship from Dispersions

Now we need to decide if there is a cause-effect relationship from data set X to Y. We use the relative dispersion given by Rd metric to help on this decision. Thus, let us suppose that partitions in Y are obtained according to ranges in X. So, if the relative dispersion of the values of a generic partition P in Y is lower than the relative dispersion of the overall time series Y, then we say that X causes Y. On the other side, if these relative dispersion are similar we say that X doesn't cause Y. Thus, we state the following null hypothesis:

$\mathcal{H}_0$: *Considering that P is a generic partition of data set Y, obtained according to Algorithm 13.1, then $Rd(P)$ is close to $Rd(Y)$.*

We use the Pearson chi-square statistic to test $\mathcal{H}_0$. So,

$$X^2 = \sum_{i=1}^{i=k} \frac{(O_i - E_i)^2}{E_i}. \tag{13.4}$$

In (13.4), $k = 2$ since there are two different cases we have to deal with: case 1 is associated to the number of partitions having a relative dispersion that is lower or equal to the relative dispersion of the overall data set Y, that is when $Rd(P) \leq Rd(Y)$, being P a generic partition; case 2 is associated to the number of partitions where $Rd(P) > Rd(Y)$. Thus, O_1 stands for the frequency of the observations associated to case 1, and E_1 means the expected frequency for case 1 considering $\mathcal{H}_0$. O_2 and E_2 are obviously associated to case 2. As an example, let us suppose that Algorithm 13.1 returned 8000 partitions from a data set Y, from where there are 7500 such that for each one, $Rd(P) \leq Rd(Y)$. Considering $\mathcal{H}_0$, we would expect about $8000/2$ partitions where $Rd(P) \leq Rd(Y)$, and the same number for those partitions where $Rd(P) > Rd(Y)$. Then $O_1 = 7500$, $O_2 = 500$, $E_1 = 4000$, $E_2 = 4000$ and $X^2 = (7500 - 4000)^2/4000 + (500 - 4000)^2/4000 = 6125$.

Thus, by using any cumulative chi-square distribution table, for a level of significance of α, we *reject* $\mathcal{H}_0$ iff:

$$X^2 > \chi^2_{df}(\alpha) \tag{13.5}$$

where df is the number of degrees of freedom, now given by $df = k - 1 = 1$. We have been using the level of significance of $\alpha = 0.05$. For these values, the critical value for X^2 is $\chi^2_1(0.05) = 3.84$. Notice that this test must be done only when the total number of observations is large enough. According with authors in [12], it has been found that this number must be at least, say, four or five times the number of cells. In our case the number of *cells* are two ($k = 2$) and the number of observations (partitions) are at least some tens, sometimes hundreds or thousands. So, tests ruled by (13.5) can be taken as valid.

We verified that whenever $\mathcal{H}_0$ was rejected, there was really a cause-effect relationship from X to Y. Although, in theory, $\mathcal{H}_0$ can be rejected because O_2 is much larger than E_2; not because O_1 is much larger than E_1. Thus, in order to prevent this kind of wrong decisions, we state that:

- *There is a cause-effect relationship from data set X to data set Y, iff $\mathcal{H}_0$ is rejected and $AvgRd(\mathcal{P}) < Rd(Y)$.*

$\mathcal{P}$ is the set of the partitions in Y returned by Algorithm 13.1 and $AvgRd(\mathcal{P})$ stands for the mean of the Rd values of all partitions in Y, which is given by

$$AvgRd(\mathcal{P}) = \frac{1}{\|\mathcal{P}\|} \sum_{P_i \in \mathcal{P}} Rd(P_i). \tag{13.6}$$

13.3.4 The Principal Cause-Effect Relationship

During our experiments, we noticed that it sometimes happens that when there is a cause-effect relationship from data set X to Y, there is also another one from Y to X. In fact, both data sets may affect each other, such as, for example a time series containing air temperatures and another one containing atmospheric pressure, with both measurements taken at the same time for the same place. Although, most of the times, there is a, say, *principal* direction when compared to the opposite direction which we may call the *secondary* direction. We can determine which one is the principal direction and how dominant it is.

Thus, be $\mathcal{P}_r$ the set of partitions in data set Y obtained according to small ranges in data set X, following Algorithm 13.1; and be $\mathcal{P}_l$ the set of partitions in X obtained according to small ranges in Y, following the same algorithm. Now let $Left = AvgRd(\mathcal{P}_l)$ and let $Right = AvgRd(\mathcal{P}_r)$. The principal direction is given by:

$$Dir = \frac{Left - Right}{\max(Left, Right)}. \tag{13.7}$$

Dir values range from −1 to +1. As an example, if *Dir* = 0.99 it means that the principal direction is from data set *X* to data set *Y*, and this direction is dominant; if *Dir* = −0.99 it means that the principal direction is dominant from *Y* to *X*; if *Dir* is close to zero it means that there is no principal direction.

13.4 Results

In order to test our approach, we applied it to some pairs of time aligned time series. For each pair, cause-effect relationships were checked, as well as the principal direction of the relationship. For each pair we kept parameter *p* in Algorithm 13.1 as small as possible, providing that partitions had at least 2 elements. Although, in future work further research will be done to consider other criteria; then *p* will be automatically chosen for each case. Concerning α, the choice of this parameter will affect the rejection/acceptance of $\mathcal{H}_0$ (see test in (13.5)): a very small value (ex.: $\alpha = 0.001$) may cause some loss of sensibility to detect existing cause-effect relationships ($\chi_1^2(0.001) = 10.83$); on the other hand, a high value (ex.: $\alpha = 0.1$) may detect false cause-effect relationships ($\chi_1^2(0.1) = 2.71$). We have been using $\alpha = 0.05$, a very popular value in statistical applications ($\chi_1^2(0.05) = 3.84$).

For reasons of space, we used each table to show two pairs of time series. The first pair to be tested was obtained from University of Oulu, Finland, http://cosmicrays.oulu.fi/: a time series for parameter *Neutron Monitor* and another one for *Solar Radio Flux*. Thus, the left part of Table 13.1 shows a small part of this pair of aligned daily time series. For this pair, we set *p* = 1%; then, cause-effect relationships from *Neutron Monitor* to *Solar Radio Flux* was detected, as well as the reverse one, that is *Solar Radio Flux* to *Neutron Monitor*. However, the second one was identified as the principal and very dominant cause-effect relationship by (13.7), since *Dir* = −0.993. This result is coherent as scientists know that *Solar Radio Flux* influences *Neutron Monitor* values. The existence of the secondary relationship just means that the values in *Solar Radio Flux* are not completely independent of the values in *Neutron Monitor*.

The second pair to be tested was obtained from the ACE Science Center, http://www.srl.caltech.edu/ace/ASC/index.html: a time series for parameter *Hydrogen (H_S1)* and another one for *Neutron Flux*. The right part of Table 13.1 shows part of these time series. For this pair, we kept *p* = 1%. Cause-effect relationships were detected for both directions. However, the direction from *Hydrogen(H_S1)* to *Neutron Flux* was identified as the principal and very dominant one, as *Dir* = 0.997. This was recognized as a coherent result too.

The left side of Table 13.2 shows part of the available data related to IBM stocks in 2009: http://finance.yahoo.com. In the right side of Table 13.2 we present a pair built from the left side: a time series containing the absolute values of the difference between the *Close* price and the previous day's *Close* price (*PrvClose*); and another one containing the *Volume* of transactions for each day. We had to use *p* = 3% here, in order to obtain partitions having at least 2 elements. Then, cause-effect

Table 13.1 Two pairs of time aligned time series from the astrophysics domain: Neutron Monitor and Solar Radio Flux; and Hydrogen (H_S1) and Neutron Flux

Neutron Monitor	Solar Radio Flux	Hydrogen H_S1	Neutron Flux
6449	77.5	4.18E+03	93.625
6433	75.5	8.42E+02	93.458
6430	76.9	2.56E+03	93.674
6404	76.9	1.67E+03	94.039
6450	76.1	3.39E+03	94.167
6383	75.7	2.52E+03	94.204
6388	75.6	5.09E+03	94.165
6403	73.6	2.56E+03	94.054
6411	75.1	1.68E+03	93.699
6406	72.9	8.42E+02	93.678
6450	74.0	2.52E+03	93.610
6453	72.8	8.48E+02	93.207
6468	73.1	2.52E+03	93.131
6405	71.7	8.48E+02	93.229
⋮	⋮	⋮	⋮

Table 13.2 IBM stock prices for 2009

Date	Open	High	Low	Close	Volume	$|Close - PrvClose|$	Volume
2009-12-31	132.41	132.85	130.75	130.90	4223400	1.67	4223400
2009-12-30	131.23	132.68	130.68	132.57	3867000	0.72	3867000
2009-12-29	132.28	132.37	131.80	131.85	4184200	0.46	4184200
2009-12-28	130.99	132.31	130.72	132.31	5800400	1.74	5800400
2009-12-24	129.89	130.57	129.48	130.57	4265100	0.57	4265100
.	.	.	.	.	.	0.07	4127600
.	.	.	.	.	.	1.28	5535500
.	.	.	.	.	.	0.74	4772500
.	.	.	.	.	.	0.51	9106600
.	.	.	.	.	.	1.31	5909500
.	.	.	.	.	.	0.22	6372500
.	.	.	.	.	.	1.44	7862600
.	.	.	.	.	.	0.25	5201300
.	.	.	.	.	.	0.34	6597200
⋮	⋮	⋮	⋮	⋮	⋮	⋮	⋮

Table 13.3 Minimum Daily Air Temperature and Precipitation

Minim Daily Air Temperature	Precipitation	Random 1	Random 2
6.0	0.1	1.70990730576	3.27963949203
6.0	3.2	6.75568659158	7.03206162118
7.5	3.2	2.36628948303	2.49647126814
7.5	0.0	1.59258546664	1.76795665048
4.9	0.0	6.70111556089	0.233526916995
6.2	3.6	8.36871360842	4.39915014498
4.2	10.1	8.21712564928	1.27329866438
1.7	0.7	4.44562604687	4.80382818154
3.0	1.8	5.86858487717	5.20883064475
3.9	7.0	0.587672998956	6.4485330581
2.5	0.5	1.34406834937	7.78643170293
1.4	0.5	6.22154597205	5.25530285293
−0.5	2.5	9.56123292807	5.37890969541
0.9	1.7	6.10342111996	6.48056520001
$\vdots$	$\vdots$	$\vdots$	$\vdots$

relationships were detected for both directions. The principal one was detected from $|Close - PrvClose|$ to *Volume*, but not as dominant as the cases presented before, as $Dir = 0.870$. This result was recognized as coherent: according to people who deal with stock market, if there is a significant change in the price from yesterday to today, then stock market get nervous, that is, volume of transactions increases.

The fourth pair we tested was formed by: a time series containing *Minimum Daily Air Temperature* in degrees Celsius; and another one having *Precipitation* values in millimeters. These data were obtained from the Daily Temperature and Precipitation Data for 223 Former-USSR Stations, file f.20674.dat, at http://cdiac.ornl.gov/ftp/ndp040/. The left side of Table 13.3 shows part of the pair we built, having data from 1936 to 2001. For this pair, we had to use $p = 2\%$. Again, cause-effect relationships were detected for both directions. However, cause-effect relationship from $Minimum Daily Air Temperature$ to $Precipitation$ was identified as the principal one, as $Dir = 0.917$. We think this is another coherent result as rain is affected by minimum air temperature.

The last pair we tested was built from two random sets of numbers generated from 1 to 10. Thus, we formed two data sets with 20 000 lines. The right side of Table 13.3 shows part of it. For this pair, we used $p = 1\%$. This time cause-effect relationships were not detected for any directions and $Dir = 0.0312$. We think this is a coherent result because data sets are independent.

Last test was made for the problematic data set (b) of Fig. 13.1. No cause-effect relationship was detected. We think this is correct.

In Algorithm 13.1 we also used the arithmetic mean instead of the median. Results were similar and not conclusive. Future work will research on this detail.

13.5 Conclusion

This paper presents an approach for mining causality from non-categorical numerical data in pairs of data series. This is an alternative to other approaches using probabilistic models, since it is not always clear how to categorize numerical data. The core idea of this approach is based on the assumption that a parameter represented in a data set X determines another parameter Y if, for all possible small ranges of values in X domain, there are small dispersions in the corresponding Y domain.

In order to assess the dispersion in data sets' partitions, we propose in this paper the Rd metric, which measures the relative dispersion and overcomes the limitations of the similar metric, *Coefficient of variation*.

By using this approach, cause-effect relationships were detected from several pairs of time series and it was possible to identify the principal direction of the cause-effect relationship and how dominant it was by comparison with the secondary direction. In our tests, all these principal directions showed to be coherent with our knowledge about these parameters.

This is also an alternative to approaches that, by using regression models, are dependent on the quality of the regression functions to conclude about the existence of cause-effect relationships.

The prototype of this approach was built in Python. Due to its simplicity, detection of cause-effect relationships are very fast even for long time series. Thus, we believe we can contribute to easily detect other unknown cause-effect relationships in other contexts. In future work, we want to extend this approach to detect cause-effect relationships from a group of parameters affecting another parameter, still in the context of non-categorical numerical data.

References

1. Polanski, A., Polanska, J., Jarzab, M., Wiench, M., Jarzab, B.: Application of Bayesian networks for inferring cause-effect relations from gene expression profiles of cancer versus normal cells. Math. Biosci. **209**(2), 528–546 (2007)
2. Pearl, J.: Causality: Models, Reasoning, and Inference. Cambridge University Press, Cambridge (2000)
3. Pearl, J.: Causal inference. J. Mach. Learn. Res. **6**, 39–58 (2010)
4. Srikant, R., Agrawal, R.: Mining quantitative association rules in large relational tables. In: Proc. of the ACM-SIGMOD 1996 Conference on Management of Data, Montreal, Canada, June (1996)
5. Cerny, Z.: WWW support for applications of system LISp-Miner. Master thesis, University of Economics Prague (2003), 81 pages (in Czech)
6. Bayardo, R.J. Jr., Agrawal, R., Gunopulos, D.: Constraint-based rule mining in large, dense databases. Data Min. Knowl. Discov. J. **4**(2/3), 217–240 (2000)
7. Cao, L.: In-depth behavior understanding and use: the behavior informatics approach. Inf. Sci. **180**, 3067–3085 (2010)
8. Bonnefon, J., Neves, R.S., Dubois, D.: H. Prade. Background default knowledge and causality ascriptions. In: Brewka, G., Coradeschi, S., Perini, A., Traverso, P. (eds.) Proc. of the 17th

European Conference on Artificial Intelligence (ECAI'06), Riva del Garda, Italy, 29 Aug–01 Sep 2006, pp. 11–15. IOS Press, Amsterdam (2006)

9. Granger, C.W.J.: Investigating causal relations by econometric models and cross-spectral methods. Econometrica **37**(3), 424–438 (1969). Published by: The Econometric Society

10. Swanson, N.R., Granger, C.W.J.: Impulse response functions based on a causal approach to residual orthogonalization in vector autoregressions. J. Am. Stat. Assoc. **92**, 357–367 (1997)

11. Chu, T., Glymour, C.: Search for additive nonlinear time series causal models. J. Mach. Learn. Res. **9**, 967–991 (2008)

12. Lindgren, B.W.: Statistical Theory, 3rd edn. MacMillan, New York (1976)

Chapter 14
A Fast Algorithm for Mining High Utility Itemsets

Show-Jane Yen, Chia-Ching Chen, and Yue-Shi Lee

Abstract Frequent itemset mining generates frequently purchased itemsets, which only considers the presence of an item in a transaction database. However, a frequent itemset may not be the itemset with high value. High utility itemset mining considers both of the profits and purchased quantities for the items, which is to find the itemsets with high utility for the business. The previous approaches for mining high utility itemsets first apply frequent itemset mining algorithm to find candidate high utility itemsets, and then scan the whole database to compute the utilities of these candidates. However, these approaches need to take a lot of time to generate all the candidate high utility itemsets, scan the whole database and search from a large number of candidate high utility itemsets to compute the utilities of these candidates. Therefore, the previous approaches are very inefficient.

In this paper, we present an efficient algorithm for mining high utility itemsets. Our algorithm is based on a tree structure in which a part of utilities for the items are recorded. A mechanism is proposed to reduce the mining space and make our algorithm can directly generate high utility itemsets from the tree structure without candidate generation. The experimental results also show that our algorithm significantly outperforms the previous approaches.

14.1 Introduction

In this section, we first introduce some preliminaries for mining high utility itemsets [7], which may refer to behaviors [3]. Let $I = \{i_1, i_2, \ldots, i_m\}$ be the set of all the items. An itemset X is a subset of I and the length of X is the number of items contained in X. An itemset with length k is called a k-itemset. A transaction database $D = \{T_1, T_2, \ldots, T_n\}$ contains a set of transactions and each transaction has a unique transaction identifier (TID). Each transaction contains the items purchased in this transaction and their purchased quantities. The purchased quantity of item i_p in a transaction T_q is denoted as $o(i_p, T_q)$. The utility of item i_p in T_q is

S.-J. Yen (✉) · C.-C. Chen · Y.-S. Lee
Department of Computer Science and Information Engineering, Ming Chuan University, Taoyuan County, Taiwan
e-mail: sjyen@mail.mcu.edu.tw

L. Cao, P.S. Yu (eds.), *Behavior Computing*,
DOI 10.1007/978-1-4471-2969-1_14, © Springer-Verlag London 2012

Table 14.1 A transaction database

TID	A	B	C	D	E	F	TID	A	B	C	D	E	F
T_1	1	0	3	0	0	1	T_6	0	0	5	0	7	0
T_2	0	4	0	5	0	0	T_7	0	10	0	3	0	0
T_3	7	2	5	7	0	0	T_8	0	2	2	0	3	0
T_4	0	1	0	0	4	0	T_9	8	1	0	5	0	0
T_5	2	0	0	9	0	1	T_{10}	0	5	2	3	0	0

Table 14.2 The profit table

Item	A	B	C	D	E	F
Profit($)	7	2	5	1	10	13

$u(i_p, T_q) = o(i_p, T_q) \times s(i_p)$, in which $s(i_p)$ is the profit of item i_p. The utility of an itemset X in T_q is the sum of the utilities of the items contained in $X \subseteq T_q$, which is shown in expression (14.1). If $X \not\subseteq T_q$, $u(X, T_q) = 0$. The utility of an itemset X in D is the sum of the utilities of X in all the transactions containing X, which is shown in expression (14.2). An itemset X is a high utility itemset if the utility of X in D is no less than user specified *minimum utility* (MU).

$$u(X, T_q) = \sum_{i_p \in X \subseteq T_q} u(i_p, T_q) \tag{14.1}$$

$$u(X) = \sum_{X \subseteq T_q \in D} u(X, T_q) \tag{14.2}$$

For example, Table 14.1 is a transaction database, in which each integer number represents the purchased quantity for an item in a transaction. Table 14.2 is the profit table which records the profit for each item in Table 14.1. Suppose the minimum utility MU is 100. The utility of itemset $\{C, E\}$ in Table 14.1 is $u(\{C, E\}) = (5 \times 5 + 7 \times 10) + (2 \times 5 + 3 \times 10) = 135 \geq 100$. Therefore, the itemset $\{C, E\}$ is a high utility itemset.

For mining frequent itemset [1], all the subsets of a frequent itemset are frequent, that is, there is a downward closure property for frequent itemsets. However, the property is not available for high utility itemsets, since a subset of a high utility itemset may not be a high utility itemset. For example, itemset $\{C, E\}$ is a high utility itemset in Table 14.1, but its subset $\{C\}$ is not a high utility itemset.

Therefore, Liu et al. [7] proposed a Two-Phase algorithm for mining high utility itemsets. They defined the *transaction utility* TU for a transaction and the *transaction weighted utility* TWU for an itemset X, which are shown in expression (14.3) and (14.4), respectively.

$$tu(T_q) = \sum_{i_p \in T_q} u(i_p, T_q) \tag{14.3}$$

$$twu(X) = \sum_{X \subseteq T_q \in D} tu(T_q) \tag{14.4}$$

If the TWU for an itemset is no less than MU, then the itemset is a *high transaction weighted utility itemset* (HTWUI). According to expression (14.4), the TWU for an itemset X must be greater than or equal to the utility of X in D. Therefore, if X is a high utility itemset, then X is a HTWUI. All the subsets of a HTWUI are also HTWUIs. Therefore, there is a downward closure property for HTWUIs. The first phase for the Two-Phase algorithm [7] is to find all the HTWUIs which are called candidate high utility itemsets by applying Apriori algorithm [1]. Two-Phase algorithm scans the database again to compute the utilities for all the candidate high utility itemsets and finds high utility itemsets in the second phase.

However, in the first phase, Two-Phase algorithm needs to repeatedly scan the database and search from a large number of candidate HTWUIs to generate candidate high utility itemsets. If the minimum utility is small, a huge number of candidate high utility itemsets will be generated. In the second phase, Two-Phase algorithm needs to scan the large database again and search from a huge number of candidate high utility itemsets, which would significantly degrade the mining performance.

In order to reduce the number of database scans, Ahmed et al. proposed HUC-Prune algorithm [2], which applies FP-Growth algorithm [5] to generate candidate high utility itemsets and then scan the database and search for the candidate high utility itemsets to compute the utilities for these candidates. Although HUC-Prune only scans database three times, it still needs to generate candidate high utility itemsets, scan the whole database and search for the candidates to find high utility itemsets.

In this paper, we propose an efficient algorithm *AM (Adsorptive Mining)* for mining high utility itemsets. Our AM algorithm only needs to scan database twice to construct a tree structure which is called an AM-Tree (Adsorptive Mining Tree). Our algorithm can generate high utility itemsets directly from the tree structure without candidate generation. We also propose an estimation method to estimate the largest utility among the supersets of an itemset, which is much less than the TWU for an itemset. Unlike TWU which needs to sum the utilities of all the items in a transaction, our estimation method only considers the utilities of some items in a transaction, which will be used in the mining process, such that the mining space can be reduced and the mining performance can be improved.

Fig. 14.1 The constructed AM-Tree for the transaction database in Table 14.1

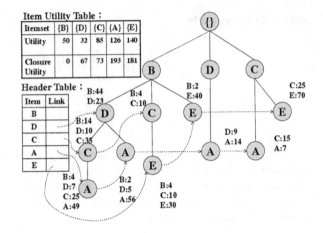

Item Utility Table :

Itemset	{B}	{D}	{C}	{A}	{E}
Utility	50	32	85	126	140
Closure Utility	0	67	73	193	181

Header Table :

Item	Link
B	
D	
C	
A	
E	

14.2 Our Algorithm

In this section, we present our AM algorithm and a running example is used to explain our algorithm. For each node in an AM-Tree, except the null root node, an item is associated. For the AM-Tree, all the child nodes of the root node do not record any information. For each node N which is not a child node of the root node in the AM-Tree, AM needs to record the accumulated utility for every item on the path from root to node N, and a link which points to the next node (not including the child node of the root) with the same associated item as node N. Figure 14.1 is the constructed AM-Tree for the transaction database in Table 14.1. In the following, we describe how to construct an AM-Tree.

For the first database scan, AM computes the TWU and the support count for each item in the database. The items whose TWUs are no less than MU are ordered by their support counts in descending order. For the items with the same support counts, they are ordered by alphabetic order. We use the function $order(x)$ to denote the order of item x. For each transaction in the database, AM sorts the items in the transaction according to the order of the items and deletes the items whose TWU are less than MU. By the way, AM computes the utility of each item in this transaction and puts this transaction into the AM-Tree. In order to explain how to construct an AM-Tree, we have the following definitions and lemmas. In the following, we suppose that item x_{min} is the first order among the items in itemset X.

Definition 14.1 The *transaction closure section* $TCS(T_q, X)$ for an itemset X in a transaction $T_q \supseteq X$ is the union of X and all the items in T_q, whose order are less than x_{min}, which is shown in expression (14.5).

$$TCS(T_q, X) = \{i \mid i \in T_q \wedge order(i) < order(x_{min})\} \cup X \qquad (14.5)$$

For example, suppose the items in transaction $T_q = \{A, B, E, F, J, M, N\}$ are ordered. $TCS(T_q, \{E, N\}) = \{A, B, E, N\}$.

Definition 14.2 The *transaction closure utility* $TCU(T_q, X)$ for an itemset X in transaction T_q is the sum of the utilities of all items in $TCS(T_q, X)$, which is shown in expression (14.6).

$$TCU(T_q, X) = \begin{cases} \sum_{i_p \in TCS(T_q, X)} u(i_p, T_q), & \text{if } TCS(T_q, X) \neq X \\ 0, & \text{if } TCS(T_q, X) = X \end{cases} \qquad (14.6)$$

Definition 14.3 The *closure utility* $CU(X)$ of an itemset X in a transaction database D is the sum of the transaction closure utilities for itemset X in all the transactions in D, which is shown in expression (14.7).

$$CU(X) = \sum_{X \subseteq T_q \in D} TCU(T_q, X) \qquad (14.7)$$

If $CU(X) \geq MU$, then itemset X is a *high closure utility itemset*.

Definition 14.4 A *closure superset* of an itemset X is $X \cup \{y\}$, in which y is an item $\notin X$ and the order of y is less than x_{min}. The *closure subset* of an itemset X is $X - \{x_{min}\}$.

AM extends the length of a high closure utility itemset to find longer high utility itemsets by extending the itemset into its closure supersets. If $TCS(T_q, X) = X$, then there is no closure superset for X in T_q, that is, itemset X cannot be extended in T_q. Therefore, the transaction closure utility $TCU(T_q, X) = 0$.

Lemma 14.1 *The closure subset of a high utility itemset must be a high closure utility itemset.*

We can first determine which itemsets are high closure utility itemsets, and then extend the high closure utility itemsets to obtain longer high closure utility itemsets, since only the closure supersets of a high closure utility itemset may be the high utility itemsets, according to Lemma 14.1. Therefore, the closure utility of an itemset X must be greater than or equal to the utility of a closure superset of X.

Lemma 14.2 *The closure subset of a high closure utility itemset must be a high closure utility itemset.*

From Lemma 14.2, we can see that there is a downward closure property for high closure utility itemsets, that is, if an itemset is not a high closure utility itemset, then all the closure supersets of this itemset are also not high closure utility itemsets. Therefore, AM uses an item utility table to register the closure utility and the utility for each extended itemset. In the following, we describe the construction of item utility table and AM-Tree.

Initially, all the entries of item utility table are zero and AM-Tree has only a null root. When the first transaction $T_1 = \{i_1, i_2, \ldots, i_n\}$ puts into AM-Tree, AM creates a child node i_1 for the root node and the child node i_{j+1} ($\forall j, 1 \leq j \leq n-1$) for node i_j, and records k (item:utility) pairs $\{i_1 : u(i_1, T_1), i_2 : u(i_2, T_1), \ldots, i_k : u(i_k, T_1)\}$ on node i_k ($\forall k, 2 \leq k \leq n$). By the way, AM registers the utility and transaction closure utility for each item in T_1 in the item utility table. In order to speed up the computation of transaction closure utility, AM computes $TCU(T_1, \{i_j\})$ ($2 \leq j \leq n$) by using $TCU(T_1, \{i_{j-1}\})$. Therefore, $TCU(T_1, \{i_2\}) = u(i_1, T_1) + u(i_2, T_1)$ and $TCU(T_1, \{i_k\}) = TCU(T_1, \{i_{k-1}\}) + u(i_k, T_1)$ ($k \geq 3$). According to Definition 14.2, $TCU(T_1, \{i_1\}) = 0$, since there is no closure superset of $\{i_1\}$ in T_1, that is, we cannot extend $\{i_1\}$ to generate longer high utility itemsets in T_1.

For each transaction $T_j = \{j_1, j_2, \ldots, j_m\}$, AM sequentially puts each item in T_j into AM-Tree. If there is no child node j_1 of the root node, then the method to put T_j into AM-Tree is the same as that of T_1. Otherwise, AM continues to check if there is a child node j_2 of node j_1 in the AM-Tree and so on. If there is a child node j_{k+1} of node j_k ($1 \leq k \leq m-1$) in the AM-Tree, then AM adds the utilities of $j_1, \ldots, j_k$ and j_{k+1} in T_j to the $k+1$ (item:utility) pairs recorded on node j_{k+1}, respectively. Otherwise, AM creates a child node j_{k+1} of node j_k and records $k+1$ (item:utility) pairs $\{j_1 : u(j_1, T_j), j_2 : u(j_2, T_j), \ldots, j_k : u(j_k, T_j), j_{k+1} : u(j_{k+1}, T_j)\}$ on node j_{k+1}. For each item j_p ($1 \leq p \leq m$), AM adds the $u(j_p, T_j)$ and $TCU(T_j, \{j_p\})$ to the utility and closure utility for j_p registered in item utility table, respectively. The method for computing $TCU(T_j, \{j_p\})$ is the same as that of $TCU(T_1, \{j_p\})$.

After processing all the transactions in the database, the item utility table and AM-Tree are constructed. The utility $u(x)$ and closure utility $CU(x)$ for each item x are recorded in the item utility table. If $u(x) \geq$ MU, then itemset $\{x\}$ is a length-1 high utility itemset. If $CU(x) \geq$ MU, then itemset $\{x\}$ is a length-1 high closure utility itemset. According to Lemma 14.1 and Lemma 14.2, only the closure supersets of a high closure utility itemset may be the high utility itemsets. We can use a high closure utility itemset to generate longer high closure utility itemsets and longer high utility itemsets. Therefore, for each length-1 high closure utility itemset $\{x\}$, AM retrieves the (item:utility) pairs recorded on each node with item x through the link corresponding to the item x in the Header Table of the AM-Tree. For each set of (item:utility) pairs recorded on a node x, the pair (x:utility) is put on the first pair and the other pairs are ordered by the original order of the items, which form a record of the conditional pattern base for $\{x\}$. For a pair ($y : u$) ($y \neq x$), itemset $\{x\}$ can be extended to a length-2 itemset $\{xy\}$, which is a closure superset of $\{x\}$. AM constructs a conditional AM-Tree for $\{x\}$ according to the conditional pattern base for $\{x\}$. For each record $\{x : u_x, i_1 : u_1, \ldots, i_n : u_n\}$ in the conditional pattern base, AM puts the record into the conditional AM-Tree for $\{x\}$. The method to construct a conditional AM-Tree is similar to the AM-Tree construction. The difference between them is that AM puts the record starting at the root node of the conditional AM-Tree, that is, the root node contains itemset $\{x\}$. Besides, the utility of $\{x\}$ on a path needs to be added to each node on the path except the child node of root node $\{x\}$. By the way the utility of itemset $\{xi_j\}$ ($1 \leq j \leq n$) and the closure utility of itemset $\{xi_j\}$ ($2 \leq j \leq n$) are also computed and recorded on the item utility table

Table 14.3 TWU and support count for each item

Item	A	B	C	D	E
TWU	219	293	282	243	181
Support count	4	7	5	6	3

Fig. 14.2 The constructed item utility table and AM-Tree after processing transaction T_1

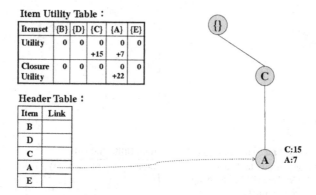

Item Utility Table :

Itemset	{B}	{D}	{C}	{A}	{E}
Utility	0	0	0 +15	0 +7	0
Closure Utility	0	0	0	0 +22	0

Header Table :

Item	Link
B	
D	
C	
A	
E	

for $\{x\}$. After constructing conditional AM-Tree and item utility table for $\{x\}$, if the utility of $\{xi_j\}$ is no less than MU, then $\{xi_j\}$ is a length-2 high utility itemset. If the closure utility of itemset $\{xi_j\}$ is no less than MU, that is, $\{xi_j\}$ is a high closure utility itemset, then AM constructs a conditional AM-Tree and an item utility table for itemset $\{xi_j\}$, and generates longer high utility itemsets recursively until there is no high closure utility itemset generated. All the high utility itemsets with item x will be generated.

We use Table 14.1 to illustrate our AM algorithm and MU is set to be 100. For the first database scan, AM computes the TWU and support count for each item in Table 14.1 and deletes the items whose TWU is less than MU, which is shown in Table 14.3 .The items which are ordered by their support counts are B, D, C, A and E. After that, AM constructs an AM-Tree and item utility table. For the first transaction T_1 in Table 14.1, the set of all the items and their utilities in T_1 is $\{C : 15, A : 7\}$. AM creates the child node C for the null root and the child node A for the node C, and the items and their utilities in T_1 (i.e., $C : 15, A : 7$) are recorded on node A. The utilities of $\{C\}$ and $\{A\}$ in T_1 and the transaction closure utility of $\{A\}$ in T_1 (i.e., $TCU(T_1, \{A\}) = u(C, T_1) + u(A, T_1) = 15 + 7 = 22$) are also registered in the item utility table, which are shown in Fig. 14.2. Note that $TCU(T_1, \{C\}) = 0$ according to Definition 14.2. The set of the items and their utilities in T_2 is $\{B : 8, D : 5\}$. Since there is no child node B for the root node, the child node B for the root node and the child node D for the node B are created. The utilities of items B and D in T_2 are recorded on node D. The closure utility of $\{D\}$ in T_2 (i.e., $TCU(T_2, \{D\}) = u(B, T_2) + u(D, T_2) = 8 + 5 = 13$) and the utilities of B and D in T_2 are registered in the item utility table, which are shown in Fig. 14.3.

For the third transaction T_3 in Table 14.1, the set of items and their utilities in T_3 is $\{B : 4, D : 7, C : 25, A : 49\}$. AM does not create nodes B and D, since there has been child node B for the root node and child node D for node B

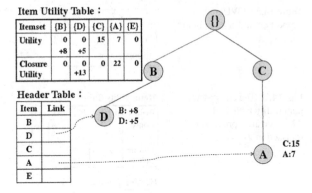

Fig. 14.3 The constructed item utility table and AM-Tree after processing transaction T_2

Item Utility Table:

Itemset	{B}	{D}	{C}	{A}	{E}
Utility	0 +8	0 +5	15	7	0
Closure Utility	0	0 +13	0	22	0

Header Table:

Item	Link
B	
D	
C	
A	
E	

D: B: +8, D: +5

C:15
A:7

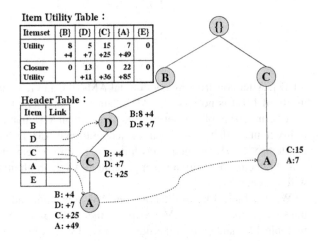

Fig. 14.4 The constructed item utility table and AM-Tree after processing transaction T_3

Item Utility Table:

Itemset	{B}	{D}	{C}	{A}	{E}
Utility	8 +4	5 +7	15 +25	7 +49	0
Closure Utility	0	13 +11	0 +36	22 +85	0

Header Table:

Item	Link
B	
D	
C	
A	
E	

D: B:8 +4, D:5 +7

C: B: +4, D: +7, C: +25

A: B: +4, D: +7, C: +25, A: +49

C:15
A:7

in the AM-Tree. AM adds the utilities of B and D in T_3 to the utilities of B and D recorded on node D, respectively. Since there is no child node C for the node D, AM creates child node C for node D and child node A for the node C, and records the utilities of B, D and C in T_3 and the utilities of B, D, C and A in T_3 on the two nodes C and A, respectively. The transaction closure utilities of D, C and A in T_3 (i.e., $TCU(T_3, \{D\}) = u(B, T_3) + u(D, T_3) = 4 + 7 = 11$, $TCU(T_3, \{C\}) = TCU(T_3, \{D\}) + u(C, T_3) = 11 + 25 = 36$, $TCU(T_3, \{A\}) = TCU(T_3, \{C\}) + u(A, T_3) = 36 + 49 = 85$) and the utilities of items B, D, C and A in T_3 are also added to the closure utilities and the utilities of these items in item utility table, respectively, which are shown in Fig. 14.4. Similarly, AM processes transactions T_4–T_{10} in Table 14.1 by using the same way. The constructed AM-Tree and item utility table are shown in Fig. 14.1.

From the item utility table in Fig. 14.1, we can see that itemset $\{A\}$ and $\{E\}$ are length-1 high utility itemsets, since their utilities are no less than MU. AM continues

Fig. 14.5 The constructed item utility table and conditional AM-Tree for itemset $\{A\}$

Item Utility Table :

Itemset	{AB}	{AD}	{AC}
Utility	111	140	96
Closure Utility	0	123	85

Header Table :

Item	Link
B	
D	
C	

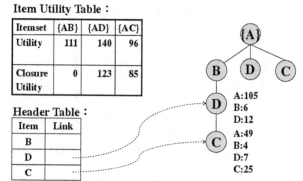

A:105
B:6
D:12

A:49
B:4
D:7
C:25

Fig. 14.6 The constructed item utility table and conditional AM-Tree for itemset $\{AD\}$

Item Utility Table :

Itemset	{ADB}
Utility	123
Closure Utility	0

Header Table :

Item	Link
B	

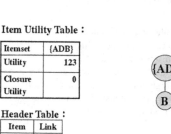

to construct conditional AM-Trees for itemsets $\{A\}$ and $\{E\}$ since itemsets $\{A\}$ and $\{E\}$ are high closure utility itemsets.

For the construction of the conditional AM-Tree for $\{A\}$, AM retrieves the (item: utility) pairs recorded on each node with item A through the corresponding link of item A in the header table. After moving the pair with item A to the first pair for each node, the generated conditional pattern base for $\{A\}$ is $\{(A : 49, B : 4, D : 7, C : 25), (A : 56, B : 2, D : 5), (A : 14, D : 9), (A : 7, C : 15)\}$. AM constructs conditional AM-Tree for $\{A\}$ from the conditional pattern base.

For the first record of $\{A\}$'s conditional pattern base, AM constructs a path $A \rightarrow B \rightarrow D \rightarrow C$ with root node $\{A\}$, and records $\{A : 49, B : 4, D : 7\}$ on node D and $\{A : 49, B : 4, D : 7, C : 25\}$ on node C. The utilities of itemsets $\{AB\}$, $\{AD\}$ and $\{AC\}$ in the first record are $49 + 4 = 53$, $49 + 7 = 56$ and $49 + 25 = 74$, respectively. The closure utilities of itemsets $\{AD\}$ and $\{AC\}$ in the first record are $49 + 4 + 7 = 60$ and $60 + 25 = 85$. Note that the closure utility of $\{AB\}$ is zero according to Definition 14.2. After processing the four records, the conditional AM-Tree and item utility table for itemset $\{A\}$ are shown in Fig. 14.5. From the item utility table in Fig. 14.5, we can see that both itemsets $\{AB\}$ and $\{AD\}$ are high utility itemsets and only the itemset $\{AD\}$ is a high closure utility itemset. Therefore, AM continues to constructs conditional AM-Tree and item utility table for itemset $\{AD\}$, which are shown in Fig. 14.6. From the item utility table in Fig. 14.6, the utility of itemset $\{ADB\}$ is 123, which is a high utility itemset. Because there is no high closure utility itemset generated, all the high utility itemsets with item A are $\{AB\}$, $\{AD\}$ and $\{ADB\}$.

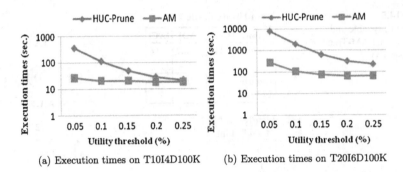

(a) Execution times on T10I4D100K   (b) Execution times on T20I6D100K

Fig. 14.7 The execution times for HUC-Prune and AM algorithm on synthetic datasets

Table 14.4 The number of the candidate high utility itemsets generated by HUC-Prune and the number of high utility itemsets for every utility threshold on the two synthetic datasets

Threshold	T10I4D100K		T20I6D100K	
	HTWUI	HUI	HTWUI	HUI
0.05%	82162	6805	740213	37362
0.10%	23185	1016	228026	4838
0.15%	7337	625	73093	989
0.20%	3055	443	35331	488
0.25%	1946	376	19445	352

14.3 Experimental Results

In this section, we evaluate the performance of our AM algorithm and compare AM with the most efficient algorithm HUC-Prune [2]. Our experiments are performed on Intel (R) CoreTM 2 Duo P8400 2.4 GHz CPU with 3 GB RAM and running on Windows XP. HUC-Prune and AM are implemented in JAVA language.

We generate two synthetic datasets T10I4D100K and T20I6D100K by using IBM Synthetic Data Generator [6], in which T is the average length of the transactions, I is the average size of maximal potentially frequent itemsets and D is the total number of the transactions. The number of distinct items is set to 1000. For the profit of each item, we use the log Normal Distribution [2, 7] and set the range of the profits between 0.01 and 10. The purchased quantity for an item in a transaction is randomly set to the number between 1 and 10.

For the experiments on the synthetic datasets, we set the utility threshold from 0.05% to 0.25%, which is the ratio of the minimum utility MU to the total utility in the dataset, to evaluate the execution times for the two algorithms HUC-prune and AM. Figures 14.7(a) and 14.7(b) show the execution times for the two algorithms on the datasets T10I4D100K and T20I6D100K, respectively. From Fig. 14.7, we can see that our AM algorithm significantly outperforms HUC-Prune. The performance gap increases as the utility threshold decreases, since the number of the candidate high utility itemsets increases for HUC-Prune. Table 14.4 shows the numbers of the generated candidate high utility itemsets for HUC-Prune and the generated high

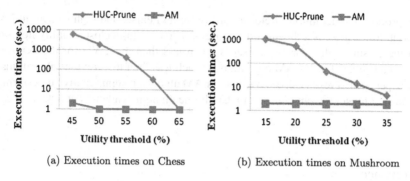

(a) Execution times on Chess   (b) Execution times on Mushroom

Fig. 14.8 The execution times for the two algorithms on real datasets

utility itemsets for each utility threshold on the two synthetic datasets, in which a huge number of candidates generated for HUC-Prune when the utility threshold is low. Therefore, HUC-Prune has to take a lot of time to scan the whole database to compute the utility for each candidate. The execution times for AM slightly decrease as the utility threshold increases, since AM generates high utility itemsets directly from AM-Tree without generating any candidate and scanning the database again. Therefore, AM is more stable than HUC-Prune.

Besides, we also evaluate the performances of HUC-Prune and our AM algorithm on the two real datasets Chess and Mushroom which were downloaded from FIMI Repository [4]. There are 3196 transactions on the dataset Chess, in which the average length of the transactions is 37 and the number of the distinct items is 76. For Mushroom dataset, the average length of the transactions is 23, the number of distinct items is 119 and the number of transactions is 8124.

Figures 14.8(a) and 14.8(b) show the execution times for the two algorithms on datasets Chess and Mushroom, respectively. From the experiments, we can see that AM also significantly outperforms HUC-prune and the performance gap is much larger than the experiments on the synthetic datasets, since few high utility itemsets are generated but a large number of candidates are generated for HUC-prune. For AM algorithm, since there is no candidate which needs to be generated and our closure utility for an itemset is much less than TWU, high utility itemsets can be generated in a short time.

14.4 Conclusions

Most of the previous approaches for mining high utility itemsets need two phases: The first phase is to generate candidate high utility itemsets. The second phase is to find the high utility itemsets by scanning the whole database and searching from the large number of candidates. In this paper, we propose an efficient algorithm AM for mining high utility itemsets. Our AM algorithm only needs one phase to directly generate high utility itemsets from our AM-Tree without candidate generation. We also propose the definition of closure utility for an itemset and high closure utility itemset. Since there is a downward closure property for high closure utility itemset and the closure utility for an itemset is much less than the TWU for the itemset, the mining space for our AM algorithm is much less than the previous algorithms. The experimental results also show that our AM algorithm significantly outperforms the current most efficient algorithm HUC-Prune on both synthetic datasets and real datasets.

References

1. Agrawal, R., Srikant, R.: Fast algorithms for mining association rules. In: Proc. of 20th International Conference on Very Large Databases, Santiago, Chile, September, pp. 487–499 (1994)
2. Ahmed, C.F., Tanbeer, S.K., Jeong, B.S., Lee, Y.K.: An efficient candidate pruning technique for high utility pattern mining. In: Proc. of the 13th Pacific-Asia Conference on Knowledge Discovery and Data Mining, April, pp. 749–756 (2009)
3. Cao, L.: In-depth behavior understanding and use: the behavior informatics approach. Inf. Sci. **180**, 3067–3085 (2010)
4. Goethals, B.: Frequent itemset mining implementations repository. http://fimi.cs.helsinki.fi/
5. Han, J., Pei, J., Yin, Y.: Mining frequent patterns without candidate generation. In: Proc. of Int. Conf. on Management of Data (ACM SIGMOD), May, pp. 1–12 (2000)
6. IBM: IBM synthetic data generator, http://www.almaden.ibm.com/software/quest/Resorces/index.shtml
7. Liu, Y., Liao, W.K., Choudhary, A.: A fast high utility itemsets mining algorithm. In: Proc. of the 1st Int. Workshop on Utility-Based Data Mining, August, pp. 90–99 (2005)

Chapter 15
Individual Movement Behaviour in Secure Physical Environments: Modeling and Detection of Suspicious Activity

Robert P. Biuk-Aghai, Yain-Whar Si, Simon Fong, and Peng-Fan Yan

Abstract Secure physical environments such as government, financial or military facilities are vulnerable to misuse by authorized users. To protect against potentially suspicious actions, data about the movement of users can be captured through the use of RFID tags and sensors, and patterns of suspicious behaviour detected in the captured data. This chapter presents four types of suspicious behavioural patterns, namely temporal, repetitive, displacement and out-of-sequence patterns, that may be observed in such a secure physical environment. We model the physical environment and apply algorithms for the detection of suspicious patterns to logs of RFID access data. Finally we present the design and implementation of an integrated system which uses our algorithms to detect suspicious behavioural patterns.

15.1 Introduction

In the wake of increased terrorist and criminal activity over the past decade, the security of physical environments has become an increasingly important topic. In many parts of the world the use of video surveillance technology has become widespread for detecting security breaches [19]. Moreover, electronic and information technology has been used to restrict access to physical environments. For example, smartcard-based access control systems have been used over the past decade to automate the identification and authentication of access to restricted physical environments such as buildings, rooms, etc. More recently, RFID (radio frequency

R.P. Biuk-Aghai (✉) · Y.-W. Si · S. Fong · P.-F. Yan
Data Analytics and Collaborative Computing Group, Department of Computer and Information Science, Faculty of Science and Technology, University of Macau, Macau, China
e-mail: robertb@umac.mo

Y.-W. Si
e-mail: fstasp@umac.mo

S. Fong
e-mail: ccfong@umac.mo

P.-F. Yan
e-mail: franciswing@163.com

L. Cao, P.S. Yu (eds.), *Behavior Computing*,
DOI 10.1007/978-1-4471-2969-1_15, © Springer-Verlag London 2012

identification) has enjoyed quick and widespread adoption in the security domain. RFID allows a person or object to be tagged with a unique identifier that can be wirelessly sensed when the RFID tag enters the range of an RFID sensor.

The low cost of RFID equipment coupled with the convenience of a wireless mode of operation and a fast detection rate makes this technology particularly suited for security applications. In 2005, the US Department of Homeland Security (DHS) announced the distribution of 40,000 RFID-based access cards to its employees and contractors to control access to both physical environments and computer systems. Other US federal agencies also are making use of similar technology to strengthen the security of their physical environments, and this technology is being adopted by governments and private agencies around the world.

Using RFID technology allows the physical access of people to secure areas to be controlled. Moreover, given enough sensors in a secure environment, it also allows the movement of people within the environment to be tracked. Current use of this technology, however, is mainly restricted to disallow unauthorized access. Once a person has gained access to a secure physical environment, the actions of that person within that environment are usually not further monitored other than detecting outright breaches of security, e.g. through video surveillance. It is possible, however, that a given person within a secure environment behaves in a way that does not constitute an outright security breach, but that could be considered suspicious behaviour. Other security problems could arise if data from a valid RFID tag is surreptitiously obtained (RFID sniffing) and used to create a clone of the RFID tag which can then be used in RFID spoofing, replay attacks, or denial of service [7, 17]. If such suspicious behaviour could be detected, security personnel could be alerted to monitor the suspicious person closely to determine whether a security breach is about to be committed.

Extensive research on intrusion detection systems (IDS) for computer networks, which covers suspicious access detection (SAD), has paralleled the fast proliferation of Internet development and penetration. On the other hand, IDS and SAD for physical access security became an important worldwide concern in recent years after the September 11 disaster in the USA. Considering the characteristic of suspicious access detection, there are certain similarities in its application both in the digital and physical realm. The required techniques of SAD for the physical realm could be based on the ones developed for the digital realm. In computers, activities can be captured easily and comprehensively, resulting in large amounts of activity data. Data analysis and mining algorithms can be applied to this data to discover abnormal and suspicious activities among the considerable volume of data. Recent development of RFID technology enables tiny contact-less tags for physical object tracing and tracking. Practical implementation of object movement identification and registration becomes feasible, simple and convenient.

The research reported here has developed techniques for extracting and analysing information on suspicious patterns of movement from people's access logs in physical SAD, and developed algorithms, methods, and tools for analysis and visualization of data related to physical object movements, especially user behaviour patterns that are suspected to be security threats. The derived result can be used for early warning of suspicious activities in a closely monitored environment equipped with multiple sensors.

Behaviour informatics has emerged as a recent research area concerned with identifying human activities that have the potential to impact the operation of business, government or public life, including terrorist and criminal activity [5, 6]. A detailed understanding of human behaviour through empirical behaviour modeling and subsequent analysis has the potential to uncover harmful patterns of behaviour [4]. Given only limited and possibly unbalanced sources of data on activities the challenge is to derive potentially harmful actions that may result from combinations of these activities. The successful identification of threats can then be used to take appropriate action such as further investigation and action to counter threats. Behaviour informatics has found application, during the past decade, in the area of Intrusion Detection Systems (IDS) for computer networks and applications. Various data mining techniques have been applied and proven to be effective, including association and frequent episode [11, 13], meta learning [13], classification [11] and clustering [15].

Research about location sensing of people or objects using radio frequency identification technology has been conducted recently. LANDMARC [14] is a location sensing prototype system that uses active RFID tags for locating objects inside buildings. Isoda et al. [9] proposed a user activity assistance system that employs a state sequence description scheme to describe the user's contexts. Willis and Helal [21] proposed a navigation and location determination system for the blind using an RFID tag grid. Leong et al. [12] developed a logical mathematical model to formulate a knowledge base of suspicious and irregular actions. Based on this model, they proposed a real-time suspicious access pattern detection prototype which allows rapid alert and reaction to irregular behaviour.

Based on the concepts and methodologies described above, we have modelled the physical environment developed an intrusion detection model for physical environments. Given the lack of availability of secure access event data, we have developed an access event generator for physical environments. The remainder of this chapter is structured as follows. In Sect. 15.2, we outline four types of suspicious patterns we detect. In Sect. 15.3, we give an overview of the system we developed, and in Sect. 15.4 we discuss the design and implementation of our simulated access event generator. In Sect. 15.5 we discuss related work, and finally draw conclusions in Sect. 15.6.

15.2 Suspicious Pattern Detection

Here we describe the semantics of four suspicious patterns and corresponding method for detecting these using concrete algorithms. Our proposed techniques detect a person's suspicious behaviour by analysing movement patterns and identifying potential security threats in a secure physical environment. Suspicious behaviour consists of a collection of suspicious patterns. Each of these patterns is a sequence

Table 15.1 Parameters for detection model

Parameter	Description
$event_i$	ith access event
cid	$cid = cardID(event_i)$, access card ID of ith access event
AP_i	$AP_i = accessPoint(event_i)$, access point of ith access event
$repThreshold$	Normal maximum allowable number of repeated accesses
$repAccMinDuration$	Normal minimum allowable duration for a sequence of repeated accesses

of actions performed by a person that may be completely legitimate when the level of analysis is a single event. However, when these events are combined over time and viewed together as a sequence they give rise to certain kinds of suspicion. The exact definition of the suspicious movement of people usually varies from one environment to another, and subjectively depends on the security requirements of each different situation. Given an existing physical environment with surveillance sensors installed, access events are captured and stored in a database together with related access right policies. Our detection functions access this data and evaluate it against administrator-defined thresholds for detection of suspicious patterns using concrete methods. We define following four suspicious patterns:

1. *Temporal pattern*: an unusually long period of stay by a person in a given area.
2. *Repetitive pattern*: unusual repetitive accesses within a given period of time.
3. *Displacement pattern*: consecutive accesses to distinct but distant neighbouring locations within an unusually short period of time.
4. *Out-of-sequence pattern*: consecutive accesses in an undefined sequence.

To detect these patterns in collected data, the following algorithms can be used in an existing physical environment that has surveillance sensors installed. Parameters for the detection algorithms are presented in Table 15.1.

Detection of Temporal Pattern Let $timeStamp(AP_i, cid)$ be the function which returns the timestamp of the ith detected access point of the person holding card cid. Let $location(AP_i)$ be the function which returns the location of the ith access point, and let $maxStay(loc, cid)$ be the function which retrieves the predefined maximum duration that the person holding card cid is allowed to stay at the location loc. We define the algorithm for detecting temporal patterns as follows:

for all new detected $event_i$ **do**
  $t_{pre} = timeStamp(AP_{i-1}, cid)$
  $t_{cur} = timeStamp(AP_i, cid)$
  $t = t_{cur} - t_{pre}$
  $t_{max} = maxStay(location(AP_{i-1}), cid)$
  if $t_{max} < t$ **then**
     pattern = "Temporal"
  else

pattern = "Normal"
end if
end for

Detection of Repetitive Pattern The detection of the repetitive pattern focuses on access events detected from a pair of access points (sensors) installed at two opposite sides of a door or entrance. In addition, two conditions must hold for a repetitive pattern: (1) the total number of repeated accesses should be greater than the predefined threshold, and (2) the total time spent during the repeated accesses must be shorter than the minimum allowable duration for a sequence of normal repeated accesses. First, the system derives the total number of repeated accesses from the last detected access event. For instance, two repeated accesses are detected from the sequence $AP_{i-4} \rightarrow AP_{i-3} \rightarrow AP_{i-2} \rightarrow AP_{i-1} \rightarrow AP_i$, where AP_i is the ith detected access point. Note that $AP_i = AP_{i-2} = AP_{i-4}$, and $AP_{i-1} = AP_{i-3}$. Let $repAccessCount(AP_n)$ be the function which counts the total number of repetitive accesses for access point AP_n. For the example of the above sequence, $repAccessCount(AP_i)$ is equal to 2. Let $timeSpent(AP_x, AP_y)$ be the function that returns the time spent by the person when accessing point y after accessing x. Therefore, the total time spent by the person for the previous access sequence can be denoted as $timeSpent(AP_{i-4}, AP_i)$. Based on these functions, we define the algorithm for detecting repetitive access patterns as follows:

for all new detected $event_i$ do
  if $(repAccessCount(AP_i) \geq repThreshold)$ and
  $(timeSpent(AP_i, AP_{2(repAccessCount(AP_i))})) < repAccMinDuration$ then
    pattern = "Repetitive"
  else
    pattern = "Normal"
  end if
end for

Detection of Displacement Pattern Let $minMove(AP_{i-1}, AP_i)$ be the function which returns the minimum time required to travel from $(i - 1)$th access point to ith access point. We define the algorithm for detecting displacement patterns as follows:

for all new detected $event_i$ do
  $t_{pre} = timeStamp(AP_{i-1}, cid)$
  $t_{cur} = timeStamp(AP_i, cid)$
  $t = t_{cur} - t_{pre}$
  $t_{min} = minMove(AP_{i-1}, AP_i)$
  if $t < t_{min}$ then
    pattern = "Displacement"
  else
    pattern = "Normal"
  end if
end for

Detection of Out-of-Sequence Pattern   A pattern is considered to be out-of-sequence when it is detected that a person attempts consecutive accesses to two distinct locations whereby the second location is unreachable from the first one. Let $isNeighbor(AP_{i-1}, AP_i)$ be a Boolean function which returns true if AP_i can be reached from AP_{i-1}. We define the algorithm for detecting out-of-sequence patterns as follows:

> **for all** new detected $event_i$ **do**
>> **if** $isNeighbor(AP_{i-1}, AP_i)$ **then**
>>> pattern = "Normal"
>> **else**
>>> pattern = "Out-of-sequence"
>> **end if**
> **end for**

Using the above four detection algorithms, a security system may decide to raise an alarm when a suspicious access pattern is detected. However, in some situations a sequence of access events may not be considered suspicious as its degree of suspicion does not exceed pre-defined threshold values. For instance, the total number of repeated accesses by a person may not exceed the limit and hence the system may not raise the alert. In such cases, the system may not be able to detect cases of slight suspicion. The prediction of future possible suspicious access patterns would be a straightforward extension of our algorithms.

15.3 System Design Overview

We have designed an integrated system for the capture of RFID sensor data, generation of simulated physical access data, training of our detection model, and real-time detection of suspicious access patterns. This system design consists of five modules arranged in three layers, as shown in Fig. 15.1.

15.3.1 System Layers

Data Layer   This layer consists of several databases. The Physical Environment Database stores data defining the physical environment, such as building layout, access point location etc. The Access Events Database stores the data about RFID access events, including real data obtained from RFID sensors installed in the physical environment, or data generated by our simulator component. The Mining Database stores mined models and detection rules. It also records parameters for training the models. The Security XML Registry stores the security requirements.

Application Server Layer   This layer consists of several applications for defining the physical environment, generating simulated access event data, training detec-

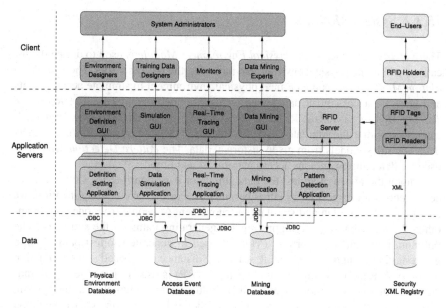

Fig. 15.1 System structure

tion models, real-time monitoring and tracking of suspicious patterns, and mining patterns from the captured access event data. For ease of use by non-expert security personnel, these applications are designed with graphical user interfaces. They comprise offline applications for creating detection models, and online applications for processing real-time access event and identifying suspicious patterns based on the models from the offline applications.

Client Layer   The client layer comprises two main parts. One part consists of the RFID holders, i.e. the persons whose movements are monitored through our system and who each hold an RFID tag that is the source of data when sensed by the RFID sensors placed in the environment. The other part consists of the users of our system's applications who are system administrators in charge of different aspects of the whole system's operation: the Environment Designer is responsible for defining the physical environment by setting environment-related parameters; the Training Data Designer is responsible for designing the simulation of physical access events by setting probability-related parameters; the Monitor is responsible for monitoring the real-time animation of real-time access events or simulated access events and looking for suspicious patterns among the users' actions; the Data Mining Expert is responsible for defining useful and efficient models/algorithms for detecting suspicious patterns.

15.3.2 System Modules

There are five modules: The *Physical Environment Module* is used to define the locations within the physical environment including the RFID sensors. The *Data Simulation Module* utilizes the defined physical environment to simulate access events and generate access event data based on user defined parameters. The *Real-time Tracing Module* visualizes RFID holders' actions in the physical environment. It can also visualize the simulated and historic data. The *Data Mining Module* extracts detection rules/models from given historical data. The *Pattern Detection Module* is used for detecting suspicious access patterns in real time.

Among the above modules, the data simulator is responsible for simulating access events with respect to specified parameters. For the sake of simplification, we adopt a fixed floor plan and simulate the movement of people from one area to another. We have devised two algorithms for generating paths for the simulator. The optimum path finding algorithm (VOP) is used to generate shortest paths from a given starting point to a target point, where the distance is regarded as the evaluation measure. A Random Path finding algorithm (VRP) is used to generate a path randomly so that random movement of RFID holders can be mimicked in the system. *User portion* (the ratio of users behaving suspiciously) and *probability* (likelihood of a certain kind of suspicious action) parameters are used to describe the ratio of the suspicious pattern access events to be generated during the simulation.

15.4 Prototype System Implementation

Based on the design from Sect. 15.3, we have implemented a prototype system. This section briefly introduces our implementation. Due to space limitations, we only illustrate some of its many functions here. Initially the Environment Designer defines characteristics of the physical environment. An example of the physical layout of a given environment is shown in Fig. 15.2, here of a portion of the US White House. The environment consists of corridors, rooms, doors, passage ways, etc. Labelled with numbers at each door and on some of the walls are locations of RFID sensors that are installed in the physical environment. The Environment Designer records all the relevant information about areas (rooms, corridors), entrance ways, connections between areas, locations of RFID sensors etc. The environment design also includes the definition of parameters related to suspicious access events, illustrated in Fig. 15.3.

In the example of Fig. 15.3 the environment designer is adjusting the minimum traversal time between an access point (RFID sensor) and one of its neighbours, as used in the detection of Displacement patterns. Given the information about the environment and the defined pattern parameters, the pattern detection is able to determine when a suspicious pattern has occurred, as explained in Sect. 15.2 above.

Once the environment is in operation, access events are captured from all connected RFID sensors and stored in the Access Events Database. Below is a sample

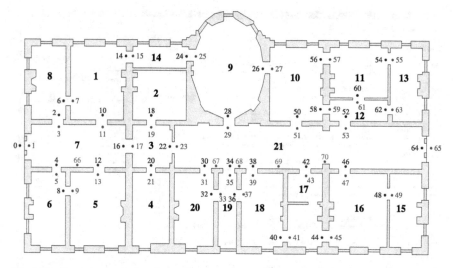

Fig. 15.2 Example physical environment layout: US White House

AccessPointID	NeighborID	MinTime	AvgTime
0	0	0	7
0	1		2	7
1	0	2	7
1	1	0	7
1	3	10	43
1	4	10	42
1	66	14	65
2	2	0	7
2	3	2	7
2	6	7	25
3	1	10	43
3	2	2	7
3	3	0	7
3	4	12	53
3	66	13	59
4	1	10	43

Displacement Rule Parameter Setting

Fig. 15.3 Parameter settings for displacement pattern at a given RFID sensor location

of some raw access event data, showing access point IDs, access card IDs and timestamps:

```
66 23 2009-05-05 12:20:07
66 78 2009-05-05 12:21:01
19 71 2009-05-05 12:21:18
```

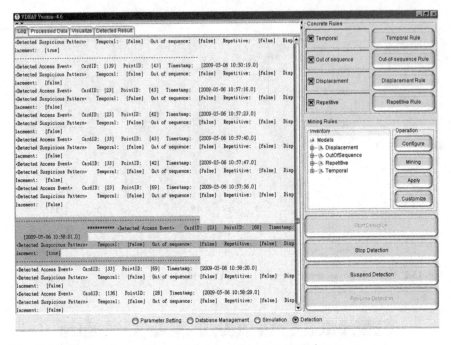

Fig. 15.4 Detection of a suspicious pattern among the access event data

Once the pattern detection is in progress, it will search for and display any suspicious patterns found in the access event data. Figure 15.4 shows an example of a detected Displacement pattern, with a panel for rule selection and configuration in the right of the same window. This allows rules to be customized at run-time in response to observed behavioural patterns.

We have evaluated our detection algorithms using our own simulated data. Not surprisingly, the detection works flawlessly. A more meaningful evaluation would use real data from an actual secure physical environment. However, given the sensitive nature of these environments and the data captured from them, we have to date not had the opportunity to get access to such data and thus evaluate our algorithms more fully. We welcome collaboration with any organization that would be interested in applying our research to their security-related data.

15.5 Related Work

The closest related work is in the area of behaviour informatics. It focuses on building a model of human activity given empirical data, and then analyzing sequences of activity that could constitute harmful actions [4–6]. An application of behaviour informatics that is closely related to the research presented here is the area of intrusion detection systems (IDS). An IDS is designed as software or hardware for detecting unwanted attempts at accessing, manipulating, or disabling of computer systems, mainly through a computer network such as the Internet. In recent years a large number of IDS have been developed to address specific needs [3]. The most commonly used models for current IDS are host-based, network-based, and protocol-based IDS. In host-based IDS, there is a unique host used to detect the intrusion by analysing data packets that travel through that host. This host comprises an agent which identifies intrusions by analysing system calls, application logs, file-system modifications and other host activities and state. OSSEC [18] is an example of host-based IDS, as it performs log analysis, integrity checking, windows registry monitoring, rootkit detection, time-based alerting and active response. In network-based IDS, a computer network intrusion detection system (NIDS) is usually installed by connecting to a hub, network switches or network taps, and is an independent platform which keeps track of network traffic data. The data from the computer network is monitored against a database and the NIDS flags those which seem to be suspicious. The audit data from single or multiple hosts are also used to detect intrusion signs. Snort [2] is an example of NIDS that performs packet logging and real-time traffic analysis on IP networks. Protocol-based IDS (PIDS) [20] usually consists of a system or agent located at the very front end of a server to monitor and analyse the protocol which is used to communicate between a connected device and the server. PIDS monitors the dynamic behaviour or states of the protocol. Depending on the requirement, two or more types of IDS are combined together to construct a hybrid intrusion detection system.

The majority of IDS use either anomaly or misuse detection models. The principle of the anomaly detection model is to look for anomalous behaviour or deviations from the predefined baseline. Although this model is effective in detecting unknown intrusions and new exploits, anomaly detection can result in a high false positive rate. For example, Qiao et al. [16] have discussed an anomaly intrusion detection method based on HMM. The intrusion detection system monitors the call trace of a UNIX privileged process, and passes it to a HMM to obtain state transition sequences. Preliminary experiments prove the state transition sequences can distinguish normal actions and intrusion behaviour in a more stable and simple manner. The misuse detection model has knowledge of suspicious patterns of behaviour and looks for activities that violate the standard policies. Misuse detection models have a lower false positive rate. Kumar and Spafford describe a generic model of matching that can be usefully applied to misuse intrusion detection [10]. Their model is based on Coloured Petri Nets.

IDS can collect a large amount of data without sufficient means to merge the data so as to extract the context for detecting attacks. Intellitactics Security Manager [8] allows users to prioritize and prevail across the full range of security threats in real time. It can capture and monitor real-time event activity and translate event codes into easy to understand terms. It can analyse complex security situations with customizable web-based reports, correlate data and prioritize threats. In another case, audit data analysis and mining (ADAM) [1] IDS used tcpdump to build profiles of rules for classification. ADAM adopts data mining technology to detect intrusions, including the combination of association rule mining and classification methodologies. Lee et al. [11] developed a data mining framework for building an intrusion detection model, which consists of programs for learning classifiers, association rules for link analysis and frequent episodes for sequence analysis. Portnoy [15] proposed an intrusion detection model with unlabelled data using clustering (unsupervised learning). The model can detect a large number of intrusions while keeping the false positive rate reasonably low.

15.6 Conclusion

In this chapter, we describe a model for detecting suspicious patterns within a large volume of access events in secure physical environments. We have defined four types of suspicious patterns that may occur in common physical access environments, namely Temporal, Repetitive, Displacement and Out-of-sequence, respectively. Using characteristics of each type of pattern we have defined algorithms for detecting these among a large set of logged access event data. Our presented integrated system allows the definition of a secure physical environment's features, the configuration of parameters related to suspicious patterns, and the detection of these patterns in collected data. For training purposes, an integrated simulator can generate large volumes of realistic access data. The use of our presented algorithms and system design can be of great use in providing an additional level of security to large physical environments in which the use of video surveillance alone is not sufficient to determine whether a sequence of valid actions performed by its users can be considered legitimate in the context of the user and location concerned. Our work is thus of particular relevance to the use in military installations, government facilities and other high-security locations. We welcome contact by organizations wishing to apply our techniques in their secure physical environments.

Acknowledgement   This research was funded by the Research Committee, University of Macau under grant number RG076/04-05S/BARP/FST.

References

1. Barbara, D., Couto, J., Jajodia, S., Popyack, L., Wu, N.: ADAM: Detecting intrusions by data mining. In: IEEE Workshop on Information Assurance and Security, pp. 11–16. IEEE Press, New York (2001)
2. Beale, J., Foster, J.C., Posluns, J., Russell, R., Caswell, B.: Snort 2.0 Intrusion Detection. Syngress, Rockland (2003)
3. Brandenburg University of Technology: Intrusion detection systems list and bibliography. http://www-rnks.informatik.tu-cottbus.de/en/node/209 (2004)
4. Cao, L.: In-depth behavior understanding and use: The behavior informatics approach. Inf. Sci. **180**, 3067–3085 (2010). doi:10.1016/j.ins.2010.03.025
5. Cao, L., Zhao, Y., Zhang, C.: Mining impact-targeted activity patterns in imbalanced data. IEEE Trans. Knowl. Data Eng. **20**, 1053–1066 (2008). doi:10.1109/TKDE.2007.190635

6. Cao, L., Zhao, Y., Zhang, C., Zhang, H.: Activity mining: From activities to actions. Int. J. Inform. Technol. Decis. Mak. **7**(02), 259–273 (2008). doi:10.1142/S0219622008002934
7. Cook, D.J., Holder, L.B.: Graph-based data mining. IEEE Intell. Syst. **15**(2), 32–41 (2000)
8. Intellitactics, Inc.: Intellitactics security manager. http://www.intellitactics.com/int/products/securitymanager.asp (2009)
9. Isoda, Y., Kurakake, S., Nakano, H.: Ubiquitous sensors based human behavior modeling and recognition using a spatio-temporal representation of user states. In: 18th International Conference on Advanced Information Networking and Applications, pp. 512–517. IEEE Press, New York (2004)
10. Kumar, S., Spafford, E.H.: A pattern matching model for misuse intrusion detection. In: 17th National Computer Security Conference, pp. 11–21 (1994)
11. Lee, W., Stolfo, S.J., Mok, K.W.: A data mining framework for building intrusion detection models. In: IEEE Symposium on Security and Privacy, pp. 120–132. IEEE Press, New York (1999)
12. Leong, A., Fong, S., Siu, S.: Smart card-based irregular access patterns detection system. In: IEEE International Conference on e-Technology, e-Commerce and e-Service, pp. 546–553. IEEE Press, New York (2004)
13. Li, Q.H., Xiong, J.J., Yang, H.B.: An efficient mining algorithm for frequent pattern in intrusion detection. In: International Conference on Machine Learning and Cybernetic, pp. 138–142. IEEE Press, New York (2003)
14. Ni, L.M., Liu, Y., Lau, Y.C., Patil, A.P.: LANDMARC: Indoor location sensing using active RFID. In: IEEE International Conference on Pervasive Computing and Communications, p. 407. IEEE Comput. Soc., Los Alamitos (2003)
15. Portnoy, L.: Intrusion detection with unlabeled data using clustering. Undergraduate thesis, Data Mining Lab, Department of Computer Science, Columbia University (2000)
16. Qiao, Y., Xin, X.W., Bin, Y., Ge, S.: Anomaly intrusion detection method based on HMM. IET Electron. Lett. **38**(13), 663–664 (2002)
17. Thornton, F., Haines, B., Das, A., Campbell, A.: RFID Security. Syngress, Rockland (2006)
18. Trend Micro, Inc.: OSSEC manual. http://www.ossec.net/main/manual (2009)
19. US Department of Justice: CCTV: Constant cameras track violators. Natl. Inst. Justice J. **249**, 16–23 (2003)
20. Wikipedia contributors: Protocol-based intrusion detection system. http://en.wikipedia.org/wiki/Protocol-based_intrusion_detection_system (2009)
21. Willis, S., Helal, S.: A passive RFID information grid for location and proximity sensing for the blind user. Technical report, University of Florida (2004)

Chapter 16
A Behavioral Modeling Approach to Prevent Unauthorized Large-Scale Documents Copying from Digital Libraries

Evgeny E. Ivashko and Natalia N. Nikitina

Abstract There are many issues concerning information security of digital libraries. Apart from traditional information security problems there are some specific ones for digital libraries. In this work we consider a behavioral modeling approach to discover unauthorized copying of a large amount of documents from a digital library. Supposing the regular user has interest in semantically related documents, we treat referencing to semantically unrelated documents as anomalous behavior that may indicate attempt of unauthorized large-scale copying. We use an adapted anomaly detection approach to discover attempts of unauthorized large-scale documents copying. We propose a method for constructing classifiers and profiles of regular users' behavior based on application of Markov chains. We also present the results of experiments conducted within development of a prototype digital library protection system. Finally, examples of a normal profile and an automatically detected anomalous session derived from the real data logs of a digital library illustrate the suggested approach to the problem.

16.1 Introduction

Over the last decade, digital libraries have been contributing significantly to the production and spreading of scientific knowledge. Free access to results of research in various fields is the key to the further development of science. Establishment and maintainance of collections of digital documents consume a lot of tangible and intangible resources. Moreover, often the further development of a digital library is placed in direct dependence on traffic, therefore popularity, of the information re-

This work was supported by Kaspersky Lab grant as part of the "Program of support for innovative projects".

E.E. Ivashko (✉) · N.N. Nikitina
Institute of Applied Mathematical Research, Karelian Research Centre of the RAS, 11, Pushkinskaya Street, Petrozavodsk, Karelia, 185910, Russia
e-mail: ivashko@krc.karelia.ru

N.N. Nikitina
e-mail: nikitina@krc.karelia.ru

L. Cao, P.S. Yu (eds.), *Behavior Computing*,
DOI 10.1007/978-1-4471-2969-1_16, © Springer-Verlag London 2012

source. Thereby, construction and operation of a digital library increase importance of the problems of information security that are concerned with the following:

1. Integrity of data that are being processed, transmitted and stored in the digital library;
2. Availability of all digital information resources that are allowed for users;
3. Confidentiality of personal and personified data of the digital library users.

As a rule, at the present time the information security issues of digital library protection mainly deal with the problems of access differentiation (construction of discretionary or mandatory access rules) and data exchange with authorized users (e.g., [12]). When publishing the digital documents, the problem of copyright observance remains relevant as well. The methods for its solution include, in particular, various technical measures related to limiting the spread and use of electronic copies of digital documents (e.g., [1, 15]).

However, apart from traditional solutions of digital data protection that are already technically mature in general, further challenges of information security emerge. Such is the task of protection of a digital library against unauthorized large-scale documents copying. Under the large-scale documents copying here we assume obtaining digital copies of all or a large part of documents from a digital library without consent of its owners. This problem is practically significant. In case an abuser copies a large amount of digital documents, they may inflict damage to operation of a digital library. Among the possible harmful actions is creation of a clone website of the copied digital library which may be used, for example, to get profit bypassing interests of the copyright holders.

The importance of the problem of protecting digital libraries against large-scale copying can be illustrated by a copyright infringement case recently brought by three major publishers against Georgia State University for allowing students "systematic, widespread, and unauthorized copying and distribution of a vast amount of copyrighted works" [2]. Typically, most publishers prohibit within their license agreements with digital libraries the widespread copying of their materials, performed either automatically or manually. Examples of straightly prohibited actions include, but are not limited to, downloading of entire journal issues (e.g., [11]).

To detect an attempt of large-scale documents copying we propose to use the anomaly detection approach based on behavior modeling, which is an important topic in behavior computing [4]. In Sects. 16.2 and 16.3 we describe the approach in detail. In Sect. 16.4 we present results of experiments conducted within development of a prototype digital library protection system against unauthorized large-scale documents copying.

16.2 The Anomaly Detection Approach

The basic idea behind the research presented in this work is as follows. If a user refers to services of a digital library she solves the tasks relevant for her. When referring to various digital documents, she assumes as part of her goal some subjective semantic connections between the documents of their interest. A student in search of material for the essay on the history of mathematics would scan through the biographies of Poincare and H. Minkowski. A specialist in microbiology would be interested in the newly discovered microorganisms and their characteristics. A philosopher would refer to the works of F. Nietzsche and A. Schopenhauer. However, there is hardly a user which would be simultaneously interested in the biographies of the great mathematicians, classical works of the philosophers and contemporary achievements in the field of microbiology.

Apparently, the situation when the same user during a short time interval is interested in highly specialized areas of physics, art history, genetics, etc., is anomalous. We suppose that interest in heterogeneous, or semantically unrelated, documents is anomalous and may indicate an attempt to copy a large amount of diverse documents for the purposes connected with copyright infringement.

In order to detect such anomalies in the users' actions automatically, we have to specify the semantic links between the documents. However, in general it is a complicated problem. We establish the semantic links by analyzing the history of user's queries to the digital library documents and use the anomaly detection approach to identify users whose behavior is suspect to be the unauthorized large-scale copying.

Anomaly detection approach is based on the assumption that an intrusion may be seen as a deviation from normal, ordinary or expected behavior of the user, and can be detected by comparing the sequence of user's actions with some given profile of their behavior. While designing a system that implements an anomaly detection approach, we have faced the following challenges:

1. Construction of a normal profile of user's behavior;
2. Design of a classifier that allows to distinguish between anomalous and normal sequences of references to documents;
3. Delimitation of the boundary values of the classifier characteristics in order to reduce the probability of misclassification;
4. Periodic update of the normal behavior profile.

As human behavior is inherently non-deterministic, the probabilistic approaches to modeling and predicting user's actions are the most common. In stochastic models of user's behavior the probability of observing a certain action is estimated based on the history of user's actions.

For example, the multinomial model [14, 16] disregards the order of user's actions in the history. In the first-order Markov models [7, 8, 14] the new action depends on the history only through the last observed action. The maximum entropy model [14] allows to construct the set of probability functions and choose the one with the highest information entropy, estimating long-term dependence of the new action on the whole history.

Other solutions such as Bayesian methods [3, 6] have also been used in modeling and predicting user's behavior.

To model the normal user's behavior we use the generalized Markov-chain-based method. While working with a digital library, the user downloads or views the single documents. This sequence of references to digital documents is being recorded in the log file. The profile of the normal user's behavior is represented by a Markov chain built on the basis of the log file. In the next section we describe a formal method proposed in [10] for constructing a Markov chain that models user's behavior and a classifier that detects anomalities. The technique adapted to detect an attempt of unauthorized large-scale documents copying is also described in [9].

16.3 Model Description

This section provides a detailed description of the construction procedure of a Markov chain modeling the normal behavior of the user, as well as an algorithm for constructing a behavior classifier which is used to detect anomalous activities of the user.

16.3.1 Behavioral Model

Let us define as a *trace* the sequence of user's references to (or downloads of) documents of the digital library during a single work session. Let Σ denote the set of unique identifiers of the documents (in the following examples we will assume that the identifiers are denoted by symbols $a, b, c, \ldots$) extended by a special null symbol $\emptyset$. Thus, a trace is a sequence of symbols of the alphabet Σ. Let Σ^* denote a set of all possible finite traces over Σ.

To construct the Markov chain there must be set a "window size" parameter w. A state in the Markov chain is associated with a subtrace of length w over Σ, i.e. each state is a sequence of w symbols from Σ. The pair (s, s') denotes a transition from state s to state s' in the Markov chain. Each state and transition are also associated with the counter.

The operation $shift(\sigma, x)$ shifts the trace σ left and appends the symbol x at the end of the trace, e.g., $shift("aba", c) = "bac"$. The operation $next(\sigma)$ returns the first symbol of the trace σ and left shifts σ by one position, e.g., $next("abcd")$ returns a symbol and updates the argument to "bcd".

The algorithm of Markov chain construction is the following. The initial state of the Markov chain is associated with the trace of the length w consisting of null symbols, e.g., if $w = 3$, the initial state is associated with the trace $[\emptyset, \emptyset, \emptyset]$. We need a training suite $T_{tr} \in \Sigma^*$ of the history of user's actions. For each trace from T_{tr} the following steps should be implemented until all symbols from σ are scanned:

1. Let $c = next(\sigma)$.
2. Set $\langle nextstate \rangle = shift(\langle currentstate \rangle, c)$.
3. Increase counters for the state $\langle currentstate \rangle$ and the transition $(\langle currentstate \rangle, \langle nextstate \rangle)$.
4. Update $\langle currentstate \rangle$ to be $\langle nextstate \rangle$.

Fig. 16.1 Example of a Markov structure

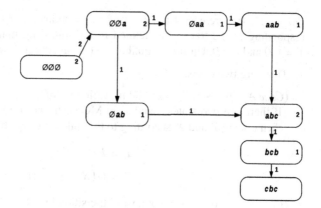

After all the traces in T_{tr} have been processed, each state and transition are associated with a positive integer—the counter. The probability of transition from state s to state s' ($P(s, s')$) is set to be $\frac{N(s,s')}{N(s)}$, where $N(s, s')$ and $N(s)$ are the counters associated with the transition (s, s') and the state s, respectively. By this procedure the Markov chain is constructed.

By construction P is a valid measure, i.e. the following equality holds for all states s:

$$\sum_{s' \in succ(s)} P(s, s') = 1 \qquad (16.1)$$

where $succ(s)$ denotes the set of successors of s.

Figure 16.1 shows an example of a Markov structure built upon the set $T_{tr} = \{aabc, abcbc\}$ with window size $w = 3$. The Markov chain constructed by the algorithm upon this set denotes the semantic relations between documents and represents a profile of the normal behavior. The training suite T_{tr} contains user activity sessions, namely sequences of documents downloaded by separate users in certain periods of time (e.g., during twelve hours).

16.3.2 Classifier

Having a model of the normal behavior, we need a classifier that would distinguish between anomalous and normal sequences of references to the documents.

Given a trace $\sigma \in \Sigma$, $|\sigma|$ denotes the length of the trace. Given a trace σ and a positive integer $i \le |\sigma|$, α_i and $\alpha[i]$ denote the prefix consisting of the first i symbols and the i-th symbol respectively.

Consider a trace $\alpha \in \Sigma^*$. Let the initial trace β_0 be $\sigma(s_0)$, i.e., the trace associated with the initial state s_0. The trace after scanning the first symbol $\alpha[1]$ is $\beta_1 = shift(\beta_0, \alpha[1])$. The trace β_k obtained after scanning the k-th symbol is recursively defined as $shift(\beta_{k-1}, \alpha[k])$. Hence, a trace α defines a sequence of traces

$\beta_0, \ldots, \beta_m$ (where each trace β_i is of length w and $m = |\alpha|$). A metric $\mu(\alpha)$ corresponding to the trace α is defined in the following manner. Initially, $X = 0.0$, $Y = 0.0$ and $i = 0$. Until i is equal m we execute the following steps:

1. There are two cases.

   (*Case A:*) $\beta_i \rightarrow \beta_{i+1}$ is a valid transition in MC
   If there are two states s and s' in MC such that $\sigma(s) = \beta_i$ and $\sigma(s') = \beta_{i+1}$, then update Y and X according to the following equations:

   $$Y = F(Y, s, (s, s')),$$
   $$X = G(X, s, (s, s')).$$

   (*Case B:*) β_i or β_{i+1} is not a valid transition in MC
   If $\beta_i \rightarrow \beta_{i+1}$ is not a valid transition in MC, then we update Y and X according to the following equations:

   $$Y = Z(Y),$$
   $$X = L(X).$$

2. Increment i to $i + 1$.

The metric $\mu(\alpha)$ is defined as $\frac{Y}{X}$ at the end of the procedure just described. A classifier over the symbol set Σ is a function $f : \Sigma^* \rightarrow \{0, 1\}$. Given the metric $\mu(\alpha)$, we can construct a classifier in the following manner:

$$f(\alpha) = \begin{cases} 1, & \mu(\alpha) \geq r, \\ 0, & \text{otherwise.} \end{cases} \tag{16.2}$$

Here r is a given threshold. Thus, the trace α is classified as anomalous if the metric is above the threshold r.

The metric μ is parametrized by the functions F, G, Z and L. Different choices of the functions lead to different classifiers. Some common functions are described in [10].

16.4 Description of the Experiments

For the experiments there has been developed a software system that preprocesses input data files, constructs the profile of normal behavior and examines user activity sessions to detect abnormalities.

16.4.1 The Source Data

The source data for the experiments has been a log file of access to documents of the Digital Library of the Republic of Karelia [5] during the period from Septem-

```
24.***.***.62 - - [03/Jul/2005:08:55:26 +0400] "GET /pagepdf.shtml?
id=315&cType=1 HTTP/1.1" 200 251
"http://elibrary.karelia.ru/book.shtml?levelID=014&id=315&cType=1"
"Mozilla/4.0 (compatible; MSIE 6.0; Windows NT 5.1; SV1; .NET CLR
1.0.3705)"
195.***.***.194 - - [03/Jul/2005:12:05:46 +0400] "GET /pagepdf.shtml?
id=707&cType=1 HTTP/1.0" 200 251
"http://elibrary.karelia.ru/book.shtml?levelID=017008&id=707&cType=1"
"Mozilla/4.0 (compatible; MSIE 6.0; Windows NT 5.1; SV1; .NET CLR
1.1.4322)"
85.***.***.133 - - [03/Jul/2005:12:22:37 +0400] "GET /pagepdf.shtml?
id=252&cType=1 HTTP/1.1" 200 251
"http://elibrary.karelia.ru/book.shtml?levelID=009&id=252&cType=1"
"Mozilla/4.0 (compatible; MSIE 6.0; Windows NT 5.1)"
85.***.***.133 - - [03/Jul/2005:12:22:38 +0400] "GET /pagepdf.shtml?
id=267&cType=1 HTTP/1.1" 200 251
"http://elibrary.karelia.ru/book.shtml?levelID=009&id=267&cType=1"
"Mozilla/4.0 (compatible; MSIE 6.0; Windows NT 5.1)"
```

Fig. 16.2 Fragment of the log file

ber 2004 to January 2011 inclusive. In total, the digital library contains over two thousand documents.

The log file of access to digital documents is written in the Common Log Format [13]. Each request to the server is logged in a single line consisting of fields separated by spaces. The log file looks like shown in Fig. 16.2. The fields that are relevant for the experiments are highlighted. The first highlighted field represents the IP address of the user's computer, the second contains the timestamp of the registered action, the third is a query string including the identifier of the requested document, the fourth contains the status of the server's response.

16.4.2 Preprocessing of the Source Data

To construct a profile of normal behavior there was carried out a preprocessing of the source data. We considered a user's session as a sequence of all document requests from this user with the time interval between two requests not exceeding twelve hours. In total there had been recorded queries from over 5500 different users. We discarded uninformative sessions containing less than five queries to the digital library. Finally, 10393 user activity sessions were derived from the log file.

16.4.3 Profile of the Normal Behavior

According to the model introduced in [10], to construct the normal profile we need two data suites: the training suite (the data that are assumed to be normal) and the

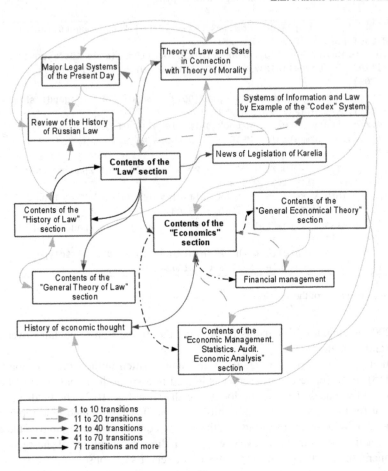

Fig. 16.3 Example (fragment) of the normal profile

testing suite (for selecting the optimal parameters). We divided user activity sessions derived from the log file into two equal parts to use as these data suites.

Figure 16.3 shows an example (fragment) of the normal profile illustrating semantic connections between documents automatically identified based on analysis of users' behavior in the digital library. Apparently, the certain digital documents demonstrate the large number of transitions between them. For example, in the profile such documents as "Memorial Book of the Olonets Province in 1867", "The Olonets Province: List of Localities According to the 1873" and "The Olonets Collection. Issue 3 (1894)" (all in Russian) are strongly tied. The names of these documents clearly state their semantic proximity, which confirms the original thesis about the identification of semantic links between documents based on analysis of users' behavior.

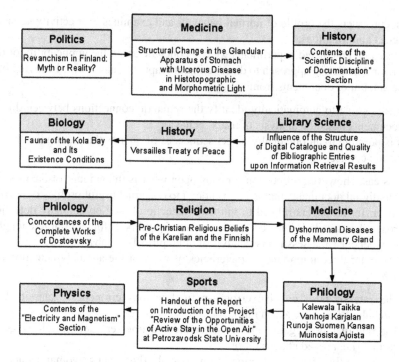

Fig. 16.4 Example of the anomalous session

16.4.4 The Classifier and the Anomalous Behavior

According to the model, the classifier is used to detect in the user activity session a significant deviation from the normal behavior represented by the Markov chain.

One of the most typical anomalous sessions that were detected automatically is shown in Fig. 16.4. In the header of each document requested by the user there is a title of the section which the document belongs to in the digital library. There is no doubt that such diversity in choice of documents and sections is not a typical behavior of the user. Detection of such anomalous sessions is the objective of the approach considered in this work. The example of an anomalous session was obtained with the following choice of the parameters of the classifier: $F(Y, s, (s, s')) = Y + 10^{-6}$; $G(X, s, (s, s')) \equiv 1$; $Z(Y) = Y + 0.1$; $L(X) \equiv 1$; $r = 1.0$.

16.5 Conclusion

This work presents the results of a series of experiments conducted to test the applicability and define characteristics of the anomaly detection approach to discovering unauthorized large-scale documents copying from a digital library. For the experiments there had been developed the software system that preprocesses input data files, constructs the profile of normal behavior and examines user activity sessions to detect abnormalities.

Based on conducted experiments we can conclude about the applicability of the behavioral modeling approach to discovering attempts of unauthorized copying of a large amount of documents from a digital library:

- it is possible to automatically identify the semantic connections between digital documents based on analysis of users' behavior;
- it is possible to automatically detect sequences of queries that contradict the semantic connections.

In this case, however, a question remains open what is the amount of data on access to digital documents that is large enough to construct useful profiles of normal behavior. In the papers related to intrusion detection based on anomaly detection approach it is usually indicated that such data should be sufficient (large enough), but any persuasive assessments (analytical or empirical) are absent.

There are three important characteristics of work of the anomaly detection approach:

1. The rate of false positive errors—normal actions erroneously classified as anomalous (i.e. false alarm rate);
2. The rate of false negative errors—anomalous actions erroneously classified as normal (i.e. missed attacks rate);
3. The mean time before detection of an anomaly (how many anomalous actions will be observed before generating an alarm).

The correlation of these three characteristics depends on the classifier, i.e. the choice of the functions F, G, Z and L, and the threshold r. However, the question of choice of the optimal classifier is beyond the scope of this work.

The final purpose of our research is to develop a system of protection against unauthorized large-scale documents copying. The system will be able to complement existing in the digital library means of copy protection (for example, limitation of the number of documents that a user can access in a short time, and license agreements that guarantee the rights of the digital library holders). In this case, detection of anomalous behavior in user activity may be the reason for temporarily blocking the user access to resources of the digital library with requirement for further investigation.

References

1. Cox, I.J., Miller, M.L., Bloom, J.A.: Digital Watermarking. The Morgan Kaufmann Series in Multimedia and Information Science. Morgan Kaufmann, San Mateo (2002)
2. Copyright Lawsuit against Georgia State University: http://ourgeorgiahistory.com/ogh/Copyright_Lawsuit_against_Georgia_State_University. Cited 7 June 2011
3. D'Ambrosio, B., Altendorf, E., Jorgensen, J.: Probabilistic relational models of on-line user behavior. In: Proceedings of the WebKDD-2003 Workshop on Webmining as a Premise to Effective and Intelligent Web Applications, Washington, DC, pp. 9–16 (2003)

4. Cao, L.: In-depth behavior understanding and use: the behavior informatics approach. Inf. Sci. **180**, 3067–3085 (2010)
5. Digital Library of the Republic of Karelia: http://www.elibrary.karelia.ru. Cited 7 June 2011
6. Hassan, M.T., Junejo, K.N., Karim, A.: Bayesian inference for Web surfer behavior prediction. In: Proceedings of ECML/PKDD Discovery Challenge Workshop (2007)
7. Hogg, T., Lerman, K.: Stochastic models of user-contributory Web sites. In: Proceedings of the 3rd International Conference on Weblogs and Social Media (2009)
8. Hu, Y., Zincir-Heywood, A.N.: Modeling user behaviors from FTP server logs. In: Proceedings of the 4th Annual Communication Networks and Services Research Conference (2006). doi:10.1109/CNSR.2006.36
9. Ivashko, E.: The defensive system against unauthorized documents-copying of the digital libraries development. In: Proceedings of the Ninth Russian Conference on Digital Libraries, pp. 300–306 (2007) (in Russian)
10. Jha, S., Tan, K., Maxion, R.A.: Markov chains, classifiers, and intrusion detection. In: Proceedings of the 14th IEEE Computer Security Foundations Workshop, 0206 (2001). doi:10.1109/CSFW.2001.930147
11. JSTOR service: Terms and conditions of use. http://www.jstor.org/page/info/about/policies/terms.jsp. Cited 7 June 2011
12. Koulouris, A., Kapidakis, S.: Access and reproduction policies of the digital material of seven national libraries. In: Proceedings of the Fifth Russian Conference on Digital Libraries, pp. 35–44 (2003)
13. Log Files—Apache HTTP Server: http://httpd.apache.org/docs/current/logs.html. Cited 7 June 2011
14. Pavlov, D., Manavoglu, E., Pennock, D., Lee Giles, C.: Collaborative Filtering with Maximum Entropy. IEEE Intell. Syst. **19**(6) (2004). doi:10.1109/MIS.2004.59
15. Wang, J.-H., Chang, H.-C., Hsiao, J.H.: Protecting digital library collections with collaborative Web image copy detection. In: Buchanan, G., Masoodian, M., Cunningham, S.J. (eds.) Digital Libraries: Universal and Ubiquitous Access to Information. Springer, Heidelberg (2008)
16. Wang, Y.: A multinomial logistic regression modeling approach for anomaly intrusion detection. Comput. Secur. (2005). doi:10.1016/j.cose.2005.05.003

Chapter 17
Analyzing Twitter User Behaviors and Topic Trends by Exploiting Dynamic Rules

Luca Cagliero and Alessandro Fiori

Abstract Everyday online communities and social networks are accessed by millions of Web users, who produce a huge amount of user-generated content (UGC). The UGC and its publication context typically evolve over time and reflect the actual user interests and behaviors. Thus, the application of data mining techniques to discover the evolution of common user behaviors and topic trends is becoming an appealing research issue. Dynamic association rule mining is a well-established technique to discover correlations, among data collected in consecutive time periods, whose main quality indexes (e.g., support and confidence) exceed a given threshold and possibly vary from one time period to another.

This Chapter presents the DyCoM (DYnamic COntext Miner) data mining system. It entails the discovery of a novel and extended version of dynamic association rules, namely the dynamic generalized association rules, from both the content and the contextual features of the user-generated messages posted on Twitter. A taxonomy over contextual data features is semi-automatically built and exploited to discover dynamic correlations among data at different abstraction levels and their temporal evolution in a sequence of tweet collections.

Experiments, performed on both real Twitter posts and synthetic datasets, show the effectiveness and the efficiency of the proposed DyCoM framework in supporting user behavior and topic trend analysis from Twitter.

17.1 Introduction

The widespread usage of online communities and social networks allows analysts to investigate Web user's behaviors and interests through the analysis of the user-generated content (UGC). The huge amount of content published on the Web (posts,

L. Cagliero (✉) · A. Fiori
Dipartimento di Automatica e Informatica, Politecnico di Torino, Corso Duca degli Abruzzi, 24, 10129 Torino, Italy
e-mail: luca.cagliero@polito.it
url: http://dbdmg.polito.it

A. Fiori
e-mail: alessandro.fiori@polito.it

L. Cao, P.S. Yu (eds.), *Behavior Computing*,
DOI 10.1007/978-1-4471-2969-1_17, © Springer-Verlag London 2012

tags) by online community or social network users may be exploited for marketing purposes to discover the actual user interests, opinions, and behaviors. Furthermore, the context in which the social content is published (e.g., the publication date, time, and place) may be useful for profiling main Web user's activities and personalizing services and promotions. However, since the user-generated content and contextual data continuously evolve over time, the discovery of the most relevant recurrences and their temporal evolution is definitely a challenging task. This prompts the need of novel and more effective data mining algorithms to support the knowledge discovery process from data coming from Web communities.

A number of approaches applied data mining techniques to UGC. They focused on building formal representations of domain-specific social knowledge [18], improving service recommendation based on social annotations or preferences [21, 27, 33], and discovering most significant user behaviors and interests [22, 25]. A significant research effort has been devoted to analyzing Twitter UGC. For instance, TwitterMonitor [25] extracted contextual knowledge from Twitter streams to detect most common topic trends, while, in [9], information retrieval techniques are exploited to discover topic trends and support analyst decision-making. To discover relevant correlations among data and their historical evolution across consecutive time periods, a well-founded approach is the extraction of dynamic association rules [2]. Dynamic rules are correlations, in the form of association rules [1], whose main quality indexes (i) exceed a given threshold, and (ii) may change from one time period to another. The history of the main pattern quality indexes reflects the most relevant temporal correlation changes.

In this chapter we present the DyCoM (Dynamic Context Miner) data mining system. It focuses on discovering and representing higher level correlations among data posted by users on the Twitter microblogging Web site by means of an extended version of dynamic association rules, namely the dynamic generalized association rules. By exploiting the Twitter Application Programming Interface (API), DyCoM retrieves both the tweet content (i.e., the textual messages) and their contextual features (e.g., publication date, time, place). Based on the values of one of the most peculiar tweet contextual features (e.g., the publication date), tweets are first partitioned in an ordered sequence of tweet collections. The textual content is tailored to a relational data schema [31] to enable the association rule mining process. Next, a taxonomy over contextual data features is semi-automatically built, by means of Extraction, Transformation, and Loading (ETL) processes, and exploited to drive the rule mining process from the sequence of tweet collections. Generalized association rules are extracted by aggregating, according to the taxonomy, lower level data items into higher level ones. The discovered correlations provide a higher level view of the analyzed data since they also include feature values (e.g., places, timestamps) at different levels of abstraction (e.g., regions or nations, time intervals or days). DyCoM integrates the discovery of generalized association rules in the context of dynamic rule mining. The discovered patterns compactly represent higher level correlations and their temporal evolution across a sequence of tweet collections. To the best of our knowledge, the integration of generalized association rules in the context of dynamic association rule mining has never been investigated so far.

Finally, the DyCoM framework allows effectively querying the extracted patterns, based on either their schema or content, to discover recurrences in Web user behaviors and topic trends. For instance, significant correlations regarding specific topics may be pointed out in consecutive days at different geographical granularity levels, e.g., within a certain city or its corresponding region or nation.

This chapter is organized as follows. Section 17.2 compares our work with related approaches. Section 17.3 presents the architecture of the DyCoM framework and describes its main blocks. Section 17.4 assesses the effectiveness of DyCoM in extracting hidden information from Twitter posts as well as describes possible use-cases of interests for the DyCoM framework. Finally, Sect. 17.5 draws conclusions and presents future developments of this work.

17.2 Related Works

A huge amount of UGC is posted by Web users in different data formats (e.g., posts, tags, videos). Both the resource content and its context of publication are suitable for driving the data mining and knowledge discovery process. A particular attention is paid by both industrial and academic researchers to the analysis of the user behaviors, opinions, and interests for marketing purposes. Thus, the usage of social network data to discover the evolution of most notable user behaviors [8] and topic trends has become an appealing research issue.

A parallel effort has been devoted to discovering temporal correlations among data by exploiting well-established data mining techniques, e.g., dynamic association rule mining. However, the application of more advanced association rule mining algorithms to discover higher level dynamic correlations has never been investigated so far.

In the following we present, in separate sections, most recent related works concerning data mining from online communities, dynamic association rule mining, and generalized association rule mining.

17.2.1 Data Mining from Online Communities

In the last years, the characterization of online communities and social networks through data mining techniques has been largely investigated. For instance, an interesting research direction is the discovery of most relevant online community user behaviors [7, 11, 20]. In [7] the authors investigated user activities (e.g., universal searches, message sending, and community creation) by means of clickstream data analysis. Differently, in [11] the characteristics of the lifetime of UGC are investigated by empirically studying workloads coming from three popular knowledge-sharing online social networks, i.e., a blog system, a social bookmark sharing network, and a question answering social network. Finally, in [20] patterns relevant for characterizing both user and group behaviors are identified. To this aim, it exploited a graph-based representation of the user connections. Results showed that online communities may be represented, for less than their half, as a giant component in which users are spread across isolated communities, while the majority of the users are outside the giant component and are typically joint by means of star connections. A parallel issue is the investigation of the evolution of the structure of social networks and online communities. For instance,authors in [23] exploited a maximum a posteriori (MAP) estimation to discover the community structure evolution by considering both the observed network and the prior distribution given by the historical community structures.

A number of approaches exploited machine learning and data mining algorithms to figure out relevant social knowledge from the online community user-generated content. For instance, in [33], the development of novel recommendation systems focused on improving the quality of personalized promotions is investigated, while the works presented in [5, 22, 34] aimed at categorizing user-generated content by means of classification and/or clustering techniques. Finally, semantic data coming from social networks is also exploited to improve the performance of query engines [6, 15]. A relevant effort has been devoted to supporting knowledge discovery from Twitter by means of data mining techniques. For instance, in [9] the authors addressed the problem of trend pattern detection to discover users that contribute towards the discussion of particular newsworthy topics. Differently, in [25] the authors presented TwitterMonitor, a context-based system to discover significant topic trends from Twitter streams. Similarly, this Chapter also presents a data mining system to perform knowledge discovery from Twitter. Unlike [9, 25] it exploits both the content of the contextual information associated with Twitter posts to perform user behavior and topic trend analysis. Furthermore, it also analyzes the evolution of the discovered patterns across consecutive time periods.

17.2.2 Dynamic Association Rule Mining

Association rule mining [1] is a widely used exploratory data mining technique to discover correlations that (i) frequently occur in the analyzed data, and (ii) hold in most cases. A number of research papers investigated the dynamic evolution of association rules across consecutive time periods. Active data mining [2] focused on representing and querying the history pattern of discovered association rule quality indexes by incrementally updating a common rule base. Similarly, other works addressed the discovery of segment-wise or point-wise periodicities in time-related datasets [14] and the anomalous time-related frequent pattern detection in network traffic analysis [26].

A parallel research issue has been devoted to discovering most relevant pattern changes over time. To address this issue, a statistical evaluator based on the chi-square test of independence [24], the discovery and selection of the emerging patterns [10], and the application of fuzzy approaches to rule change evaluation [3] have been exploited. However, the application of the traditional association rule mining process may entail discarding relevant but rare knowledge due to the pushing of a minimum support constraint, i.e., a minimum frequency of occurrence of the extracted patterns, into the mining process. This Chapter proposes a data mining system that exploits a semi-automatically generated taxonomy to drive the extraction of higher level correlations among timestamped data collected from Twitter. To address this issue, it discovers generalized association rules and their history, in terms of the evolution of their main quality indexes (e.g., support and confidence [1]), from a sequence of tweet collections.

17.2.3 Generalized Association Rule Mining

A significant research effort has been devoted to the design and development of novel algorithms to efficiently discover association rules including items at different abstraction levels. The first attempt to address generalized association rule mining has been done in [28, 29] in the context of market basket analysis. More specifically, in [28] the authors proposed Cumulate, a generalized association rule mining algorithm that generates itemsets by considering, for each item, its parents in the hierarchy. Hence, candidate frequent itemsets are generated by exhaustively evaluating the taxonomy and, thus, producing a large amount of redundant patterns. The generalization process has been performed by exploiting a traditional rule mining process on an extended source dataset in which transactions include all the possible generalizations of the relative data items. Indeed, a postprocessing step is needed to prune redundant patterns (i.e., itemsets containing higher level items belonging to the same attribute).

Several optimization strategies to perform a more efficient generalized association rule extraction process has been proposed [4, 12, 13, 16]. For instance, in [16] a faster support counting is provided by exploiting the TID intersection computation, which is common in algorithms designed for the vertical data format [35]. Differently, in [12, 13] an optimization based on a top-down hierarchy traversal is proposed. It identified in advance itemsets which cannot be frequent in the dataset by means of the Apriori principle. The discovery of interesting multiple-level association rules is driven by a level-dependent multiple support threshold enforcement when itemsets belonging to the same abstraction level are extracted. Differently, in [4] the authors proposed a support-driven itemset generalization approach, in which rare yet relevant knowledge discarding is prevented by triggering its corresponding generalization. Generalizations that cover only frequent descendants are not extracted as they are deemed redundant for analyst decision making. DyCoM integrates both a traditional (exhaustive) approach to generalized itemset mining [28] and a more recent one [4], in the dynamic generalized rule mining process.

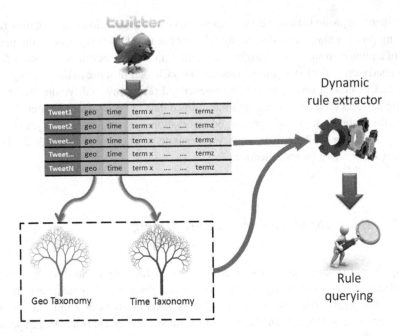

Fig. 17.1 The DyCoM framework

17.3 The DyCoM Framework

DyCoM (DYnamic COntext Miner) is a data mining framework focused on support-ing the discovery of relevant correlations among the textual content and the publi-cation context of messages posted on Twitter. To address this issue, it investigates the evolution of most significant patterns hidden in a sequence of tweet collections. Figure 17.1 reports the DyCoM framework architecture. In the following, its main blocks and functionalities are briefly described.

User-Generated Content and Context Data Retrieval   This block entails the re-trieval, preprocessing, and categorization of messages (tweets) posted by Web users on Twitter. The retrieved data is tailored to a relational data schema which includes both content features (i.e., the most relevant keywords) and contextual features (e.g., the geographical location). Furthermore, tweets are partitioned into a sequence of subsets based on the value its most peculiar features (e.g., tweet collections posted in consecutive days).

Taxonomy Generation over Contextual Data   Taxonomies over the tweet con-textual features are semi-automatically generated. More specifically, Extraction, Loading, and Transformation (ETL) processes are exploited to aggregate values of lower level contextual features (e.g., the GPS coordinate) into their higher level ag-gregations (e.g., the city and the region).

Dynamic Rule Extractor   This block aims at discovering the evolution of the significant higher level correlations hidden in the sequence of tweet collections. It discovers generalized association rules whose main quality indexes (i.e., support and confidence) exceed a given threshold and possibly vary from one interval to another. The generalization is performed by evaluating the previously generated taxonomies.

Rule Querying   The extracted dynamic rules are queried to efficiently retrieve the information of interest based on either their content or schema. To ease the domain expert analysis' task, the resulting dynamic rules are ranked based on the value of their main quality indexes (e.g., support and confidence) in some user-specified tweet collections of increasing level.

A more detailed description of the main DyCoM framework blocks is presented in the following sections.

17.3.1 User-Generated Content and Context Data Retrieval

Twitter (http://twitter.com) is one of the most popular microblogging and social networking services. Textual messages posted by Twitter users (i.e., the tweets) are at most 140 characters long and publicly visible by default. This block addresses the retrieval of tweets posted on Twitter by means of the Search Application Programming Interfaces (APIs). Data is returned by Twitter APIs in the JSON format (Java Script Object Notation), which is an XML-based standard for client-server data exchange. The tweet textual content is enriched by several contextual feature values (e.g., publication place coordinates, city, date, hour). Some of them are peculiar characteristics of the context in which tweets are posted by users (e.g., the source location GPS coordinates), while others are just high level aggregations of the previous ones (e.g., the city). Both the textual message keywords and the contextual feature values are modeled as data items.

Definition 17.1 (Item) Let t_i be a label, called attribute, which describes a data feature. Let Ω_i be the discrete domain of attribute t_i. An item $(t_i, value_i)$ assigns the value $value_i \in \Omega_i$ to attribute t_i.

In the case of continuous attributes, the value range is discretized into intervals and the intervals are mapped to consecutive positive integers. The items represent either the textual message content, e.g., (text, "This is a message by Obama"), or a contextual feature value, e.g., (Date, 2010-10-10). A tweet could be represented as a set of items, called record, as stated in the following definition.

Definition 17.2 (Record) Let $\mathcal{T} = \{t_1, t_2, \ldots, t_n\}$ be a set of attributes and $\Omega = \{\Omega_1, \Omega_2, \ldots, \Omega_n\}$ the corresponding domains. A record r is a set of items that contains at most one item for each attribute in $\mathcal{T}$. Each record is characterized by a level l.

The level l associated with each tweet is an integer non-negative number that identifies the collection, in a sequence of tweet collections, to which the record (tweet) belongs to. For instance, if tweets are partitioned based on their daily submission hour and labeled in order of increasing time interval, tweets submitted at 3 a.m. are characterized by level 3 as they are included in the third tweet collection. A set of records (tweets) all characterized by a common level l, is called relational tweet collection of level l.

Definition 17.3 (Relational Tweet Collection of Level l) Let $\mathcal{T} = \{t_1, t_2, \ldots, t_n\}$ be a set of attributes and $\Omega = \{\Omega_1, \Omega_2, \ldots, \Omega_n\}$ the corresponding domains. A relational tweet collection D_l is a collection of records, where each record r is characterized by level l.

Since data retrieved by Twitter is not compliant with a relational tweet collection, a preprocessing phase is needed. Data is tailored to a common relational data schema and preprocessed by means of a data cleaning process. Data cleaning discards useless and redundant information and correctly manages missing values. For each tweet, the geographical information, the publication date and time stamp are selected. The tweet textual content is preprocessed by removing stopwords, numbers, and links. Furthermore, a stemming algorithm based on Wordnet [17] is applied to reduce words to their base form, i.e., the stem. The preprocessed messages are tailored to a relational data schema, by selecting distinct words belonging to the BOW data representation, and included in the corresponding relational tweet collection (cf. Definition 17.3). Each record at a certain level represents a Twitter message and is composed of (i) the stemmed terms of the textual content and (ii) its publication context features. This data representation will be exploited to address dynamic generalized association rule mining from the sequence of tweet collections.

17.3.2 Taxonomy Generation over Contextual Data

A taxonomy is a hierarchical knowledge representation that defines is-a relationships between concepts and their instances (i.e., the data items). Taxonomies are exploited to categorize objects and provide additional information about data. This block addresses the semi-automatic taxonomy generation over the contextual feature values belonging to any relational tweet collection. The generated taxonomy will be exploited to drive the generation of the generalized itemsets. To aggregate attribute values into higher level concepts, we define the concept of aggregation tree, i.e., an aggregation hierarchy built over the domain of one attribute of the relational tweet collection.

Definition 17.4 (Aggregation Tree) Let t_i be an attribute and Ω_i its domain. An aggregation tree AT_i is a tree representing a pre-defined set of aggregations over values in Ω_i. AT_i leaves are all the values in Ω_i. Each non-leaf node in AT_i is an aggregation of all its children. Node $\bot$ aggregates all values for attribute t_i.

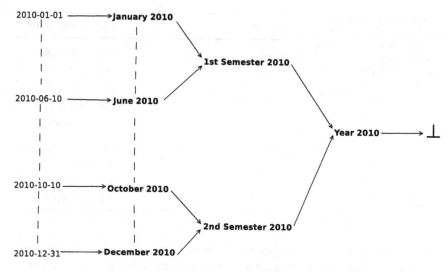

Fig. 17.2 A portion of an aggregation tree over the date attribute

We define a taxonomy as a set of aggregation trees built over distinct data attributes.

Definition 17.5 (Taxonomy) Let $\mathcal{T} = \{t_1, t_2, \ldots, t_n\}$ be a set of attributes and $\rho = \{AT_1, \ldots, AT_m\}$ a set of aggregation trees defined on $\mathcal{T}$. A taxonomy $\Gamma \subseteq \rho$, is a set of aggregation trees.

Despite a taxonomy may potentially include more aggregation trees over the same attribute, for the sake of simplicity, in the following we will consider only taxonomies that contain at most one aggregation tree $AT_i \in \rho$ for each attribute $t_i \in \mathcal{T}$.

A portion of an example aggregation tree built over the *date* attribute is reported in Fig. 17.2. Taxonomies over contextual features (e.g, spatial and temporal information) are derived by means of aggregation functions based on a hierarchical model. The hierarchical model represents the relationships between different levels of aggregation. Similarly to what usually done in data warehousing [19], the information is extracted by means of Extraction, Transformation and Load (ETL) processes, called here aggregation functions. For instance, in the relational tweet representation, aggregations functions may define either associations among different contextual attributes (e.g., *City* $\Rightarrow$ *State*) or aggregations over a singular contextual attributes (e.g., *Date* $\Rightarrow$ *Semester*) which could be obtained by simply parsing the corresponding attribute domain values.

Given a set of aggregation functions over UGC features, the context taxonomy generation block allows semi-automatically building a taxonomy. It associates with each item the corresponding set of generalizations organized in a hierarchical fashion. For instance, consider a temporal contextual feature (e.g., *Month*) that represents a higher level knowledge abstraction of another contextual feature (e.g., *Date*).

Table 17.1 Aggregation functions used for the taxonomy generation over the temporal and the spatial contextual features	Data features	Aggregation function
	Temporal	$Date \Rightarrow WeekDay$
		$Date \Rightarrow Month$
		$Month \Rightarrow Year$
		$Time \Rightarrow Hour$
		$Hour \Rightarrow TimeSlot$
	Spatial	$GPSCoordinates \Rightarrow Id$
		$Id \Rightarrow Place$
		$Place \Rightarrow Region$
		$Region \Rightarrow State$

A conceptual hierarchy of aggregations may be devised by mapping the two attribute domains by means of the corresponding aggregation function (e.g., $Date \Rightarrow Month$). Consider again the *Date* attribute and its high level aggregation *Semester*. Although the corresponding higher level attribute does not exist yet, the corresponding mapping may be simply derived by parsing the lower level *Date* domain values (e.g., 2010-10-10) and generating upper level concepts (e.g., *2nd Semester* 2010) according to the corresponding aggregation function (i.e., $Date \Rightarrow Semester$). In Table 17.1 the aggregation functions exploited in the experiments (see Sect. 17.4) over temporal and spatial contextual data features are resumed. However, the DyCoM system allows easily integrating different and more complex aggregation functions as well.

Given a taxonomy Γ, we formalize the concept of generalized item as a item $(t_i, expression_i)$ such that $expression_i$ is a non-leaf node in some $AT_i \in \Gamma$.

Definition 17.6 (Generalized Item) Let t_i be an arbitrary attribute, Ω_i its domain, and AT_i be an aggregation tree defined on values in Ω_i. A generalized item $(t_i, expression_i)$ assigns the value $expression_i$ to attribute t_i. $expression_i$ is a non-leaf node in AT_i which defines an aggregation value over values in Ω_i. $leaves(expression_i) \subseteq \Omega_i$ is the set of items whose values are leaf nodes descendant of $expression_i$ in AT_i.

The support of a generalized item $(t_i, expression_i)$ in a relational tweet collection D_l is the (observed) frequency of $leaves(expression_i)$ in D_l.

17.3.3 Dynamic Association Rule Mining

This block focuses on analyzing the dynamic evolution of the tweet content and publication context. Correlations hidden in each tweet collection may be effectively discovered by means of association rule mining [1]. The traditional association rule mining problem is commonly addressed by means of a two-step process: (i) frequent itemset mining, driven by a minimum support threshold and (ii) association rule generation, driven by a minimum confidence threshold. To also consider the evolution of the discovered patterns across a sequence of tweet collections, a well-established approach is the discovery of dynamic association rules [2]. Dynamic rules are association rules whose main quality indexes (i.e., support and confidence) may change from one time period to another.

The DyCoM framework addresses dynamic association rule mining by exploiting the previously generated taxonomies to extract generalized association rules and analyze their temporal evolution across consecutive time periods. Generalized rules allow discovering dynamic correlations at different abstraction levels and preventing the discarding of relevant but rare knowledge. In the following, we both introduce preliminary definitions and separately addressed the dynamic generalized itemset and association rule mining steps.

Preliminary Definitions To formally state the dynamic generalized association rule mining problem, we preliminary introduce the concepts of not generalized and generalized itemset.

Definition 17.7 ((Generalized) Itemset) Let T be a set of attributes, Ω the corresponding domain, and Γ a taxonomy defined on values $value_i \in \Omega_i$. A *not generalized itemset* is a set of items $(t_k, value_k)$ in which each attribute t_k may occur at most once. A *generalized itemset* is an itemset that includes at least a generalized item $(t_k, value_k)$ such that $value_k \in \Gamma$.

For instance, $\{(Place, New York), (date, October\ 2010)\}$ is a (generalized) itemset of length 2 (i.e., a generalized 2-itemset). A (generalized) itemset covers a given record (tweet) r of level l, i.e., $r \in D_l$ (cf. Definition 17.2), if all its (possibly generalized) items $x \in X$ are either (i) included in r, or (ii) ancestors of items $i \in r$ (i.e., $\exists i \in leaves(x)|\ i \in r$). The support of a (generalized) itemset X in a relational tweet collection D_l of level l is given by the number of tweets $r \in D_l$ covering X divided by the cardinality of D_l. A descendant of an itemset represents its knowledge at a lower aggregation level.

Definition 17.8 (Generalized Itemset Descendant) A (generalized) itemset X is a descendant of a generalized itemset Y if (i) X and Y have the same length and (ii) for each item $y \in Y$ there exists at least an item $x \in X$ that is a descendant of y.

Consider the itemset $\{(Place, New York), (date, 2010\text{-}10\text{-}10)\}$. According to the aggregation tree reported in Fig. 17.2, it is an example of descendant of the generalized itemset $\{(Place, New York), (date, October\ 2010)\}$.

A generalized association rule is an implication $X \Rightarrow Y$, where X and Y are disjoint generalized or not generalized itemsets.

Definition 17.9 (Generalized Association Rule) Let A and B be two (generalized) itemsets such that $attr(A) \cap attr(B) = \emptyset$, where $attr(X)$ is the set of attributes belonging to itemset X. A generalized association rule is represented in the form $A \Rightarrow B$, where A and B are the body and the head of the rule respectively.

Generalized association rules are usually characterized by support and confidence quality indexes. The rule support *sup* is the support of the (generalized) itemset $A \cup B$, while the rule confidence *conf* is given by $\frac{sup(A \cup B)}{sup(A)}$ (i.e., the rule strength).

Dynamic Generalized Itemset Mining   Dynamic generalized itemsets are generalized itemsets whose main quality indexes may vary from one time period to another and exceed a given threshold in at least one of them. A commonly used approach [1] is to constrain the itemset mining process by means of a minimum support threshold.

Definition 17.10 (Dynamic Generalized Itemset) Let $D_1, D_2, \ldots, D_n$ be a sequence of n relational tweet collections of increasing level and *min_sup* a minimum support threshold. Let $G = \{g_1, g_2, \ldots, g_k\}$ be the set of generalized itemsets whose support value is equal to or exceeds *min_sup* for some D_i where $i \in \{1, 2, \ldots, n\}$. A dynamic generalized itemset is a pattern represented in the form $g_i [s_1, s_2, \ldots, s_n]$ where $g_i \in G$ and $[s_1, s_2, \ldots, s_n]$ is a vector of support values. s_i is the support of g_i in the i-th tweet collection D_i.

For instance, suppose that D_1, D_2, D_3 is a sequence of three relational tweet collections of increasing levels 1, 2, and 3. By enforcing a minimum support threshold equal to 1%, an example of dynamic generalized itemset is: $\{(Place, New York), (date, October 2010)\}$ [1.5%, 2%, 1%]. It states that the itemset $\{(Place, New York), (date, October 2010)\}$ is frequent in all tweet collections and its support values in D_1, D_2, and D_3 are, respectively, 1.5%, 2%, and 1%.

To address the dynamic generalized itemset mining problem efficiently, i.e., without performing multiple itemset mining steps over each tweet collection followed by postprocessing, DyCoM adopts an Apriori-like algorithm, namely the DYGEN (DYnamic GENeralized itemset extractor) algorithm. Furthermore, to reduce the number of generated patterns, a tuned version of the DYGEN algorithm, namely *Tuned* DYGEN (*Tuned* DYnamic GENeralized itemset extractor) algorithm, which entails a support-driven opportunistic generalization of infrequent patterns only, is integrated in the DyCoM framework as well. More specifically, generalized itemsets that have no infrequent descendants, according to the support threshold, in a given time period are no longer extracted since their covered knowledge is already supported by the lower level descendants.

In the following section, we thoroughly describe both DYGEN and *Tuned* DYGEN algorithms.

Algorithm 17.1 DYGEN: DYnamic GENeralized itemset extractor

Input: a sequence of relational tweet collections $D_1, \ldots, D_n$ of levels $1, \ldots, n$,
a minimum support threshold min_sup, a taxonomy Γ

Output: set of dynamic generalized itemsets DGI

1: $k = 1, DGI = \emptyset$
2: C_k = set of distinct k-itemsets in $D_1, \ldots, D_n$
3: **repeat**
4:   scan $D_i \in \{D_1, \ldots, D_n\}$ and count the support $sup(c, D_i) \; \forall c \in C_k$
5:   $Gen = \emptyset$ // generalized itemset container
6:   **for all** c in C_k **do**
7:     $gen(c)$ = set of new generalizations of itemset c
8:     $gen(c) = $ taxonomy_evaluation(Γ, c)
9:     $Gen = Gen \cup gen(c)$
10: **end for**
11: **if** $Gen \neq \emptyset$ **then**
12:     count support in D_i for each itemset $gen(c) \in Gen$
13: **end if**
14: $DGI_k = \{$itemsets in $C_k \cup Gen$ that satisfy $min\_sup\}$
15: $k = k + 1$
16: $C_{k+1} = $ candidate_generation(DGI_k)
17: **until** $C_k \neq \emptyset$
18: **return** DGI

The DYnamic GENeralized Itemset Extractor Algorithm   The DYGEN algorithm takes in input a sequence of relational tweet collections $D_1, \ldots, D_n$ of increasing level (cf. Definition 17.3), a taxonomy Γ (cf. Definition 17.5), and a minimum support threshold min_sup. It produces of the set of all dynamic generalized itemsets (cf. Definition 17.10) obtained from $D_1, \ldots, D_n$ by evaluating the input taxonomy Γ and by enforcing a minimum support threshold min_sup.

Algorithm 17.1 reports the pseudo-code of the DYGEN algorithm. It adopts an Apriori-like approach to dynamic itemset mining. The DYGEN algorithm iteratively generates frequent generalized itemsets from each relational tweet collection by means of a level-wise approach. In an arbitrary iteration k, DYGEN performs three steps: (i) k-itemset generation from each the relational tweet collection D_i (lines 16), (ii) support counting and generalization of (generalized) itemsets with length equal to k (lines 3–17), and (iii) generation of candidate (generalized) itemsets of length $k + 1$ by joining k-itemsets and candidate pruning. A loop on the set of candidates of length k (lines 6–10) triggers the generalization procedure over the taxonomy Γ. Given an itemset c of level l and a taxonomy Γ, the taxonomy evaluation procedure generates a set of generalized itemsets by applying on each item $(t_j, value_j)$ of c the corresponding aggregation tree $AT_j \in \Gamma$. All the itemsets obtained by replacing one or more items in c with their generalized versions of level $l + 1$ are generated and included into the Gen set (line 9). Finally, their supports are computed by performing the dataset scans (line 12). Since the input dataset is structured, DYGEN also exploits this feature to further reduce the number of generated candidates. In particular, it does not generate candidates including two items with the same attribute label. The DYGEN algorithm ends the mining loop when the set of candidate itemsets is empty (line 17).

The *Tuned* DYnamic GENeralized Itemset Extractor Algorithm   The *Tuned* DYGEN algorithm improves the DYGEN performance by addressing the issue of reducing the number of generated dynamic generalized itemsets. To accomplish this task, it adopts a support-driven itemset generalization procedure, first proposed in [4], that lazily generalizes a higher level itemset at a certain level only if it has at least an infrequent descendant (cf. Definition 17.8).

The main modifications to the DYGEN algorithm (see Algorithm 17.1) needed to perform lazy itemset generalization are described in the following. Once the candidate itemset support counting (line 4) has been performed, the generalization procedure (line 8) is lazily invoked only on candidates that are infrequent in at least one tweet collection. This approach prevents the extraction of higher level correlations that represent only the knowledge covered by its lower level descendants.

The experimental evaluation, reported in Sect. 17.4, highlights the reduction, in terms of both the number of extracted dynamic generalized itemsets and execution time, of *Tuned* DYGEN with respect to DYGEN, evaluated on synthetic data.

The Rule Generation Algorithm   The last DyCoM mining step focuses on generating dynamic generalized association rules from the set of extracted dynamic generalized itemsets.

In the following we extend the concept of generalized association rule (cf. Definition 17.9) to a dynamic context as follows.

Definition 17.11 (Dynamic Generalized Association Rule) Let $D_1, D_2, \ldots, D_n$ be a sequence of relational tweet collections. Let *min_sup* be a minimum support threshold and *min_conf* be a minimum confidence threshold. Let $R = \{gr_1, gr_2, \ldots, gr_k\}$ be the set of generalized rules whose support and confidence are equal to or exceed the corresponding thresholds, i.e., *min_sup* and *min_conf*, for some D_i where $i \in \{1, 2, \ldots, n\}$. A dynamic generalized association rule is a pattern represented in the form $gr_i [s_1, s_2, \ldots, s_n][c_1, c_2, \ldots, c_n]$, where $gr_i \in R$ and $[s_1, s_2, \ldots, s_n]$ and $[c_1, c_2, \ldots, c_n]$ are, respectively, the support and confidence vectors, i.e., s_i and c_i are the support and confidence of gr_i in the i-th tweet collection D_i.

The rule mining algorithm generates the dynamic generalized association rules (cf. Definition 17.11) starting from the complete set of dynamic frequent (generalized) itemsets. Thus, it performs the second step of the Apriori algorithm [1]. Since confidence of rules generated from the same itemset has the anti-monotone property, candidate rules of length k are generated by merging two $(k - 1)$-length rules that share the same prefix in the rule consequent [1]. Although confidence and support are the most popular rule quality indexes [1], the DyCoM framework allows easily integrating different quality indexes as well (e.g., lift [30]).

17.3.4 Querying Rules

The block entails the selection and ranking of most valuable dynamic rules for better supporting in-depth analysis. Rule selection is constrained by either (i) the rule schema (i.e., the attributes that have to appear in the rule body or head), or (ii) the rule content. For example, the schema constraint $\{(Keyword, *)\} \rightarrow \{(Time, *)\}$ selects all 2-length dynamic rules that include, respectively, an item characterized by attribute *Keyword* in the rule body and attribute *Time* in the rule head. The generalized rule $\{(Keyword, Sport)\} \rightarrow \{(Time, from\ 10\ p.m.\ to\ 12\ p.m.)\}$ [...] satisfies the above schema constraint. Differently, the item constraint $\{*\} \rightarrow \{(Time, from\ 10\ p.m.\ to\ 12\ p.m.)\}$ selects all rules that contain item (Time, from 10 p.m. to 12 p.m.) as rule consequent. Thus, the rule $\{(Keyword, Sport)\} \rightarrow \{(Time, from\ 10\ p.m.\ to\ 12\ p.m.)\}$ [...] satisfies the example item constraint as well.

Results of rule querying may be sorted according to confidence and support quality index values obtained in a specific tweet collection to better support in-depth analysis.

17.4 Experimental Results

A number of experiments have been conducted to evaluate the performance of the DyCoM framework. The following issues have been addressed: (i) the performance of the DYGEN and *Tuned* DYGEN mining algorithms (see Sect. 17.4.1) and (ii) the usefulness of the mined patterns in different real use cases (see Sect. 17.4.2).

17.4.1 DyCoM Performance Analysis

The DyCoM data mining system focuses on extracting dynamic generalized association rules from sequences of tweet collections by exploiting the DYGEN algorithm. However, when a large amount of rules is extracted the domain expert validation task becomes complex. To reduce the number of extracted patterns, a tuned version of the DYGEN algorithm, namely the *Tuned* DYGEN algorithm, is integrated in the DyCoM framework as well. *Tuned* DYGEN exploits a recently proposed support-driven approach to itemset generalization, i.e., GENIO (GENeralized Itemset DiscOverer) [4], which generates a higher level itemset only if it has at least an infrequent descendant (cf. Definition 17.8).

In this section, we compared the performance of the DYGEN and *Tuned* DYGEN algorithms by addressing the following issues: (i) the impact of both support and confidence thresholds on the number of extracted dynamic generalized rules and (ii) the extraction time.

Impact of Support and Confidence Thresholds We analyzed the performance of DYGEN and *Tuned* DYGEN by comparing the number of dynamic generalized rules extracted by the two algorithms. To this aim, a set of experiments was performed on synthetic datasets generated by means of the TPC-H generator [32]. The TPC-H data generator consists of a suite of business oriented ad-hoc queries. The queries and the data populating the database have been chosen to have broad industry-wide relevance. By varying the scale factor parameter, datasets with different sizes are generated. We generated a dataset starting from the lineitem table by setting a scale factor equal to 0.075 (i.e., around 450,000 records). To partition the whole dataset in three distinct time-related data collections we queried the source data by enforcing different constraints on the shipping date value (attribute *ShipDate*). More specifically, we partitioned lineitems shipped in the three following time periods: [1992-01-01, 1994-02-31], [1994-03-01, 1996-05-31], and [1996-06-01, 1998-12-01]. For the sake of brevity, we will denote the corresponding datasets as data-1, data-2, and data-3 in the rest of this section.

Since the minimum support and confidence thresholds enforced, respectively, during the itemset and rule mining steps significantly affect the number of extracted patterns, we performed different mining sessions, for all combinations of algorithms and mining constraints. In Figs. 17.3(a) and 17.3(b) we plotted the number of dynamic (generalized) rules mined by DYGEN and *Tuned* DYGEN from data-1, data-2, and data-3, by varying the support and confidence thresholds. To test the DYGEN and *Tuned* DYGEN algorithms, we considered the generated datasets in increasing order of shipment date interval. For both algorithms, the number of mined dynamic (generalized) rules significantly increases for lower minimum support and confidence values (e.g., 1%). When higher support thresholds are enforced, most of the lower level patterns are infrequent and the number of pruned generalized patterns is limited. When lowering the support threshold, some of them become frequent and a number of generalized patterns covering only frequent knowledge are discarded at each time period as they are deemed not relevant for domain expert analysis. Indeed, the pruning effectiveness of the *Tuned* DYGEN algorithm with respect to DYGEN becomes more and more significant at lower support thresholds. For most of the settings of minimum support and confidence value, the reduction, in terms of the number of extracted dynamic rules, achieved by *Tuned* DYGEN against DYGEN is at least 5%.

DyCoM Extraction Time We also analyzed, on synthetic datasets, the time spent by DyCoM in each dynamic generalized rule mining phase, i.e., frequent dynamic generalized itemset mining, and dynamic generalized association rule generation. To perform this analysis, we exploited the same TPC-H lineitem tables data-1, data-2, and data-3 presented in the previous section. As expected, the majority of the DyCoM extraction time is devoted to the itemset mining phase. For all tested datasets, support, and confidence threshold values the rule generation step never takes more than 10% of the whole extraction time.

In Fig. 17.4, we reported the execution time spent by the DYGEN and the *Tuned* DYGEN algorithms in dynamic generalized itemset mining, by varying the minimum support threshold. The extraction time is mainly affected by the cost of the

Fig. 17.3 Number of generated dynamic generalized rules extracted from data-1, data-2, and data-3. Number of time periods = 3. Taxonomy height = 3

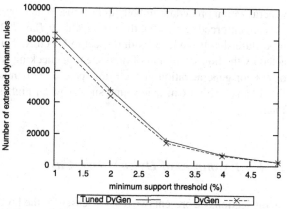

(a) Impact of the support threshold. Minimum confidence threshold = 50%

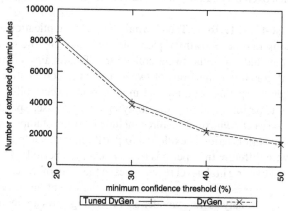

(b) Impact of the confidence threshold. Minimum support threshold = 3%

Fig. 17.4 Dynamic generalized itemset extraction time by varying the minimum support threshold. data-1, data-2, and data-3 datasets

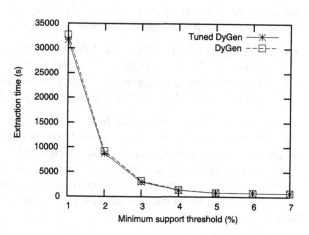

generalization procedure in both algorithms. When lower minimum support thresholds are enforced (e.g., 1%), the time spent by *Tuned* DYGEN in the generalization procedure slightly reduces with respect to DYGEN. This effect is partially counteracted by the higher amount of time spent in checking whether an itemset is eligible or not for generalization at each time period. This makes the *Tuned* DYGEN and DYGEN algorithm extraction time similar when higher support thresholds are enforced.

17.4.2 Examples of DyCoM Use-Cases

In this section, we present two real use-cases for the DyCoM system that are focused on user behavior and topic trend analysis. For each use-case, some examples of discovered dynamic generalized association rules are reported as well.

Use-Case 1: Topic Trend Analysis   This application scenario enables analysts to look into newsworthy topic trends by analyzing the temporal evolution of correlations hidden in the tweet collections. To this aim, analysts may follow this steps: (i) crawling collections of tweets posted at consecutive time periods, (ii) dynamic generalized association rule mining from tweet collections in order of increasing time periods, and (iii) rule querying based on user-specified constraints.

Tweet contextual features include the time stamp at which messages are posted. This information is exploited to partition tweets into disjoint collections associated with different time periods (e.g., 1-day time period). The DYGEN algorithm allows discovering high level recurrences in the form of generalized association rules. They may represent unexpected trends in the evolution of relevant tweet topics. For instance, analysts may wonder how breaking news are matter of contention on Twitter. To delve into the impact of breaking news coming from the United States Capitol, We first collected tweets whose submission time is uniformly distributed in the range [2011/03/22, 2011/03/24]. Then, we categorizing them based on their submission date. For example, the following dynamic generalized rules are extracted by enforcing a minimum support threshold equal to 1%.

(A) {(Keyword, Obama), (Keyword2, Libya)} → {(Place, Washington, D.C.)}
       (sup = [1.3%, 0.3%, 0.3%], conf = [100%, 83%, 85%])
(B) {(Keyword, Obama), (Keyword2, Libya)} → {(Place, Washington, D.C.)}
       (sup = [2.3%, 2.5%, 1.3%], conf = [100%, 91%, 91%])

The U.S. congress meeting, that held on March 22nd in Washington D.C., was focused on the conflict in Libya. Keywords *Obama* and *Libya*, which have been frequently posted on March, 22nd in Washington, D.C., becomes infrequent in the same location the day after. However, the extraction of the higher level correlation (B) allows figuring out that the same topic remains of interest in the U.S.A. yet.

Depending on the granularity of the selected time periods, different mining results may be achieved.

Use-Case 2: Context-Based User Behavior Analysis   This application scenario investigates the attitudes of Twitter users in posting, citing, and answering Twitter messages concerning newsworthy topics in different spatial contexts. To achieve this goal, analysts should perform the following steps: (i) crawling tweet collections posted from regions or states of interest within a given time period, (ii) dynamic generalized association rule mining from tweet collections by following a user-provided significance order, and (iii) dynamic rule querying based on user-specified constraints.

We consider again the previously collected tweets, posted during the time period [2011/03/22, 2011/03/24], and we reorganized them, based on their submission place, in: (i) a collection of tweets posted within a 2,500 km radius far from New York (i.e., lands along Eastern American coastline) and (ii) a collection of tweets posted within a 2,500 km radius far from London (i.e., North-West of Europe). We performed a dynamic generalized association rule mining session by enforcing a minimum support threshold equal to 1% and a minimum confidence threshold equal to 80% and by considering the American tweet collection first. Following the chain of spatial tweet message propagation set the analyst may discover valuable knowledge about user attitudes in Twitter service usage by looking into the history of the discovered patterns across a sequence of places of interest. For instance, the following dynamic rules are extracted:

(A) {(Keyword, Obama), (Place, Washington, D.C.)} $\rightarrow$ {(Date, 2011/03/22)}
       (sup $= [3.6\%, 2.1\%, 0.5\%]$, conf $= [100\%, 100\%, 91\%]$)
(B) {(Keyword, Obama), (Place, United Kingdom)} $\rightarrow$ {(Date, 2011/03/22)}
       (sup $= [1.1\%, 2.3\%, 1.2\%]$, [conf $= 100\%, 95\%, 95\%]$)

The dynamic generalized association rules (A) is discovered from the collection of American tweets, while rule (B) is mined from the European tweet collection. American Twitter users paid particular attention to the foreign policy undertaken by president Obama and the American Congress, which had been topic of discussion in the past meeting held in the United States Capitol Washington, D.C. (USA) on March, 22nd 2011. Furthermore, users coming from Europe are also interested in posting messages related to the same topic as the U.S. Government decisions bias the European state economies.

17.5 Conclusions and Future Works

This paper presents the DyCoM data mining system that focuses on analyzing Twitter user behaviors and topic trends by discovering and looking into dynamic correlations hidden in the Twitter user-generated content. A taxonomy over the contextual data features allows the discovery of correlations, i.e., the generalized association rules, at different abstraction levels. Two dynamic rule mining algorithm are integrated in the DyCoM system: the DYGEN and the *Tuned* DYGEN algorithm. In particular, *Tuned* DYGEN exploits a support-driven opportunistic approach that allows preventing, at each time period, the generation of redundant higher level patterns.

Experimental results show both the effectiveness and the efficiency of the DyCoM framework. Real use-cases are proposed and exploited to highlight the usefulness of the DyCoM system in both user behavior analysis and topic trend detection.

Future works will address: (i) taxonomy inference from tweet textual content, (ii) incremental updating of both the taxonomy content and the mined rule sets, and (iii) the enforcement of analyst-provided constraints into the generalized association rule mining process.

References

1. Agrawal, R., Imielinski, T., Swami, A.N.: Mining association rules between sets of items in large databases. In: SIGMOD Conference, pp. 207–216 (1993)
2. Agrawal, R., Psaila, G.: Active data mining. In: Proceedings of the 1st International Conference on Knowledge Discovery and Data Mining, pp. 3–8 (1995)
3. Au, W.-H., Chan, K.C.C.: Mining changes in association rules: a fuzzy approach. Fuzzy Sets Syst. **149**, 87–104 (2005)
4. Baralis, E., Cagliero, L., Cerquitelli, T., D'Elia, V., Garza, P.: Support driven opportunistic aggregation for generalized itemset extraction. In: 5th IEEE International Conference of Intelligent Systems, pp. 102–107 (2010)
5. Basile, P., Gendarmi, D., Lanubile, F., Semeraro, G.: Recommending smart tags in a social bookmarking system. In: Bridging the Gep Between Semantic Web and Web, vol. 2, pp. 22–29 (2007)
6. Bender, M., Crecelius, T., Kacimi, M., Michel, S., Neumann, T., Parreira, J.X., Schenkel, R., Weikum, G.: Exploiting social relations for query expansion and result ranking. In: IEEE 24th International Conference on Data Engineering Workshop, pp. 501–506 (2008)
7. Benevenuto, F., Rodrigues, T., Cha, M., Almeida, V.: Characterizing User Behavior in Online Social Networks. In: Proceedings of the 9th ACM SIGCOMM Conference on Internet Measurement Conference, pp. 49–62. ACM, New York (2009).
8. Cao, L.: In-depth behavior understanding and use: the behavior informatics approach. Inf. Sci. **180**, 3067–3085 (2010)
9. Cheong, M., Lee, V.: Integrating web-based intelligence retrieval and decision-making from the twitter trends knowledge base. In: Proceeding of the 2nd ACM Workshop on Social Web Search and Mining, pp. 1–8. ACM, New York (2009)
10. Dong, G., Li, J.: Mining border descriptions of emerging patterns from dataset pairs. Knowl. Inf. Syst. **8**, 178–202 (2005)
11. Guo, L., Tan, E., Chen, S., Zhang, X., Zhao, Y.E.: Analyzing patterns of user content generation in online social networks. In: Proceedings of the 15th ACM SIGKDD International Conference on Knowledge Discovery and Data Mining, pp. 369–378. ACM, New York (2009)
12. Han, J., Fu, Y.: Discovery of multiple-level association rules from large databases. In: Proceedings of the International Conference on Very Large Data Bases, pp. 420–431 (1995)
13. Han, J., Fu, Y.: Mining multiple-level association rules in large databases. IEEE Trans. Knowl. Data Eng. **11**(5), 798–805 (2002)
14. Han, J., Gong, W., Yin, Y.: Mining segment-wise periodic patterns in time-related databases. In: Proceedings of International Conference on Knowledge Discovery and Data Mining, pp. 214–218 (1998)
15. Heymann, P., Ramage, D., Garcia-Molina, H.: Social tag prediction. In: Proceedings of the 31st Annual International ACM SIGIR Conference on Research and Development in Information Retrieval, pp. 531–538. ACM, New York (2008)

16. Hipp, J., Myka, A., Wirth, R., Güntzer, U.: A New Algorithm for Faster Mining of Generalized Association Rules, pp. 74–82. Springer, Berlin (1998)
17. Hovy, E., Lin, C.Y.: Automated text summarization in SUMMARIST (1999)
18. Kasneci, G., Ramanath, M., Suchanek, F., Weikum, G.: The YAGO-NAGA approach to knowledge discovery. ACM SIGMOD Record 37(4), 41–47 (2009)
19. Kimball, R., Ross, M., Merz, R.: The Data Warehouse Toolkit: The Complete Guide to Dimensional Modeling. Wiley, New York (2002)
20. Kumar, R., Novak, J., Tomkins, A.: Structure and evolution of online social networks. In: Proceedings of the 12th ACM SIGKDD International Conference on Knowledge Discovery and Data Mining, pp. 611–617 (2006)
21. Li, Q., Wang, J., Chen, Y.P., Lin, Z.: User comments for news recommendation in forum-based social media. Inf. Sci. (2010)
22. Li, X., Guo, L., Zhao, Y.E.: Tag-based social interest discovery. In: Proceeding of the 17th International Conference on World Wide Web, pp. 675–684 (2008)
23. Lin, Y.-R., Chi, Y., Zhu, S., Sundaram, H., Tseng, B.L.: Analyzing communities and their evolutions in dynamic social networks. ACM Trans. Knowl. Discov. Data 3, 8.1–8.31 (2009)
24. Liu, B., Hsu, W., Ma, Y.: Discovering the set of fundamental rule changes. In: Proceedings of the Seventh ACM SIGKDD International Conference on Knowledge Discovery and Data Mining, pp. 335–340. ACM, New York (2001)
25. Mathioudakis, M., Koudas, N.: TwitterMonitor: trend detection over the twitter stream. In: Proceedings of the 2010 International Conference on Management of Data, pp. 1155–1158. ACM, New York (2010)
26. Qin, M., Hwang, K.: Frequent episode rules for Internet anomaly detection. In: Proceedings of Third IEEE International Symposium on Network Computing and Applications, pp. 161–168 (2004)
27. Shepitsen, A., Gemmell, J., Mobasher, B., Burke, R.: Personalized recommendation in social tagging systems using hierarchical clustering. In: Proceedings of the ACM Conference on Recommender Systems, pp. 259–266 (2008)
28. Srikant, R., Agrawal, R.: Mining generalized association rules. In: VLDB, pp. 407–419. Morgan Kaufmann, San Mateo (1995)
29. Srikant, R., Vu, Q., Agrawal, R.: Mining association rules with item constraints. In: KDD, pp. 67–73 (1997)
30. Tan, P.N., Kumar, V., Srivastava, J.: Selecting the right interestingness measure for association patterns. In: Proceedings of the Eighth ACM SIGKDD International Conference on Knowledge Discovery and Data Mining, p. 41 (2002)
31. Tan, P.N., Steinbach, M., Kumar, V., et al.: Introduction to Data Mining. Pearson, Addison Wesley, Boston (2006)
32. TPC-H: The TPC benchmark H. Transaction processing performance council (2009). http://www.tpc.org/tpch/default.asp
33. Xue, Y., Zhang, C., Zhou, C., Lin, X., Li, Q.: An effective news recommendation in social media based on users' preference. In: International Workshop on Education Technology and Training, vol. 1, pp. 627–631 (2009)
34. Yin, Z., Li, R., Mei, Q., Han, J.: Exploring social tagging graph for web object classification. In: Proceedings of the 15th ACM SIGKDD International Conference on Knowledge Discovery and Data Mining, pp. 957–966. ACM, New York (2009).
35. Zaki, M.J., Parthasarathy, S., Ogihara, M., Li, W., et al.: New algorithms for fast discovery of association rules. In: Proceedings of the 3rd International Conference on Knowledge Discovery and Data Mining, vol. 20 (1997)

Part IV
Behavior Applications

Chapter 18
Behavior Analysis of Telecom Data Using Social Networks Analysis

Avinash Polepally and Saravanan Mohan

Abstract In Mobile Social Network Analysis, mobile users interaction pattern change frequently and hence it is very hard to detect their changing patterns because humans posses an extremely high degree of randomness in their calling behavior. To identify regularity in such random behavior, we propose a new method using network attributes to find periodic or near periodic graphs in dynamic social networks. We try to analyze real-world mobile social networks and extract its periodicity through a simple practical and efficient method using effective network attributes of the social network. We demonstrate the applicability of our approach on real-world networks and extract meaningful and interesting periodic interaction patterns. This helps in defining targeted business models in cellular communication arena.

Abbreviations

Δ	Time snap shot rate
E	is an edge representation in the graph
K	degree of a node in social network
$\langle K \rangle$	mean degree of a node in social network (per node)
C	clustering coefficient of a node in social network
$\langle C \rangle$	mean clustering coefficient (per node)
W	weight of an edge in social network
$\langle W \rangle$	mean weight of an edge in social network (per node)

18.1 Introduction

It is well known that humans have the potential for high degree of random behavior, but still we could expect observable calling patterns from a single mobile user.

A. Polepally (✉) · S. Mohan
Ericsson R&D, Chennai, India
e-mail: avinash.polepally@gmail.com

S. Mohan
e-mail: m.saravanan@ericsson.com

L. Cao, P.S. Yu (eds.), *Behavior Computing*,
DOI 10.1007/978-1-4471-2969-1_18, © Springer-Verlag London 2012

These patterns could vary with timescales. Calling patterns could vary with daily calls made between employees, calling family, close friends, girlfriends etc. The patterns could vary accordingly on weekends and weekdays. Our aim is to find the periodicity that can learn aspects of a user's life by building simple mechanisms that can recognize any of the common structures in the user's routine calling. Learning the structure of individual's routine has already been demonstrated using other modalities like Lahiri [7] work where periodic interaction are extracted from changing graphs in a tractable manner; our contribution will be to demonstrate learning of changing social structures. We represent the telecommunication network through graph structure. Each individual is represented by a node in the network, and there is an edge between two nodes if a social interaction has occurred at any point in time between the two individuals represented by these nodes. In telecom scenario, social interactions should happen through sending text message (SMS) or making a call. Edges were weighted by the duration of voice interactions. The competitiveness in this area could be withstood with operational proficiency, technological edge and consumer/subscriber understanding [12]. The technology and proficiency of work are not considered to be bottlenecks of the present scenario. Instead, it is always important to know your customer. The consumer behavior helps us in providing better services and gain sustainable advantages in business.

Our research is mainly to extent telecom operator's business by understanding the behavior of their subscribers through social interactions which could help them in better marketing and advertising of their products. The identification of different category of consumers who are of different value to the operator helps in promoting marketing activities like up-selling, cross-selling etc. In today's extremely challenging business environment, many telecommunications operators have shifted the focus of Telecom BI from consumer acquisition to consumer retention. Interestingly, as mobile penetration is increasing and even approaching saturation, it has been estimated that it is much cheaper to retain an existing consumer than to acquire a new one. To maintain profitability, telecom service providers must control churn, i.e. the loss of subscribers who switch from one operator to another [5]. There are many models that predict churn, but to prevent churn it is very important to understand the behavior of the consumers and act likely to provide promotional offers that favor their behavior to retain them on their network. The operator must offer the right incentives, and adopt right marketing strategies, and place network assets appropriately to protect its consumers.

Retrieving information from call detail record data can provide major business insights for designing such strategies. In mobile social network which is established by calls made between mobile users, the dynamics of the network change at very high and random rate. Due to frequent changing patterns of social interactions it is very difficult to cut-off a timeline for identifying evolving communities and changing behavior patterns of the mobile users [10]. In telecom business it is important to identify the community structure and see how they evolve which helps the operators in marketing and advertising.

Fig. 18.1 The figure shows three actors behavior with corresponding colors. Bottom point = "consumer 1", Left point = "consumer 2", Right point = "consumer 3"

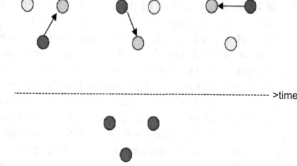

18.2 Preliminaries and Related Work

Lahri's [7] work regarding periodic mining of social networks shows the computational complexity of enumerating all periodic subgraphs. They have shown that there are at most $O(T2\ln T\sigma)$ closed periodic subgraphs at minimum support σ in a dynamic network of T time steps. Furthermore, they have described a polynomial time, parameter-free, one-pass algorithm to mine all periodic subgraphs, including a 'jitter' heuristic for mining subgraphs that are not perfectly periodic. They have also proposed a new measure, *purity*, for ranking mined subgraphs according to how perfectly periodic a subgraph is.

Many other works related to subgraph mining were try to handle dynamic social networks depend on time snapshot i.e., the time span when the social network should be captured to observe the dynamic changes [2, 3, 9, 10] which depend on time snapshot rate to study the changing dynamics. In this paper we find the periodicity of mobile social networks driven by daily rhythms of calling behavior. In Fig. 18.1, let us consider three time instances morning, afternoon, night of the three mobile users who have different behavior.

$$\text{Red} = \text{calls time usage for} > 2\text{ hrs}$$

$$\text{Green} = \text{calls made} > 1\text{ hr but} < 2\text{ hrs}$$

$$\text{Yellow} = \text{calls made} < 1\text{ hr}$$

In case if the mobile operator interested to offer a promotional scheme to their subscribers to keep them active and stay within the network. Then it would be better if a promotional scheme is made such that "consumer 1" gets some discount on calls made during mornings, "consumer 2" gets some discount on calls made during afternoons and "consumer 3" gets some discount on calls made during nights. This would be in the best interest of the consumers, the operator could also give an option to choose at what time of the day the consumer would prefer discount on calls. While looking at an operator point of view, the operator cannot satisfy all consumers else he would have to bank on low revenues. Rather he could choose to release such a promotional plan to public that would benefit maximum consumers. The operator should know which section of the day should the offer carry discounts. Here the efficient time slicing helps to make suitable decisions.

The work related to dynamic networks [6] have been analyzed by creating a sequence of network snapshots, in which edges that vary in time are aggregated over a window of fixed length. It means that all the edges that have come active in time span t_1 and t_2 would be captured as the behavior of the network in that particular window. To move beyond static structures and consider questions relating to how behavior changes over time like e.g. how the communities evolve over time or how the strength of connection between the individuals in the network changes over time [8], we are in a need of an engineering approach to represent continuous variations. The network variations pose its own set of problems, as the choice of snapshot rate Δ determines many properties of social networks. Hence, we should take a proper approach while choosing a snapshot rate Δ, incorrect choice of Δ may end up in a strong bias in resulting analysis and conclusions. Thus as suggested by Clauset and Eagle [4], the problem for calculating the periodicity is solved by capturing a natural snapshot rate that smoothes out high frequencies while trying to preserve low frequency social patterns. In our paper we have tried to capture the weight of an edge as an important network parameter in determining the periodicity.

18.3 Network Statistics

The social change analysis is to use the network attributes for static topologies, and use that information and extend to dynamic topologies [11]. A similarity statistic measure [11] is used to have an intuition on the degree of topological overlap between two sequential snapshots t_1 and t_2. Our proximity data would appear as tuple of the form (i, j, t_1, t_2) denoting that i and j are proximate to each other starting at time t_1 and ending at time t_2. In order to transform this information into a sequence of T network snapshots $\vec{N} = N^1, N^2, \ldots, N^T$ a length of time Δ is chosen and the network change is observed over different values of Δ and analyzed, Δ is called the snapshot rate, that each snapshot covers. We then simply say that $N_{i,j}^t = 1$ if and only if the nodes i and j are connected at any time between t and $t + \Delta$, $N_{i,j}^t = 0$ otherwise. The aim of the experiment is to observe what effect the free parameter Δ has on the patterns in dynamic network topology. When both Δ and the density of nodes in physical space are small, these snapshots are expected to be very sparse, each containing only a few edges.

- The average degree $\langle K \rangle$ of a node

$$\langle K \rangle = \frac{\sum_{i=1}^{n} K}{n} \tag{18.1}$$

- The mean local density of triangles $\langle C \rangle$ (also called the clustering coefficient)

$$\langle C \rangle = \frac{\sum_{i=1}^{n} C}{n} \tag{18.2}$$

- The mean weight of the edges $\langle W \rangle$ which represented duration of voice call

$$\langle W \rangle = \frac{\sum_{\forall ij} W}{n} \tag{18.3}$$

18.4 Experimental Evaluation

18.4.1 Data Set

Every day mobile call transactions are very high in volume related to any specific operators. We have extracted a data set which consists of CDR data for 8 days of one of the largest African mobile operators. The data set contains detailed information about voice calls, SMS, value added service calls etc. of subscribers. Our analysis is based on a representative region in the operator's network and all intra-region (local) calls made during the specified period. Total data set is about 5 million nodes and 400 million edges.

18.4.2 Telecom Network Representation

Telecom Network is generally referred as call graphs and each graph having many vertices and one or more connecting edges.

Vertices. Each vertex represents a mobile user and is represented with an ID [in this case it was the MSISDN (mobile number)]. There are other parameters which were represented according to the requirements.

Edges. Each edge is described with the two IDs corresponding to the two vertices that define an edge (vertices connected by the edge). Each edge is characterized by the weight that corresponds to the amount of call duration in seconds.

We include a pair of nodes A and B, if A calls B or vice-versa. We prune out edges which last for less than 5 seconds assuming them to be dropped calls. Availability of data was for only 8 days, to translate the data into a network representation that captures the characteristics of the underlying communication network, we consider an edge $E_{A,E}$ if there has been at least one call made between them. The weight $W_{A,E}$ of an edge $E_{A,E}$ is the aggregate of all calls between A to B in that observed period of time. During pre-processing, we also excluded the service numbers, e.g. the operator's help-line service number, number for retrieving voice mail and emergency numbers included in them. We also churned out landline numbers, thus concentrating only on the mobile network structure. We also pruned out calls made internationally, to concentrate on the local mobile structure.

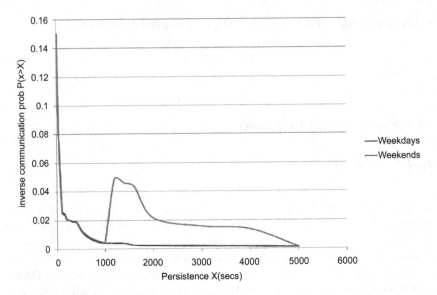

Fig. 18.2 The empirical distribution of the percentage of edges in the network for weekday and weekend during the 8 days data observation

18.4.3 Network Analysis

We conducted some initial analysis on networks; our main intention was to study how mobile socialization network persists. Since every call made has a time span attached i.e. the duration of the call made, it is interesting to analyze the persistence of call. From the starting to the ending times of each edge, we computed its temporal duration, or persistence, as $t_2 - t_1$. In Fig. 18.2, we can see the empirical distribution for these durations (in seconds) for both weekends and weekdays. The distributions are largely similar except for slight variations in the bottom tail of the curve ($x > 1000$ seconds), which looks quite obvious as weekends stay for more social communication.

$$P(X < x) = \frac{number\ of\ users\ in\ the\ sample \leq x}{n} \tag{18.4}$$

Figure 18.2 suggests an equilibrium system with quite expected and strongly regular features. The average persistence of an edge in telecom mobile network turns out to be relatively small ($\langle E \rangle = 5$ mins 52 secs) while there are several edges that persist for more than a couple of hours especially on weekends, the variation suggests that networks evolve in broad range time scales which is quite expected. We conclude that much of this regularity is due to strong periodic behavior in human cycle. The peak of the curve during weekends is quite expected which brings us to obvious conclusion of greater socialization through calls with close ones during weekends.

We tried to calculate the impact of snapshot rate Δ on the observed network dynamics, choosing a Δ larger than the natural time-scale of the topological variation of the network will cause high-frequency variations, so we need to average out the potentially obscuring variation in ordering. In general research studies of dynamic networks, the snapshot rate is often determined by the method of data collection, or chosen to guarantee a certain density of edges in each snapshot or mostly by natural intuition; here we are trying to calculate the ideal snapshot rate temporally to avoid the possible loss of subtle data due to incorrect choice and also to determine ideal snapshot for proximity networks. We have to consider the effect that represents increasingly large snapshot rate Δ on the measured network statistics. Lengthening Δ has the effect of increasing the density of edges in each snapshot; thus, we would expect the degree to increase with Δ. The mean degree could have a rise or fall with lengthening of Δ depending on number of mobile users in the network.

We note that choosing a Δ larger than the natural time-scale of the topological variation of the network will naturally cause high-frequency variations to be averaged out, thus knowing the impact of how Δ the snapshot rate might cause for network dynamics. We present an empirical formulation over calculating Δ. Here, we use our highly temporally resolved data to explore the question of what kind of artifacts can be introduced by an incorrect choice of Δ, and whether there might be a ideal choice for proximity networks. The relevant figures are drawn using MATLAB software [1].

Choosing snapshot rate of $\{\Delta = 2 \text{ hrs}; 4 \text{ hrs}; 8 \text{ hrs}; 24 \text{ hrs}\}$ we compute the mean degree $\langle K \rangle$, mean correlation coefficient $\langle C \rangle$ and mean edge weight $\langle W \rangle$ for each snapshot series N for each of the eight days. The shortest rate $\Delta = 2$ hrs shows high frequency noise overlaid on low frequency structure. It was also expected that as the snapshot rate increases the high frequency variations are averaged out. It can be noticed from Figs. 18.3, 18.4, 18.5 of mean degree $\langle K \rangle$, clustering coefficient $\langle C \rangle$ and mean voice duration (edge weight) $\langle W \rangle$ correspondingly that for $\Delta = \{1 \text{ day}, 8 \text{ hrs}, 4 \text{ hrs}, 2 \text{ hrs}\}$, most of the days look same with prominent differences between week and weekend days. Also considering the effect of using increasingly large snapshot rate Δ on the network, we observe that a large Δ would cause increase in the density of edges in each graph thus increasing the clustering coefficient and degree.

From Figs. 18.3 to 18.7, based on mean degree $\langle K \rangle$ and clustering coefficient $\langle C \rangle$ observe that as Δ grows it clearly averages the higher frequency fluctuations and we see a steady increase in the voice call duration $\langle W \rangle$ series. This is precisely what we are trying to avoid by empirically selecting the time-snapshots for dynamically evolving social networks. We can also see how the value of clustering coefficient and voice duration values increases with increase in snapshot range. The value given in Figs. 18.3 to 18.6 were averaged out and plotted against the snapshot rate, for respective measures to analyze for the growth of the above curves which were not pictorially visible.

In Fig. 18.7, the degree, clustering and voice call duration statistics averaged over the set of snapshots derived over a week, as a function of Δ shows essentially monotonic growth of the curves. This illustrates that the choice of snapshot rate

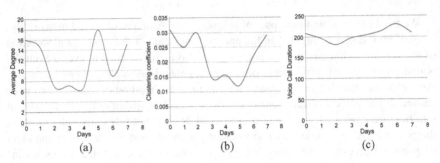

Fig. 18.3 (a) Mean degree $\langle K \rangle$, (b) clustering coefficient $\langle C \rangle$, (c) voice call duration $\langle W \rangle$ as a function of time for $\Delta = 1$ day during the week of may. Δ has a smooth curve with a few visible fluctuations on a daily basis

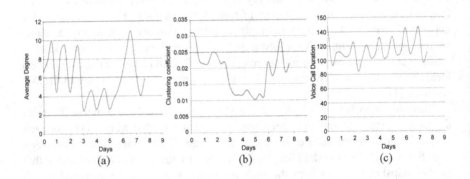

Fig. 18.4 (a) Mean degree $\langle K \rangle$, (b) clustering coefficient $\langle C \rangle$, (c) voice call duration $\langle W \rangle$ as a function of time for $\Delta = 8$ hrs during the week of may. A little more fluctuations are seen in graphs with mornings having relatively low usage

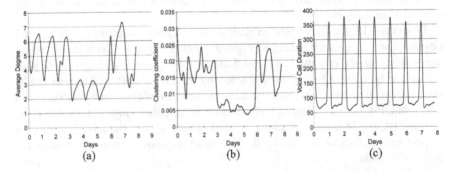

Fig. 18.5 (a) Mean degree $\langle K \rangle$, (b) clustering coefficient $\langle C \rangle$, (c) voice call duration $\langle W \rangle$ as a function of time for $\Delta = 4$ hrs during the week of may. High fluctuations are seen in graphs and distinct high usage seen in evening and night time

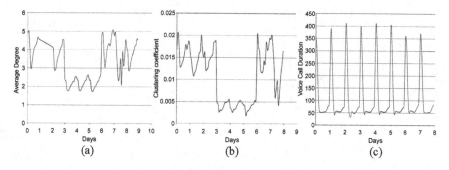

Fig. 18.6 (a) Mean degree $\langle K \rangle$, (b) clustering coefficient $\langle C \rangle$, (c) voice call duration $\langle W \rangle$ as a function of time for $\Delta = 2$ hrs Δ grows in sampling, higher fluctuations are clearly noticed

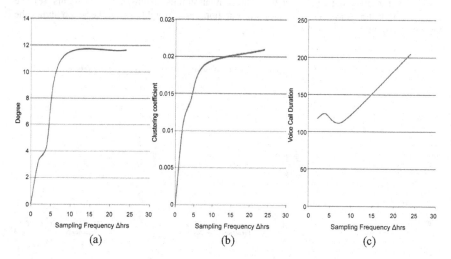

Fig. 18.7 (a) The values of network metric (mean degree $\langle K \rangle$) and (b) (clustering coefficient $\langle C \rangle$) for 8 days data as a function of snapshot rate, the value is seen increasing to the value of Δ. We can see that as early as $\Delta = 7$–8 the graph seems to flatten out. This clearly tells us that snapshot rate is around that point. (c) The value of network metric (voice call duration (edge weight $\langle W \rangle$)) for 8 days data as a function of snapshot rate Δ, the value is seen increasing to the value of Δ monotonically

completely determines the measured value of the network statistics. The shape of these curves certainly conveys some information about how to choose Δ. Thus, by working out the autocorrelation function for each of the two signals and finding the power spectra using the FFT (Fourier transform of the autocorrelation signal) we were able to roughly capture the snapshot rate of the dynamic proximity data. Figure 18.8 shows autocorrelation of degree, clustering and voice duration coefficient respectively.

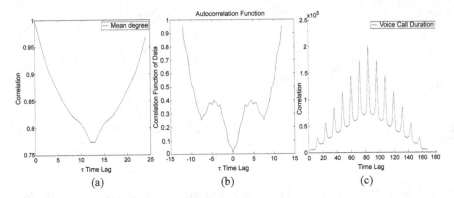

Fig. 18.8 (a) The autocorrelation for $\Delta = 2$ hrs for mean degree $\langle K \rangle$. (b) The autocorrelation for $\Delta = 2$ hrs, for clustering coefficient $\langle C \rangle$. The correlation falls to zero at $\Delta = 12.08$ hrs. (c) The autocorrelation for $\Delta = 2$ hrs for voice call duration (edge weight) $\langle W \rangle$. The graph shows the visible periodicity

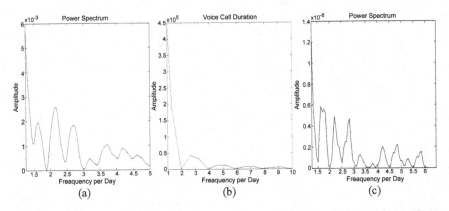

Fig. 18.9 (a) The power spectra of degree time series at $\Delta = 2$ hrs. The principle peaks are at $\Delta = \{6, 8, 10, 13\}$ hrs with some minor peaks at $\Delta = 6$ hrs. (b) The power spectra of clustering coefficient time series at $\Delta = 2$ hrs. The principle peaks are at $\Delta = \{6, 8, 10, 13\}$ hrs with some minor peaks at $\Delta = 6$ hrs. (c) The power spectra of voice duration (edge weight) time series. The principle peaks are at $\Delta = \{5, 8\}$ hrs

The spectral analysis given in Fig. 18.8 shows a strong and expected daily periodic behavior. The spectral density function can be seen in Fig. 18.9 for degree, clustering coefficient and voice call duration respectively.

Figures 18.9(a), (b) and (c) shows the power spectra for the same time series, with strong peaks in mean degree spectra, clustering coefficient and voice call duration match at roughly more than 3 times a day. Thus, the ideal snapshot rate for dynamic proximity data is the highest of these frequencies, which is 8 hrs, but a cross section of $\Delta_{ideal} = 8$ hrs at Δ vs. network statistic in Fig. 18.7 above shows that 8 hrs is

Table 18.1 Understand the Graph Behavior for 3 different cut-off values

Graph Behavior	Cut-off at $\Delta = 2$ hrs	Cut-off at $\Delta = 4$ hrs	Cut-off at $\Delta = 8$ hrs	Cut-off at $\Delta = 1$ day
Degree [Fig. 18.7(a)]	Monotonic increase	Monotonic increase	**Graph flattens**	Saturated
Clustering Coefficient [Fig. 18.7(b)]	Monotonic increase	Monotonic increase	**Graph flattens**	Saturated
Edge Weight [Fig. 18.7(c)]	Monotonic increase	Monotonic increase	Monotonic increase	Monotonic increase

quite a valid cut-off point, that us $\Delta_{ideal} = 8$ hrs is the point where the curve flattens. Choosing this value gives a cross section of the curves on 1st graph of Fig. 18.7, we get the natural average degree $\langle K \rangle = 10.01$, average clustering coefficient $\langle C \rangle = 0.0191$ and average voice call duration (edge weight) $\langle W \rangle = 127$ secs. The ideal rate that we calculate for the dynamic proximity network $\Delta_{ideal} = 8$ hrs is presumably closely related to both the length of the day and the length of the human work day. Our research is one of the valid and novel attempts to calculate approximate rate for the dynamic proximate telecom network and also we proposed a methodology for finding the periodicity in mobile networks.

18.5 Logistic Evaluation

Let's take a cut-off value at each point of the 3 graphs given in Fig. 18.7. From Table 18.1, the natural rate that we calculate for the dynamic proximity network $\Delta_{ideal} = 8$ hrs is presumably closely related to both the length of the day and the length of the human work day. Thus, the natural rate for other dynamical social systems may vary considerably, depending on the particular context of the social ties.

18.6 Applications

Using the above findings, we can easily identify the individuals, communities, and units who play central roles in different time zones. With the specific time zone, we can leverage mobile networks with many interesting findings.

- Discern information breakdowns, bottlenecks, structural holes, as well as isolated individuals, communities, and units in various time zones.
- Make out opportunities to accelerate information flows/talk-time across functional and organized communities.
- Strengthen the efficiency and effectiveness of existing, formal communication channels in weak time zones.

- Raise awareness of and reflection on the importance of informal networks and ways to enhance their usage performance.
- Leverage on network support according to distinctive times.
- Refine strategies before launching promotional offers.

18.7 Conclusion

We have proposed empirical method to capture the periodicity of dynamic networks. We have included network attributes like clustering coefficient, vertex degree, and edge weight in our analysis. We have demonstrated our approach on the huge telecom data set. The method is quite simple and better to find the periodicity. This efficiently helps us break the social network into timeslots to study the evolution of communities, core values and many other graph theory related metrics that is useful in different applications. In telecom it gives us business use case for greater understanding of the behavior of mobile users.

Acknowledgements   This work was supported by Ericsson R&D, Chennai. We would like to thank our team members and managers, R&D Head for moral support and encouragement to complete this research work. We propose our sincere thanks to Prasad Garigipati, Anand Varadarajan, and Lennart Isaksson, for their excellent support in conceptualizes our research idea.

References

1. http://www.mathworks.com
2. Berger-Wolf, T.F., Saia, J.: A framework for analysis of dynamic social networks. In: The 12th ACM SIGKDD International Conference on Knowledge Discovery and Data Mining, August 20–23, Philadelphia, PA, USA (2006)
3. Cao, L.: In-depth behavior understanding and use: the behavior informatics approach. Inf. Sci. **180**, 3067–3085 (2010)
4. Clauset, A., Eagle, N.: Persistence and periodicity in a dynamic proximity network. In: DIMACS Workshop on Computational Methods for Dynamic Interaction Networks, pp. 1–5 (2007)
5. Dasgupta, K., Singh, R., Viswanathan, B., Chakraborthy, D., Joshi, A., Mukherje, S., Nanavati, A.: Social ties and their relevance to churn in mobile telecom networks. In: The 11th International Conference on Extending Database Technology Advances in Database Technology (EDBT), France (2008)
6. Eagle, N., Pentland, A., Lazer, D.: Inferring social network structure using mobile phone data. In: The National Academy of Sciences, vol. 106(36), pp. 15274–15278 (2009)
7. Lahiri, M., Berger-Wolf, T.F.: Mining periodic behavior in dynamic social networks. In: 8th IEEE International Conference on Data Mining, December 15–19, pp. 373–382 (2008)
8. Robardet, C.: Constraint-based pattern mining in dynamic graphs. In: The Ninth IEEE International Conference on Data Mining (ICDM), Fl, USA, pp. 950–955 (2009)
9. Saravanan, M., Prasad, G., Karishma, S., Suganthin, D.: Labeling communities using structural properties. In: International Conference on Advances in Social Networks Analysis and Mining (ASONAM), Odense, Denmark, pp. 217–224 (2010)

10. Tantipathananandh, C., Berger-Wolf, T.F., Kempe, D.: A framework for community identification in dynamic social networks. In: The 13th ACM SIGKDD International Conference on Knowledge Discovery and Data Mining, August 12–15, San Jose, California, USA (2007)
11. Wasserman, S., Faust, K.: Social Network Analysis: Methods and Applications. Cambridge University Press, New York (1994)
12. Weiss, G.: Data Mining in Telecommunications, Data Mining and Knowledge Discovery Handbook: A Complete Guide for Practitioners and Researchers, pp. 1189–1201. Kluwer Academic, Dordrecht (2005)

Chapter 19
Event Detection Based on Call Detail Records

Huiqi Zhang and Ram Dantu

Abstract In this paper we propose the model of the inhomogeneous Poisson for
call frequency and inhomogeneous exponential distribution for call durations to detect events based on mobile phone call detail records. The maximum likelihood
method is used to estimate the rate of frequency and call duration. This work is
useful for enhancing homeland security, detecting unwanted calls (e.g., spam) and
commercial purposes. For validation of our results, we used actual call logs of 100
users collected at MIT by the Reality Mining Project group for a period of 8 months.
The experimental results show that our model achieves good performance with high
accuracy.

19.1 Introduction

Analyzing patterns of human behavior [15] is an area of increasing interest in a number of different applications. The automatic detection of events by studying patterns
of human behavior is one of them and has recently attracted attention. An event is
something that happens at a given point in time and at a given place. We use *event* to
refer to a large-scale activity that is unusual relative to normal patterns of behavior.
To understand such data, we often care about both the patterns of typical behavior
and detecting and extracting information from deviations from this behavior. Almost all previous approaches for event detection are based on text, website data and
video data (see related work).

In this paper we propose and investigate the inhomogeneous Poisson and inhomogeneous exponential distribution model to detect events, and we illustrate how
to learn such a model from data to both characterize normal behavior and detect
anomalous events based on call detail records. There are no contents in call detail
records that are the main difference from the text and website data and more difficult
to detect events hidden in them. We can only use information such as the time of

H. Zhang (✉) · R. Dantu
Computer Science and Engineering, University of North Texas, Denton, TX 76201, USA
e-mail: hz0019@unt.edu

R. Dantu
e-mail: rdantu@unt.edu

L. Cao, P.S. Yu (eds.), *Behavior Computing*,
DOI 10.1007/978-1-4471-2969-1_19, © Springer-Verlag London 2012

initiation of calls, number of calls in a period of time, call duration, incoming calls, outgoing calls and location. The maximum likelihood estimation is used to estimate the rates and the thresholds of the number of calls and call duration. The experimental results show that our model achieves good performance with high accuracy.

In Sect. 19.2 we briefly review the related work. In Sect. 19.3 the model is described. We perform the experiments with the actual call logs and discuss the results in Sect. 19.4. In Sect. 19.5, we conduct the validation of our model using the actual call logs. Finally, we have the conclusions in Sect. 19.6.

19.2 Related Work

There are a large amount of previous work on event detection in text, data stream and video. In [1] the authors proposed a method based on an incremental TF-IDF model and the extensions include generation of source-specific models, similarity score normalization based on document-specific averages, source-pair specific averages, term re-weighting based on inverse event frequencies, and segmentation of the documents. In [2] the authors examine the effect of a number of techniques, such as part of speech tagging, similarity measures, and an expanded stop list on the performance. In [3] the authors use text classification techniques and named entities to improve the performance. In [4] a novelty detection approach based on the identification of sentence level patterns is proposed. In [5] the authors propose a probabilistic model to incorporate both content and time information in a unified framework, which gives new representations of both news articles and news events. They did explorations in two directions because the news articles are always aroused by events and similar articles reporting the same event often redundantly appear on many news sources. In [6] the authors propose to detect events by combining text-based clustering, temporal segmentation, and graph cuts of social networks in which each node represents a social actor and each edge represents a piece of text communication that connects two actors. In [7, 8] the authors propose the conceptual model-based approach by the use of domain knowledge and named entity type assignments and showed that classical cosine similarity method fails for the anticipatory event detection task. In [9] the authors proposed the online new event detection (ONED) framework, which includes a combination of indexing and compression methods to improve the document processing rate, a resource-adaptive computation method to maximize the benefit that can be gained from limited resources, new events to be further filtered and prioritized before they are presented to the consumer when the new event arrival rate is beyond the processing capability of the consumer and implicit citation relationships to be created among all the documents and used to compute the importance of document sources.

In [10] the authors propose an iterative algorithm and use likelihood criterion to segment a time-series into piecewise homogeneous regions to detect the change points, which are equivalent to events defined by them and evaluate them with the highway traffic data. In [11] the authors use an infinite automaton in which bursts are state transitions to detect burst events in text streams and conduct the experiments with emails and research papers. In [12] the authors use suffix tree to encode the frequency of all observed patterns and apply a Markov model to detect patterns in the symbol sequence. In [13] the authors find piecewise constant intensity functions to represent continuous intensity functions using a combination of Poisson models and Bayesian estimation methods and use dynamic programming method to find them. In [14] the authors use a time-varying Poisson process model and statistical estimation techniques for unsupervised learning in the context. They applied this model to freeway traffic data and building access data.

The above approaches are to detect events for text, novel and unusual data points or segments in time-series that have either contents or are traffic data. However, none of the previous work focuses on the specific problem we study here, using the inhomogeneous Poisson and inhomogeneous exponential distribution model by studying the calling pattern based on call detail records to detect events that reflect the human activity.

19.3 Model

19.3.1 Formulation

In event detection we need to analyze and classify categorical data, either in an exploratory or in a confirmatory context. Exploratory analysis of such data often has to do with extracting relevant hidden knowledge from a large dataset. We need to develop and use robust and flexible classification methods. In this paper we use probabilistic models for the classification of variables, based on inhomogeneous Poisson process for number of calls and exponential distribution for call durations.

The observed data can be represented in a bi-dimensional matrix, where rows describe data units and columns describe categorical variables. Empirical clustering models are usually used to analyze such data.

We assume that number of calls follows inhomogeneous Poisson process and call duration follow inhomogeneous exponential distribution.

Let $N_i = \{n_{i1}, \ldots, n_{ik}\}$ be random variable for number of calls of a given day i, $D_i = \{d_{i1}, \ldots, d_{ik}\}$ be random variable for call duration for day i, $i = 1, 2, \ldots, 7$ be a day of week, 1 for Sunday, $\ldots$, 7 for Saturday. Then

$$
N = \begin{bmatrix}
n_{11} & n_{12} & \cdots & n_{1k} \\
n_{21} & n_{22} & \cdots & n_{2k} \\
\cdots & \cdots & \cdots & \cdots \\
n_{71} & n_{72} & \cdots & n_{7k}
\end{bmatrix}
$$

is the matrix of number of calls on 7 days of week and

$$D = \begin{bmatrix} d_{11} & d_{12} & \cdots & d_{1k} \\ d_{21} & d_{22} & \cdots & d_{2k} \\ \cdots & \cdots & \cdots & \cdots \\ d_{71} & d_{72} & \cdots & d_{7k} \end{bmatrix}$$

is the matrix of call duration on 7 days of week.

Then Poisson density function for day i is given by

$$P_{N_i}(N_i = n_{ij}) = \frac{e^{-\lambda_i} \lambda_i^{n_{ij}}}{n_{ij}!} \tag{19.1}$$

where λ_i is the rate (average) of number of calls for day i.

By the properties of Poisson distribution, the mean $= \lambda_i$, the variance $var = \lambda_i$ and the standard error $\sigma = \pm\sqrt{\lambda_i}$.

The exponential distribution density function of call duration for day i is given by

$$P_{D_i}(D_i = d_{ij}) = \frac{1}{\mu_i} e^{-\frac{d_{ij}}{\mu_i}} \tag{19.2}$$

where μ_i is the mean of call duration for day i.

By the properties of exponential distribution, the variance $var = \mu_i^2$ and the standard error $\delta = \pm\sqrt{\mu_i^2} = \pm\mu_i$.

Now using maximum likelihood estimates [16] to estimate the λ_i for day i. The cumulated probability distribution function is

$$P_{N_i}(N_i = n_{i1}, n_{i2}, \ldots, n_{ik} | \lambda_i) = \prod_{j=1}^{k} \frac{e^{-\lambda_i} \lambda_i^{n_{ij}}}{n_{ij}!} = \frac{e^{-k\lambda_i} \lambda_i^{\sum_{j=1}^{k} n_{ij}}}{\prod_{j=1}^{k} n_{ij}!}$$

$$\ln P_{N_i} = -k\lambda_i + (\ln \lambda_i) \sum_{j=1}^{k} n_{ij} - \ln\left(\prod_{j=1}^{k} n_{ij}\right) \tag{19.3}$$

$$\frac{d(\ln P_{N_i})}{d\lambda_i} = -k + \frac{\sum_{j=1}^{k} n_{ij}}{\lambda_i} = 0$$

$$\hat{\lambda}_i = \frac{\sum_{j=1}^{k} n_{ij}}{k}$$

For μ_i The cumulated probability distribution function of call duration is

$$P_{D_i}(D_i = d_{i1}, d_{i2}, \ldots, d_{ik}|\mu_i) = \prod_{j=1}^{k} \frac{1}{\mu_i} e^{-\frac{d_{ij}}{\mu_i}} = \frac{1}{\mu_i^k} e^{-\frac{1}{\mu_i} \sum_{j=1}^{k} d_{ij}}$$

$$\ln P_{D_i} = -k \ln \mu_i - \frac{1}{\mu_i} \sum_{j=1}^{k} d_{ij} \qquad (19.4)$$

$$\frac{d(\ln P_{D_i})}{d\mu_i} = -\frac{k}{\mu_i} + \frac{1}{\mu_i^2} \sum_{j=1}^{k} d_{ij} = 0$$

$$\hat{\mu}_i = \frac{\sum_{j=1}^{k} d_{ij}}{k}$$

The maximum likelihood estimates are used to estimate average number of calls and call duration. Next we consider the maximum average number of calls and call duration obtained for all weekday/weekend and week by week. Suppose that the m week data is used to compute the rates of number of calls and call duration for user p. Let $\hat{\lambda}_{d1}^{p}, \hat{\lambda}_{d2}^{p}, \ldots, \hat{\lambda}_{d7}^{p}$ be the rate of number of calls obtained for all weekday/weekend and $\hat{\lambda}_{w1}^{p}, \hat{\lambda}_{w2}^{p}, \ldots, \hat{\lambda}_{wm}^{p}$ be the rate of call duration obtained week by week for m weeks of user p respectively. Let $\hat{\mu}_{d1}^{p}, \hat{\mu}_{d2}^{p}, \ldots, \hat{\mu}_{d7}^{p}$ be the mean of call duration obtained for all weekday/weekend and $\hat{\mu}_{w1}^{p}, \hat{\mu}_{w2}^{p}, \ldots, \hat{\mu}_{wm}^{p}$ be the mean of call duration obtained week by week for m weeks of user p respectively.

Then the maximum means of number of calls and call duration are respectively computed by:

$$\hat{\lambda}_{max}^{p} = \max\left(\hat{\lambda}_{d1}^{p}, \hat{\lambda}_{d2}^{p}, \ldots, \hat{\lambda}_{d7}^{p}, \hat{\lambda}_{w1}^{p}, \hat{\lambda}_{w2}^{p}, \ldots, \hat{\lambda}_{wm}^{p}\right) \qquad (19.5)$$

$$\hat{\mu}_{max}^{p} = \max\left(\hat{\mu}_{d1}^{p}, \hat{\mu}_{d2}^{p}, \ldots, \hat{\mu}_{d7}^{p}, \hat{\mu}_{w1}^{p}, \hat{\mu}_{w2}^{p}, \ldots, \hat{\mu}_{wm}^{p}\right) \qquad (19.6)$$

where $\hat{\lambda}_{max}^{p}$ and $\hat{\mu}_{max}^{p}$ are the maximum likelihood estimates of number of calls and call duration for user p over the number of days specified respectively. The thresholds define the limits for all weekday/weekend and week by week. The assumption is that the calling pattern could be different. Each person has his/her own thresholds, and if the number of calls or call duration are greater than the thresholds of their own for some day, we define that there is some event in that day.

To calculate the threshold of number of calls for user p, N_{thres}^{p}, we define

$$N_{thres}^{p} = \hat{\lambda}_{max}^{p} + \hat{\sigma}_{max}^{p} \qquad (19.7)$$

where $\hat{\lambda}_{max}^{p}$ and $\hat{\sigma}_{max}^{p}$ are the maximum rate of number of calls and correspondent standard error with positive $\hat{\sigma}_{max}^{p}$.

To calculate the threshold of call duration for user p, D_{thres}^{p}, we define

$$D_{thres}^{p} = \hat{\mu}_{max}^{p} + \hat{\delta}_{max}^{p} \qquad (19.8)$$

where $\hat{\mu}_{max}^{p}$ and $\hat{\delta}_{max}^{p}$ are the maximum mean of call duration and correspondent standard error with positive $\hat{\delta}_{max}^{p}$.

Definition of an Event   A collection of call log data can be represented as

$$C = \langle (t_1, a_1, d_1, l_1), (t_2, a_2, d_2, l_2), \ldots, (t_n, a_n, d_n, l_n) \rangle,$$

where t_i is a time point, d_i is a call duration, l_i is a location and a_i is a pair of actors, caller-callee $\langle s_i, r_i \rangle$ where s_i is an actor who initiates a call at time t_i and r_i is an actor who receive a call. An event is defined as a subset $E \subset C$ of a tuple

$$E = \left\{ (t_1, a_1, d_1, l_1), (t_2, a_2, d_2, l_2), \ldots, (t_m, a_m, d_m, l_m) \right\}$$

such that either $\sum_{i=1}^{m} d_i > D_{thres}$ or $count(d_i) > N_{thres}$ defined as the above in the time period $\Delta t = t_m - t_1$.

19.3.2 Real-Life Data Sets and Parameters

Real-Life Traffic Profile   In this paper, the actual call logs are used for analysis. These actual call logs are collected at MIT [17] by the Reality Mining Project group for a period of 8 months. This group collected mobile phone usage of 100 users, including their user IDs (unique number representing a mobile phone user), time of calls, call direction (incoming and outgoing), incoming call description (missed, accepted), talk time, and tower IDs (location of phone users). These 100 phone users are students, professors and staff members. The collection of the call logs is followed by a survey of feedback from participating phone users for behavior patterns such as favorite hangout places; service provider; talk time minutes and phone users' friends, relatives and parents. We used this extensive dataset for our social group analysis and validation of 20 sample users in this paper. More information about the Reality Mining Project can be found in [17].

We use the formulas (19.3) and (19.4) to detect events based on the day of week, call frequencies and call duration.

Day of week: Everyone has his/her own schedule for working, studying, entertainment, traveling and so on. The schedule is mainly based on the day of the week.

Call frequencies: The call frequency is the number of incoming or outgoing calls in a period of time. The greater the number of incoming or outgoing calls in a period of time, the more socially close the caller and callee relationship.

Call duration: The call duration is how long both caller and callee want to talk to each other. The longer the call duration is in a period of time, the more socially close the caller and callee relationship.

Fig. 19.1 The number of incoming, outgoing and total calls per day for user3

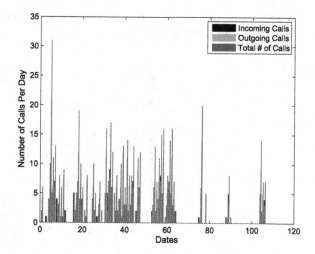

19.3.3 Computing the Thresholds of Frequency and Duration

The thresholds of frequency and duration are computed by formulas (19.7) and (19.8) based on day of week (Sunday, Monday, ..., Saturday) and week sequence (1st week, 2nd week, ...). Then the thresholds of frequency and duration are chosen to compare with the frequency and duration for each day. If the frequency or duration of some day is greater than the thresholds of frequency or duration, we define that there is an event in that day.

We used the data from the data set of four months, a semester since the communication members were relatively less changed in a semester for students.

19.4 Experiment Results and Discussion

Figures 19.1 and 19.2 show the number of calls and call duration for user3, where the x-axis indicates the days and y-axis indicates the number of calls and call duration (incoming, outgoing and total of them) respectively.

In Fig. 19.3 the x-axis is days and y-axis indicates the number of calls and call duration (incoming, outgoing and total of them) respectively, which show that there are events in these dates. From the Fig. 19.3 we may see that there are 7 event days which are the 5th, 18th, 31st , 33rd, 58th, 62nd, 76th, days during 106 days.

The experiment results of user3 and user74 as examples are listed in Table 19.1. In Table 19.1 the thresholds of number of calls and duration are calculated by maximum likelihood estimates. There are two types of events: one has location change and the other has no location change. For example, in Table 19.1, there are 17 calls, which is greater than 15 (the threshold of the number of calls) for user3 on the 33rd day and the location is on campus. We define that there is some event in that day. For

Fig. 19.2 The duration of incoming, outgoing and total calls per day for user3

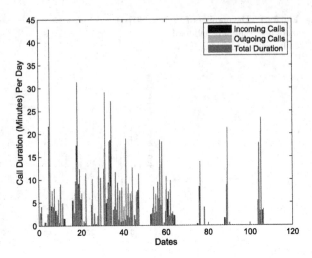

Fig. 19.3 There are events in these days for user3

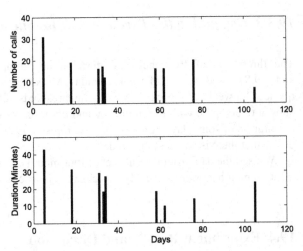

user74 on the 24th day, although there are only 2 calls, the call duration is 76.8 minutes, which is much greater than 18 minutes (the maximum rate of the call duration) and there is some event in that day.

Note: the thresholds of number of calls and duration are computed by formulas (19.7) and (19.8).

19.5 Validation

To evaluate the accuracy of our model, we used actual call logs of 100 phone users and randomly choose 20 phone users. These users include students, professors and staff members. The best way to validate the results is to contact the phone users to

Table 19.1 Event dates and locations

Users	Event	Days	# of contacts	# of calls the day	Duration (minutes)	Location	Note
User3	1	5	11	31	42.9	Visit world trade center	Both large # of calls and duration
	2	18	3	19	31.3	Visit Harvard Univ.	Both large # of calls and duration
	3	31	8	16	29.1	Visit Harvard Univ.	Large # of calls
	4	33	6	17	18.3	Campus	Large # of calls
	5	58	11	16	18	Campus	Large # of calls
	6	62	10	16	9.8	Campus	Large # of calls
	7	76	7	20	13.9	Campus	Large # of calls
User74	1	24th	1	2	76.8	Campus	Large duration
	2	26th	4	13	20.3	Visit central square	Large # of calls
	3	27th	4	8	15.6	Campus	Large # of calls
	4	28th	2	11	64.2	At home	Large # of calls
	5	46th	1	6	40.2	Visit Stateplace-New York	Large duration
	6	78th	1	2	54.4	At home	Large duration
	7	86th	2	6	82.8	At home	Large duration
	9	93rd	3	10	24.5	Visit Stateplace-New York	Large # of calls
	10	122nd	4	9	7.9	Visit Stateplace-New York	Large # of calls
	11	130th	2	7	46.8	At home	Large duration
	12	131st	2	9	53.7	Home (next day visit Cityplace-Providence StateRI)	Both large # of calls and duration
	13	153rd	2	4	60	Visit Stateplace-New York	Large duration

get feedback, but because of the privacy issues it is almost impossible to use this way. Thus we use hand labeling method to validate our model. We used the data of the four months to detect events. Note that we cannot use the data of the next four months to validate our model since the events may happen in a different time period. In order to validate our model, we hand labeled the events based on the number of calls, duration of calls in the day, history of call logs, location, time of arrivals, and other humanly intelligible factors.

Table 19.2 shows the validation results. We achieve 92% accuracy.

Table 19.2 Validation results

Users	# of events	Threshold of # of calls per day	Threshold of duration per day (minutes)	Ave. # of calls per day	Ave. duration per day (minutes)	False positive	False negative
3	7	15	30	5	5.5	0%	11%
14	10	9	82	4	40	0%	9%
15	9	14	21	6	7	0%	10%
16	11	8	24	4	8	0%	8%
21	8	11	56	5	18	0%	0%
22	4	22	60	11	20	0%	9%
29	12	10	14	4	7	0%	7%
33	9	6	43	1	8	0%	10%
35	5	21	76	10	21	0%	8%
38	15	15	67	10	29	0%	6%
39	13	13	52	5	14	0%	9%
50	9	16	75	7	31	0%	10%
57	8	8	15	3	5	0%	11%
72	13	12	70	6	23	0%	6%
74	13	7	36	2	7	0%	7%
78	13	10	76	3	13	0%	6%
83	12	10	28	5	9	0%	7%
85	14	9	24	4	9	0%	6%
88	11	7	16	3	4	0%	8%
95	8	4	10	2	4	0%	10%

Note: the thresholds of number of calls and duration are computed by formulas (19.7) and (19.8).

19.6 Conclusion

In this paper we proposed the inhomogeneous Poisson process model for detecting events based on mobile phone call detail records. We used the data from the data set of four months, a semester since the communication members were relatively less changed in a semester for students.

The maximum likelihood estimates are used to estimate average number of calls and call duration to compare with the frequency and duration for each day. If the frequency or duration of some day is greater than the thresholds of frequency and duration, we define that there is some event in that day.

The best way to validate the results is to contact the phone users to get feedback, but because of the privacy issues it is almost impossible to use this way. Thus we use our defined conditions to validate our model.

This work is useful for enhancing homeland security, detecting unwanted calls (e.g., spam), communication presence, marketing etc. The experimental results show that our model achieves good performance with high accuracy.

In our future work we plan to detail the event classification and to analyze and use some criterion to optimize the detection process.

Acknowledgement We would like to thank Nathan Eagle and Massachusetts Institute of Technology for providing us the call logs of Reality Mining dataset.

References

1. Brants, T., Chen, F.: A system for new event detection. In: Proceedings of International ACM SIGIR Conference, pp. 330–337 (2003)
2. Chen, F., Farahat, A., Brants, T.: Story link detection and new event detection are asymmetric. In: Human Language Technology Conference (HLT-NAACL) (2003)
3. Kumaran, G., Allan, J.: Text classification and named entities for new event detection. In: Proceedings of international ACM SIGIR Conference, pp. 297–304 (2004)
4. Li, X., Croft, B.W.: Novelty detection based on sentence level patterns. In: Proceedings of ACM CIKM, pp. 744–751 (2005)
5. Li, Z., Wang, B., Li, M.: A probabilistic model for retrospective news event detection. In: Proceedings of International SIGIR Conference, pp. 106–113 (2005)
6. Zhao, Q., Mitra, P.: Event detection and visualization for social text streams. In: Proceedings of International Conference on Weblogs and Social Media (ICWSM) (2007)
7. He, Q., Chang, K., Lim, E.P.: A model for anticipatory event detection. In: Proceedings of the 25th International Conference on Conceptual Modeling (ER). LNCS, vol. 4215, pp. 168–181. Springer, Berlin (2006)
8. He, Q., Chang, K., Lim, E.P.: Anticipatory event detection via sentence classification. In: Proceedings of IEEE International Conference on Systems, Man, and Cybernetics, pp. 1143–1148 (2006)
9. Luo, G., Tang, C., Yu, P.: Resource-adaptive real-time new event detection. In: Proceedings of International ACM SIGMOD Conference on Management of Data (2007)
10. Guralnik, V., Srivastava, J.: Event detection from time series data. In: Proceedings of the Fifth ACM SIGKDD International Conference on Knowledge Discovery and Data Mining, pp. 33–42. ACM Press, New York (1999)
11. Kleinberg, J.: Bursty and hierarchical structure in streams. In: Proceedings of the Eigth ACM SIGKDD International Conference Knowledge Discovery and Data Mining, pp. 91–101. ACM Press, New York (2002)
12. Keogh, E., Lonardi, S., Chiu, B.Y.: Finding surprising patterns in a time series database in linear time and space. In: Proceedings of the Eigth ACM SIGKDD International Conference on Knowledge Discovery and Data Mining, pp. 550–556. ACM Press, New York (2002)
13. Salmenkivi, M., Mannila, H.: Using Markov chain Cityplace Monte Carlo and dynamic programming for event sequence data. Knowl. Inf. Syst. **7**(3), 267–288 (2005)
14. Ihler, A., Hutchins, J., Smyth, P.: Adaptive event detection with time-varying Poisson processes. In: Proceedings of the ACM SIGKDD International Conference on Knowledge Discovery and Data Mining, pp. 207–216 (2006)

15. Cao, L.: In-depth behavior understanding and use: the behavior informatics approach. Inf. Sci. **180**, 3067–3085 (2010)
16. Harris, J.W., Stocker, H.: Maximum likelihood method. In: Handbook of Mathematics and Computational Science, p. 824. Springer, New York (1998), §21.10.4
17. Massachusetts Institute of Technology: Reality mining. http://reality.media.mit.edu/ (2008)

Chapter 20
Smart Phone: Predicting the Next Call

Huiqi Zhang and Ram Dantu

Abstract Prediction of incoming calls can be useful in many applications such as social networks, (personal, business) calendar and avoiding voice spam. Predicting incoming calls using just the context is a challenging task. We believe that this is a new area of research in context-aware ambient intelligence. In this paper, we propose a call prediction scheme and investigate prediction based on callers' behavior and history. We present Holt-Winters method to predict calls from frequent and periodic callers. The Holt-Winters method shows high accuracy. Prediction and efficient scheduling of calls can improve the security, productivity and ultimately the quality of life.

20.1 Introduction

Prediction plays an important role in various applications. Several schemes have been widely deployed for predicting weather, environment, economics, stock, market, earthquakes, flooding, network traffic and call center traffic [1–5, 7–14]. Companies use predictions of demands for making investments and efficient resource allocation. The call centers predict workload so that they can get the right number of staff in place to handle it. Network traffic prediction is used to access future network capacity requirements and to plan network development for optimum use network resources and improve quality of services. Prediction is also applied in the human behavior study [6] by combining the computer technology and social networks [10, 15, 16].

Over the past few years, there has been a rapid development and deployment of new strategic services based on the IP protocol, including Voice over IP (VoIP) and IP-based media distribution (IPTV). These services operate on private and public IP networks, and share their network with other types of traffic such as web traffic. Not

H. Zhang (✉) · R. Dantu
Computer Science and Engineering, University of North Texas, Denton, TX 76201, USA
e-mail: hz0019@unt.edu

R. Dantu
e-mail: rdantu@unt.edu

L. Cao, P.S. Yu (eds.), *Behavior Computing*,
DOI 10.1007/978-1-4471-2969-1_20, © Springer-Verlag London 2012

only will VoIP reduce communication costs and provide enhanced and more flexible communication experiences to people, but it will also pave the road for innovative, valued added, highly-personalized services. We can expect, for example, that interactive multimedia and broadcast video services will be reusing the infrastructure that is being deployed for VoIP. Such trends will result in what we can call IP-based multimedia communications infrastructure, encompassing both the equivalence of conventional phone conversations and advanced communication and content distribution services. For example, we expect VoWiFi (voice over WiFi) and mobile-TV services will be widely used in the next five years. Hence, it is very conceivable that the people will be using or wearing several communication devices simultaneously. Unwanted interruptions due to these devices can waste people's time and can be of serious impact to the productivity. On the other hand, efficient scheduling of transactions can improve the security, productivity, and ultimately quality of life.

Therefore, we need context-based services for accepting or rejecting the incoming transactions. For example, predicting the wanted and unwanted calls from customers can improve the revenues for sales people. Predicting the call volume in E911 agencies can improve the efficient deployment of the resources. Predicting the expected calls for a busy business executive (personal or business) can be very useful for scheduling a day. Match making services can use calling patterns and calling behavior for the compatibility studies [17]. Moreover, the prediction of incoming calls can be used to avoid unwanted calls and schedule a time for wanted calls. For example, the problem of spam in VoIP networks has to be solved in real time compared to e-mail systems. Compare receiving an e-mail spam at 2:00 a.m. that sit in the inbox until you open it the next morning to receiving a junk phone call that must be answered immediately. There was some work reported on telephone telepathy based on psychology. So far no scientific research has been reported in predicting the incoming calls for context-based services. Predicting of incoming calls using just the context is a challenging task. We believe that this is a new area of research. In this paper we predict calls from our social network.

Real-Life Data Sets   Every day calls on the cellular network include calls from different sections of our social life. We believe calls from family members, friends, supervisors, neighbors, and strangers. Every person exhibits a unique traffic pattern. Calls to our home from neighbors and business associates may not be as frequent as those from family members and friends. Similarly, we talk for longer periods to family members and friends compared to neighbors and distance relatives. These traffic patterns can be analyzed for inferring the closeness to the callee. This closeness represents the social closeness of the callee with the caller on the cellular communication network.

To study closeness the people have with their callers, we collected the calling patterns of 20 individuals at our university. We are in process of collecting calling patterns from 20 more people. The details of the survey are given in Dantu et al. [18]. We found that it is difficult to collect the data set because many people are unwilling to given their calling patterns due to privacy issues. Nevertheless, the collected datasets include people with different type of calling patterns and call distributions. As part of the survey, each individual downloaded two months of detailed cell phone records from his online accounts on the cellular service provider's website. Each call record in the dataset had the 5-tuple information *Call record: (date, start time, type, caller id, talk-time)* where: date is the date of communication; start time is the start time of the communication; type is the type of call, i.e., "Incoming" or "Outgoing"; caller id is the caller identifier; and talk-time is the amount of time spend by caller and the individual during the call.

We used the call records for predicting the next caller. In Sect. 20.2 we describe a prediction technique based on a statistical model. In Sect. 20.3 we proposed a prediction scheme. In Sect. 20.4 the experiment and validation results are presented. Section 20.5 is the conclusion.

20.2 A Call Prediction Model for Frequent and Periodic Callers Using Holt-Winters Method

We used both Holt-Winters and ARIMA (Autoregressive Integrated Moving Average) [19] models which are the most general class of models for predicting a time series to predict the incoming calls and compared the accuracy of the results for these two methods. We found that there was no obvious difference between them for the accuracy since Holt-Winters method is the subset of ARIMA method. Therefore we choose the Holt-Winters method to predict incoming calls since the ARIMA method become the Holt-Winters method by adjusting the parameters of it.

Holt-Winters method [19] uses mathematical recursive functions to predict the future quantitative behavior. It uses a time series model to make predictions and assumes that the future behavior will follow the same pattern as the past. In our case we have patterns corresponding to some periodicity. This means that we use a factor in the equations that uses information from past days to make a prediction of what will happen in the future. The Holt-Winters method equations are given by:

$$Y_t = \alpha \frac{x_t}{I_{t-l}} + (1 - \alpha)(Y_{t-1} - b_{t-1}) \qquad (20.1)$$

$$b_t = \gamma(Y_t - Y_{t-1}) + (1 - \gamma)b_{t-1} \qquad (20.2)$$

$$I_t = \beta \frac{x_t}{Y_t} + (1 - \beta)I_{t-l} \qquad (20.3)$$

$$P_{t+m} = (Y_t - mb_t)I_{t-l+m} \qquad (20.4)$$

where x_t is the observed value at time t; Y_t is the smoothed observation at time t; b_t is the trend smoothing at time t; I_t is the seasonal smoothing at time t. ($t - l$ is used in (20.1) to (20.4) to present the use of information from previous periods); l is the number of periods that complete 1 season; P_{t+m} is the prediction at m periods ahead; m is the number of periods ahead we want to predict; α is the overall smoothing parameter; β is the seasonal smoothing parameter; γ is the trend smoothing parameter.

α is the short term parameter. A large value of α will give a large weight to measurements very near in the past, while a small value of alpha will give more weight to measurements further in the past. γ is the trend parameter. A large value of γ will give more weight to the difference of the last smoothed observations; while a small value of γ will use information further in the past. β is the seasonal parameter. A large value of β will give more weight to the present relation between the observation and the smoothed observation, and small values of β will give more weight to past days relation between the observation and the smoothed observation.

Some values in (20.1) to (20.4) correspond to nonexistent periods of time in the case of Y_{t-1} or b_{t-1} When $t = 0$, we don't have any values for Y or b. So we need initial values. The same happen for the seasonal index I when $t < l$. The initial values of Y for the observation smoothing is assumed to be $S_0 = 0$. The initial values for the trend smoothing b is given in (20.5). This value is mainly an average of the differences in the observations of the first two periods divided by the length of a period.

$$b_0 = \frac{(x_{l+1} + x_{l+2} + \cdots + x_{l+l} - x_1 - x_2 - \cdots - x_l)}{l^2} \tag{20.5}$$

The initial values for the seasonal smoothing are given by (20.6). This value of the seasonal index is basically an average of the observed values of every period and an average of the day n of every period divided by the average of the observed values for its period.

$$I_{k-l} = \sum_{i=0}^{n} \frac{x_{k+il}}{\sum_{j=k+il}^{(i+1)l} x_j} \qquad \text{for } k = 1, 2, \ldots, l-1, l \tag{20.6}$$

To find the parameters α, β and γ we use the minimum mean square error, where the error is the difference between the prediction and the observed values at time t. The minimization equation is given by (20.7) using $0 = \alpha, \beta, \gamma = 1$.

$$\text{Min}\left(\frac{1}{N} \sum (P_t - x_t)^2\right) \tag{20.7}$$

20.3 Prediction Scheme

Holt-Winters method can be used to predict one-step-ahead value of time series, which can be extended to k-step-ahead. We can use the mathematical expression to explain what is one-step-ahead and k-step ahead prediction. Let Yt to be the time series that we want to predict its performance. The k-step ahead prediction can be defined with $Yt + k$. This means $Yt + k$ denotes the k-step prediction made at origin time t. When $k = 1$, it is one-step-ahead prediction.

We use the one-step-ahead prediction to explain the prediction scheme, and then we may extend it to be the k-step ahead prediction.

In one-step-ahead prediction scenario, we first set the prediction step to be one, which means each time we only predict value at one time unit. According to the Holt-Winters method, we have known all the parameters from the historical actual time series, and we also know the last one of the historical time series data. From the basic idea of the prediction, we use the minimum mean square error (MMSE) prediction method to predict its performance at one time unit.

To extend the horizon of time series, the k-step-ahead prediction value can be computed recursively. For example, to obtain the two-step-ahead prediction value, the one-step-ahead predicted value is computed first. Then it is used with the other lagged values to compute the two-step-ahead predicted value. This procedure is repeated to generate subsequent k predicted values. This is similar as the one-step-ahead prediction. The call prediction algorithm is presented as follows.

Step 1: Predict the number of incoming calls on the next day using the all recorded calls of a caller and find what day is that day.

Step 2: If the prediction result is not 0, compute the correspondent frequencies of the incoming calls in the 3 time intervals, that is morning (8–12 O'clock), afternoon (13–17 O'clock) and evening (18–24 O'clock) for all those weekdays/weekend using the all recorded calls of those days (say, calls of all Saturdays in a certain period of time).

Step 3: Choose the time interval with the highest, second high and third high call frequency and using all the incoming call times in that time interval to predict the time of incoming calls respectively.

20.4 Experimental Results and Validation

The following information is available for call prediction of callers.

1. The number of incoming calls per day during a certain period of time.
2. The number of incoming calls from Monday thru Sunday respectively (7 data sets).
3. Divide the total incoming call time from Monday thru Sunday (7 data sets) into 3 time intervals, that is morning (8–12 O'clock). Afternoon (13–17 O'clock) and evening (18–24 O'clock) respectively.

Because of the limited space we only choose the data set on all previous Fridays for the receiver 1 as an example to show how to predict the incoming calls on the next day which is Friday.

Figures 20.1, 20.2, 20.3 and Table 20.1 show the prediction results of the calls using our call prediction algorithm and Holt-Winters procedure, where the red dots denote the predicted values and green dots denote the observation values on all Fridays for the receiver 1.

Step 1: Predict the number of incoming calls for next day and this day is Friday. Figure 20.1 shows that there will be 1 incoming call (predicted value is 1.2) and the observation is 2 calls. In Fig. 20.1 the x-axis is the days and y-axis is the number of calls.

Fig. 20.1 The predicted and actual number of calls by caller 1

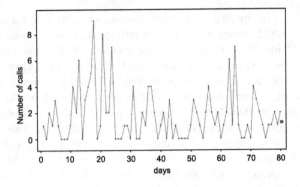

Fig. 20.2 The 1st predicted call time and actual call time of user 1 on Friday. The *red dot* indicates the predicted value and *green dots* indicate the actual values (Color figure online)

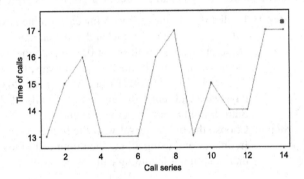

Fig. 20.3 The 2nd predicted call time and actual call time of user 1 on Friday

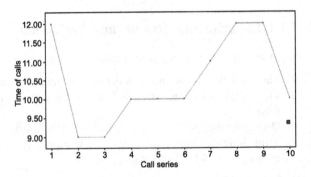

Step 2: Compute the correspondent frequencies of the incoming calls in the 3 time intervals on all previous Fridays. There are about 36% call time are in 8–12 O'clock interval, 50% call time are in 13–17 O'clock interval, and 14% call time are in 18–24 O'clock interval shown in Table 20.1 respectively.

The 13–17 O'clock time interval on all previous Fridays was selected to predict the incoming call time since there is the highest frequency in this time interval (50% of the all incoming calls in this time interval).

Next we predict the time of the incoming calls on the next day (Friday). Figure 20.2 shows that there will be an incoming call at about 17.3 O'clock which is

Table 20.1 Call frequency

Time interval	Frequency	Cumulative %
8–12	10	35.71%
13–17	14	85.71%
18–24	4	100.00%

Table 20.2 Prediction results

Receivers	Callers	Total # of days/calls	# of calls on weekdays	Prediction of # of calls	Prediction of time	Observation of # of calls	Observation of time
Receiver1	Caller1	80/148		1.2		2	
			14 (13–17 Fri.)		17.3		17
			10 (8–12 Fri.)		9.3		10
	Caller4	92/86		1.4		1	
			9 (18–24 Mon.)		23		22
	Caller2	87/181		−0.3		1	
			7 (8–12 Wed.)		11.0		11
Receiver2	Caller9	129/229		2.3		3	
			12 (18–24 Sat.)		21.0		22
	Caller10	116/218		1.5		2	
			15 (18–24 Fri.)		23.8		24
			17 (13–17 Fri.)		13.0		13
Receiver10	Caller1	84/234		2.5		4	
			10 (18–24 Sat.)		21.5		22
	Caller114	49/138		2.5		3	
			14 (18–24 Sun.)		22.3		22
	Caller3	75/109		0.9		1	
			7 (18–24 Fri.)		22.0		23
Receiver3	Caller10	116/282		2.4		2	
			29 (13–17 Fri.)		13.6		13
			11 (18–24 Fri.)		23.3		24

the predicted value (17:20 O'clock) and the observation is 17 O'clock. In Fig. 20.2 the x-axis is the call series on all Fridays (i.e. 1st call, 2nd call, ...) and y-axis is the incoming call time. Next the 8–12 O'clock time interval on Friday is computed to predict the second incoming call time. Figure 20.3 shows that the second predicted incoming call will be at about 9.3 O'clock (9:20 O'clock) and the observation is 10 O'clock.

The number of calls and corresponding call times for additional callers and receivers have also been predicted. The results are summarized as follows and shown in Table 20.2. Caller 4 for receiver 1: The number of call predicted is 1.4 and the observation is 1. The predicted call time is timeHour23Minute023 O'clock and the observation is timeHour22Minute022 O'clock. Caller 2 for receiver 1: The number of calls predicted is −0.3 and the observation is 1. This call time predicted is time-Hour11Minute011 O'clock and the observation is timeHour11Minute011 O'clock. Caller 9 for receiver 2: The number of call predicted is 2.3 and the observation is 3. This call time predicted is 21 O'clock and the observation is 22 O'clock.

Predicted calls for other receivers and frequent callers on the weekdays/weekend are shown in Table 20.2. In Table 20.2 the "80/148" indicates that there are 148 incoming calls in 80 days in the "Total # of days/calls" column and the "14(13–17 Fri.)" indicates that there are 14 incoming calls in 13–17 O'clock time interval on all previous Fridays in the "# of calls on weekdays" column.

20.5 Conclusion

The results show that this approach can be used to predict the calls and are reasonable accurate. The prediction technique proposed are preliminary and other approaches need to be considered in order to improve accuracy. The sample size is only 20, but, represents typical profiles (socially active, socially inactive, socially moderate). We are also working on the call logs of 100 phone users from MIT Reality Mining Group over a period of 8 months (between October 2004 and May 2005) [17]. We are observing similar results issues are to be addressed by future work.

References

1. Gray, A.G., Thomson, P.J.: On a family of finite moving-average trend filters for the ends of series. J. Forecast. **21**(2), 125–149 (2002)
2. Taylor, G.W.: Smooth transition exponential smoothing. J. Forecast. **23**(6), 385–404 (2004)
3. Man, K.S.: Long memory time series and short term forecasts. Int. J. Forecast. **19**(3), 477–491 (2003)
4. Zou, H., Yang, Y.: Combining time series models for forecasting. Int. J. Forecast. **20**(1), 69–84 (2004)
5. Liu, H., Hall, S.G.: Creating high frequency national accounts with state space modeling A Monte Carlo experiment. J. Forecast. **20**(6), 441–449 (2001)

6. Cao, L.: In-depth behavior understanding and use: the behavior informatics approach. Inf. Sci. **180**, 3067–3085 (2010)
7. Janacek, G., Swift, L.: Time Series: Forecasting, Simulation, Applications. Ellis Horwood, Chichester (1993)
8. Magalhaes, M.H., Ballini, R., Molck, P., Gomide, F.: Combining forecasts for natural stream flow prediction. In: IEEE Annual Meeting of the Fuzzy Information, vol. 1, pp. 390–394 (2004)
9. Guang, C., Jian, G., Wei, D.: Nonlinear-periodical network traffic behavioral forecast based on seasonal neural network model. In: IEEE International Conference on Communications, Circuits and Systems, vol. 1, pp. 683–687 (2004)
10. Young, P.C.: Non-stationary time series analysis and forecasting. J. Prog. Environ. Sci. **1**, 3–48 (1999)
11. Young, P.C., Tych, W., Pedregal, D.J.: Stochastic unobserved component models for adaptive signal extraction and forecasting. In: IEEE Signal Processing Society Workshop, vol. 8, pp. 234–243 (1998)
12. Groschwitz, N.K., Polyzos, G.C.: A time series model of long term NSFNET backbone traffic. In: IEEE International Conference on Communications, vol. 3, pp. 1400–1404 (1994)
13. Hansen, J.W., Nelson, R.D.: Neural networks and traditional time series methods: a synergistic combination in state economic forecasts. IEEE Trans. Neural Netw. **8**(4), 863–873 (1997)
14. Tych, W., Pedregal, D.J., Young, P.C., Davies, J.: An unobserved component model for multi-rate forecasting of telephone call demand: the design of a forecasting support system. Int. J. Forecast. **18**(4), 673–695 (2002)
15. Harvey, A.C.: Forecasting, Structural Time Series Models and the Kalman Filter. Cambridge University Press, Cambridge (1989)
16. Nelson, M., Hill, T., Remus, B., O' Connor, M.: Can neural networks applied to time series forecasting learn seasonal patterns: an empirical investigation. In: IEEE International Conference on System Sciences, vol. 3, pp. 649–655 (1994)
17. Eagle, N., Pentland, A., Lazer, D.: Inferring social network structure using mobile phone data. PNAS (2006, in submission)
18. Kolan, P., Dantu, R.: Socio-Technical Defense Against Voice Spamming, ACM Trans. Auton. Adapt. Syst. December (2006, accepted)
19. Bowerman, B.L.: Forecasting, time series, and regression: an applied approach (2005)

Chapter 21
A System with Hidden Markov Models and Gaussian Mixture Models for 3D Handwriting Recognition on Handheld Devices Using Accelerometers

Wang-Hsin Hsu, Yi-Yuan Chiang, and Jung-Shyr Wu

Abstract Based on accelerometer, we propose a 3D handwriting recognition system in this paper. The system is consists of 4 main parts: (1) data collection: a single tri-axis accelerometer is mounted on a handheld device to collect different handwriting data. A set of key patterns have to be written using the handheld device several times for consequential processing and training. (2) Data preprocessing: time series are mapped into eight octant of three-dimensional Euclidean coordinate system. (3) Data training: hidden Markov models (HMMs) and Gaussian mixture models (GMMs) are combined to perform the classification task. (4) Pattern recognition: using the trained HMM to carry out the prediction task. To evaluate the performance of our handwriting recognition model, we choose the experiment of recognizing a set of English words. The accuracy of classification could be achieved at about 96.5%.

21.1 Introduction

In recent years mobile devices have become popular as a result of the growth of sensor-enabled mobile devices. Users can utilize diverse digital contents anywhere, anytime due to its portability. If the mobile terminal can aware of user's current context then it could react in some appropriate manner to suit the user without the need of user interaction.

To implement the handwriting recognition system, many different techniques, such as vision-based gesture recognition, touch-based gesture recognition have been

W.-H. Hsu (✉) · Y.-Y. Chiang
Department of CSIE, Vanung University, Chungli, Taoyuan, Taiwan 320
e-mail: kimble@vnu.edu.tw

Y.-Y. Chiang
e-mail: yychiang@vnu.edu.tw

W.-H. Hsu · J.-S. Wu
Department of EE, National Central University, Chungli, Taoyuan, Taiwan 320

J.-S. Wu
e-mail: jswu@wireless.ee.ncu.edu.tw

L. Cao, P.S. Yu (eds.), *Behavior Computing*,
DOI 10.1007/978-1-4471-2969-1_21, © Springer-Verlag London 2012

utilized. In recent years, a new kind of interaction technology that recognizes users' movement has emerged due to the rapid development of sensor technology, which is an interesting topic in behavior computing [2]. An accelerometer measures the amount of acceleration of a device in motion. Analysis of acceleration signals enables three kinds of gesture interaction methods: tilt detection, shake detection and gesture recognition [1, 3–6].

Although in the literature there are already exist some approaches of using acceleration signals for gestures recognition, most work focuses on recognizing the simple gestures such as Arabic numerals [3–5], simple linear movements and direction [6]. In our work, we attempt to recognize a set of handwritten English words.

We propose a 3D handwriting recognition system in this paper. The system is consists of 4 main parts: (1) *data collection*: a single tri-axis accelerometer is mounted on a handheld device to collect different handwriting data. A set of *key patterns* have to be written using the handheld device several times for consequential processing and training. (2) *Data preprocessing*: time series are mapped into eight octant of three-dimensional Euclidean coordinate system. (3) *Data training*: HMMs and GMMs are combined to perform the classification task. (4) *Pattern recognition*: using the trained GMM to carry out the prediction task. To evaluate the performance of our handwriting recognition model, we choose the experiment of recognizing a set of English words. The accuracy of classification could be achieved at about 96.5%.

The rest of this paper is organized as follows. Theoretical backgrounds, including hidden Markov models and Gaussian mixture models, are presented in Sect. 21.2. In Sect. 21.3, we discuss the feasibility of applying HMMs to 3D handwriting recognition. The proposed recognition system is presented in Sect. 21.4. Then, the effectiveness of this scheme is demonstrated through experimental analysis in Sect. 21.5 followed by Conclusions in Sect. 21.6.

21.2 Theoretical Backgrounds

21.2.1 Hidden Markov Models

A HMM is a stochastic model used for representation an underlying stochastic process that is not observable, but could be observed through the sequence of observed symbols [7]. HMM are often used for signal processing such as speech recognition [8]. A HMM with N urns could be described via a compact notation: $\lambda = (A, B, \Pi)$ and the following notation is other symbols in HMM model.

$$N = \text{Number of states in the model,}$$

$$T = \text{length of observation sequence,}$$

$$A = a_{ij},$$

where $a_{ij} = P(i_{t+1}|i_t = i)$, the probability of being in state j at time $t + 1$ if state is i at time t.

$$B = b_j(k) = P(v_k \text{ at } t|i_t = i),$$

the probability of symbol v_k observed in state j at time t, and v_k is one of the possible observation symbols.

$$\Pi = \pi_i = P(i_1 = i)$$

the probability of being in state i at the start time ($t = 1$).

HMMs assume that the observed sequence is generated in following manner: At time 0, the model starts in one of the N states with probability π_i. A random observation value k is selected with probability $b_j(k)$ and then the model jump to next state j from current state i at time 1 with probability a_{ij}, and repeated the operations until T outputs have been generated.

There are three fundamental problems that HMMs are used to solve mainly.

1. Computing the $P(O|\lambda)$, the probability of occurrence of the observation sequence O, when given the model $\lambda = (A, B, \pi)$.
2. Choosing a state sequence I so that $P(O, I|\lambda)$, the joint probability of the observation sequence O, could be maximum.
3. Adjusting the HMMs' parameters, A, B, and π, so that $P(O|\lambda)$ or $P(O, I|\lambda)$ could be maximum.

Since a data sequence could be viewed as an observation sequence, this study could have to solve these three problems. In this study, the training phase of HMMs is as Problem 3 and the Segmental K-means algorithm and Baum-Welch algorithm are used to solve it. When the model has been trained, the Forward-Backward algorithm is utilized to calculate the likelihood value of each sequence as Problem 1.

21.2.2 Gaussian Mixture Models

Gaussian mixture models are the most statistically mature methods for clustering and density modeling. Mixture Models are a type of density model which comprises a number of component functions, for example, probability mixture model, parametric mixture model and continuous mixture model. GMMs have been successfully used for texture and color image analysis and applied to speech recognition. This study assumes that a probability distribution existed that could represent the feature statistics of a block of each patient and Gaussian mixture models are selected for this purpose.

For a D-dimensional feature vector, x, the mixture density used for the likelihood function is defined as

$$p(x|\lambda) = \sum_{i=1}^{M} \omega_i p_i(x).$$

The density function is a weighted linear combination of the M uni-model Gaussian densities, $p_i(x)$, each parameterized by a mean $D \times 1$ vector, μ_i, and a $D \times D$ covariance matrix, Σ_i, and ω_i is the mixture weight for ith component, which satisfies $\omega_i > 0$ and

$$\sum_{i=1}^{M} \omega_i = 1,$$

$$p_i(x) = \frac{1}{(2\pi)^{D/2}|\Sigma_i|^{1/2}} \exp\left[-\frac{1}{2}(x - \mu_i)^T (\Sigma_i)^{-1}(x - \mu_i)\right].$$

The complete Gaussian mixture density is denoted by

$$\lambda = \{p_i, \mu_i, \Sigma_i\}, \quad i = 1, \ldots, M.$$

For 3D handwriting sequence recognition, each HMM model is used to calculate the 3D feature vectors for pattern distribution.

21.3 The Feasibility of Applying HMMs and GMMs to 3D Handwriting Recognition

21.3.1 Data Sets and Data Preprocessing

Data is collected from an handheld device with accelerometer, such as Apple iPhone, Google HTC Phone, etc. Since acceleration signals are sampled in equal-time interval, the length of raw data is variable according to different key pattern and different input speed. Data from the accelerometer has the following attributes: time, acceleration along x-axis, y-axis, and z-axis.

We obtain three acceleration time series $\mathbf{a}_x$, $\mathbf{a}_y$, $\mathbf{a}_z$ from the previous step. In order to obtain the position time series, we can use integration calculus twice on the acceleration time series. That is, $\mathbf{v}_x = \int_{t_0}^{t_N} \mathbf{a}_x dt$ and $\mathbf{s}_x = \int_{t_0}^{t_N} \mathbf{v}_x dt$, where $\mathbf{v}_x$ and $\mathbf{s}_x$ are respectively the velocity and position time series of x-axis. The other two position time series $\mathbf{s}_y$ and $\mathbf{s}_z$ could be derived using the same method.

In order to use a single sequence instead of three to represent a *letter* of alphabet or a *word*, we have further transform the three position time series into one sequence. The method is described as follows. Suppose that

$$\mathbf{s}_x = \{s_x(1), s_x(2), \ldots, s_x(n)\},$$

$$\mathbf{s}_y = \{s_y(1), s_y(2), \ldots, s_y(n)\},$$

$$\mathbf{s}_z = \{s_z(1), s_z(2), \ldots, s_z(n)\}$$

are given, we could have a difference sequence as below

$$ds_X(t) = \{\tau(s_X(t) - s_X(t-1)) : t = 2, \ldots, n\},$$

$$X \in \{x, y, z\},$$

where $\tau : \mathbb{R} \to \{0, 1\}$ is defined as

$$\tau(x) = \begin{cases} 1, & \text{if } x \geq 0, \\ 0, & \text{if } x < 0. \end{cases}$$

Then, we can transform $ds_X(t)$, $X \in \{x, y, z\}$ into a single sequence composed of $\{0, 1, \ldots, 7\}$ as follows:

$$S(t_i) = ds_x(t_i) \cdot 2^0 + ds_y(t_i) \cdot 2^1 + ds_y(t_i) \cdot 2^2.$$

21.3.2 Performance Criteria

The classification performance can be evaluated using mis-classification rate such as *apparent error rate* and/or graphical representation tools such as the *receiver operating characteristic* (ROC) curve.

Let the training data be denoted by $Y = \{y_i : i = 1, \ldots, n\}$, the pattern y_i consisting of two parts, $y_i^T = (x_i^T, z_i^T)$, where $\{x_i : i = 1, \ldots, n\}$ are the measurements and $\{z_i : i = 1, \ldots, n\}$ are the corresponding class labels, now coded as a vector, $(z_i)_j = 1$ if $x \in$ class ω_j and zero otherwise. Let $\omega(z_i)$ be the corresponding categorical class label. Let the decision rule designed using the training data be $\eta(x; Y)$ and let $Q(\omega(z), \eta(x; Y))$ be the loss function

$$Q(\omega(z), \eta(x; Y)) = \begin{cases} 0 & \text{if } \omega(z) = \eta(x; Y), \\ 1 & \text{otherwise.} \end{cases}$$

The apparent error rate, e_A, is obtained by using the design set to estimate the error rate,

$$e_A = \frac{1}{n} \sum_{i=1}^{n} Q(\omega(z), \eta(x; Y)).$$

Another assessment tool for performance is the ROC curve. For a given classifier and an instance, there are four possible outcomes: true positive, false negative, true negative, and false positive. The true positive rate is

$$tp\ rate = \frac{\text{Positive correctly classified}}{\text{Total positives}}.$$

The false positive rate is

$$fp\ rate = \frac{\text{Negatives incorrectly classified}}{\text{Total negatives}}.$$

Table 21.1 The average likelihoods of the sequences between two symbols of '*A*', '*B*', '*C*', '*D*'

Average likelihoods	A	B	C	D
A		−45.12	−49.17	−55.89	−51.44
B		−	−50.78	−56.11	−52.78
C		−	−	−52.44	−57.20
D		−	−	−	−47.65

Additional terms associated with ROC curves are

$$\text{sensitivity} = \text{recall},$$

$$\text{specificity} = \frac{\text{True negatives}}{\text{False positives} + \text{True negatives}}$$

$$= 1 - fp\ rate.$$

Then, ROC curves are two-dimensional graphs in which *tp rate* is plotted on the *Y* axis and *fp rate* is plotted on the *X* axis. All the ROC curves pass throughout $(0, 0)$ and $(1, 1)$ points and as the separation increases the curve moves into the top left corner.

21.3.3 Experimental Results

We choose some letters of alphabet for testing the significance of HMM-based pattern matching method. In the first experiment, signals for four capital letters '*A*', '*B*', '*C*', and '*D*' are collected. For each letter, 10 patterns are collected. Then, utilizing the data pre-processing method mentioned in Sect. 21.3.1, the acceleration data are transformed into a sequence composed of $\{0, 1, \ldots, 7\}$. We name these sequences as $S_{A,i}$, $S_{B,i}$, $S_{C,i}$, and $S_{D,i}$, where $i \in \{1, \ldots, 10\}$ indicates the pattern number. Using HMM, we compute the following likelihood between two symbols:

$$HMM\text{-}likelihood(S_{X,i}, S_{X,j}), \quad X \in \{A, B, C\},$$
$$i, j \in \{1, \ldots, 10\},$$
$$i \neq j,$$
$$HMM\text{-}likelihood(S_{X,i}, S_{Y,i}), \quad X, Y \in \{A, B, C\},$$
$$X \neq Y,$$
$$i \in \{1, \ldots, 10\}.$$

The average likelihood between two letters are shown in Table 21.1. Besides, the ROC curve are shown in Fig. 21.1. The cut-off point for best Sensitivity and Specificity is 88.00. Using the cut-off, accuracy can be achieved at 82.62%. From these experimental results, we find that the HMM-based pattern matching method for 3D

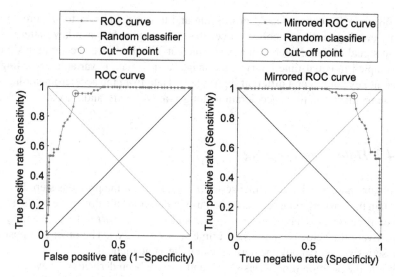

Fig. 21.1 Performance assessment using the ROC curve

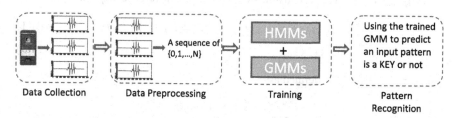

Fig. 21.2 The architecture of the proposed 3D handwriting recognition system

handwriting recognition is feasible. In next section, we will propose the system architecture for the 3D handwriting recognition.

21.4 The Proposed 3D Handwriting Recognition System

The architecture of the proposed 3D handwriting recognition system is presented as shown in Fig. 21.2 which consists of 4 main parts: data collection, data preprocessing, data training, and pattern recognition. We detail them in what follows.

21.4.1 Data Collection

A single tri-axis accelerometer is mounted on a handheld device to collect different handwriting data. A set of *key patterns* have to be written using the handheld device several times for consequential processing and training. In order to acquire an adequate training result, we collect larger than 10 samples for each key pattern. The output signal of the accelerometer is sampled at 300 Hz. Since acceleration signals are sampled in equal-time interval, the length of raw data is variable according to different key pattern and different input speed. Data from the accelerometer has the following attributes: time, acceleration along x-axis, y-axis, and z-axis.

21.4.2 Data Preprocessing

We obtain three acceleration time series a_x, a_y, a_z from the previous step. In order to obtain the position time series, we can use integration calculus twice on the acceleration time series. That is, $v_x = \int_{t_0}^{t_N} a_x dt$ and $s_x = \int_{t_0}^{t_N} v_x dt$, where v_x and s_x are respectively the velocity and position time series of x-axis. The other two position time series s_y and s_z could be derived using the same method.

While the position time series s_x, s_y, and s_z have been derived, we have to transform them into a sequence which composed of a finite set of symbols. Suppose that $\{s_X(t)\}_{t=t_0}^{t_N}$, $X \in \{x, y, z\}$ are given, we could have a difference sequence as below

$$ds_X(t) = \left\{ \tau\big(s_X(t) - s_X(t-1)\big) : t = (t_0 + 1), \ldots, t_N \right\}, \tag{21.1}$$

$$X \in \{x, y, z\}, \tag{21.2}$$

where $\tau : \mathbb{R} \to \{0, 1\}$ is defined as

$$\tau(x) = \begin{cases} 1, & \text{if } x \geq 0, \\ 0, & \text{if } x < 0. \end{cases}$$

Then, we can transform $ds_X(t)$, $X \in \{x, y, z\}$ into a sequence composed of $\{0, 1, \ldots, 7\}$ as follows:

$$S(t_i) = ds_x(t_i) \cdot 2^0 + ds_y(t_i) \cdot 2^1 + ds_y(t_i) \cdot 2^2. \tag{21.3}$$

The geometric meaning of transformation (21.3) is to mapping the difference sequence (21.1) into eight octant of three-dimensional Euclidean coordinate system.

21.4.3 Data Training

For data training, we have to prepare two sets of time series: *key patterns* and *non-key patters*. After preprocessing, we have two sets of sequences, named *KEY* and *NONKEY*. Then, we apply HMM to all pairs of sequences selected from $\{KEY \times KEY\} \cup \{KEY \times NONKEY\}$ and we get a weight value for each pair of sequences. For example,

$$KEY = \{k_1, k_2, \ldots, k_m\}$$

Table 21.2 The average likelihoods of the sequences between two words in the KEY WORD set

Average likelihoods	Kimble	Apple	Test	Nathan	Wonderful
Kimble	−59.36	−67.03	−78.24	−67.32	−66.73
Apple	−	−63.04	−77.65	−69.39	−69.09
Test	−	−	−71.45	−77.06	−75.14
Nathan	−	−	−	−58.62	−62.31
Wonderful	−	−	−	−	−51.39

and

$$NONKEY = \{nk_1, nk_2, \ldots, nk_n\},$$

any selected pair of sequences can be one of the following type:

$$(k_i, k_j), \quad i \in \{1, \ldots, m\}, \ j \in \{1, \ldots, n\}$$

or

$$(k_i, nk_j), \quad i \in \{1, \ldots, m\}, \ j \in \{1, \ldots, m\}.$$

Then, the type associated with the weight values form a set for the input of support vector classifier. Thai is, the training process is performed by HMMs and GMMs.

21.4.4 Pattern Recognition

After the step of data training, we have an HMM model. For a new input pattern, a_{0x}, a_{0y}, a_{0z}, we have to process them using the data preprocessing method and we could get a sequence $S_0(t)$. Then, the GMM could be applied. We compute the likelihood p_i of the longest common subsequence between $S_0(t)$ and a particular key word, K. Using the likelihood, the GMM model would tell us whether the new pattern is the key word K or not.

21.5 Experimental Results

To evaluate the performance of our handwriting recognition model, we choose the experiment of recognizing a set of English words. The set of English words contains {Kimble, Apple, Test, Nathan, Wonderful}. For each word, we collect at least 10 patterns from the handheld device (HTC G1 mobile phone). Table 21.2 is a statistic of the average likelihood between these words. It is easy to see that the patterns indicating the same word have larger length than the patterns indicating the different words. Then, using GMM, the accuracy of classification could be achieved at about 96.5%.

21.6 Conclusions

In this paper, we propose a handwriting recognition system based on a single tri-axis accelerometer mounted on a cell phone for human computer interaction. The system is consists of 4 main parts: (1) *data collection*: a single tri-axis accelerometer is mounted on a handheld device to collect different handwriting data. A set of *key patterns* have to be written using the handheld device several times for consequential processing and training. (2) *Data preprocessing*: time series are mapped into eight octant of three-dimensional Euclidean coordinate system. (3) *Data training*: HMMs and GMMs are combined to perform the classification task. (4) *Pattern recognition*: using the trained GMM to carry out the prediction task. The experimental results show that the accuracy of classification could be achieved at about 96.5%.

References

1. Baek, J., Yun, B.J.: A sequence-action recognition applying state machine for user interface. IEEE Trans. Consum. Electron. **54**(2), 719–726 (2008)
2. Cao, L.: In-depth behavior understanding and use: the behavior informatics approach. Inf. Sci. **180**, 3067–3085 (2010)
3. Cho, S.J., Choi, E.a.: Two-stage recognition of raw acceleration signals for 3-d gesture-understanding cell phones (2006)
4. Choi, E.-S., Bang, W.-C.A.: Beatbox music phone: Gesture interactive cell phone using tri-axis accelerometer (2006)
5. Choi, S.-D., Lee, A.S., Lee, S.-Y.: On-line handwritten character recognition with 3d accelerometer, pp. 845–850 (2006)
6. Kallio, S., Kela, J., Mantyjarvi, J.: Online gesture recognition system for mobile interactioncs, vol. 3, pp. 2070–2076 (2003)
7. Rabiner, L.R., Juang, B.H.: An introduction to hidden Markov models. IEEE ASSP Mag. **3**, 4–16 (1986)
8. Rabiner, L.R., Juang, B.H.: A tutorial on hidden Markov models and selected applications in speech recognition. Proc. IEEE **77**, 257–286 (1989)

Chapter 22
Medical Students' Search Behaviour: An Exploratory Survey

Anushia Inthiran, Saadat M. Alhashmi, and Pervaiz K. Ahmed

Abstract Medical information searching has become an integral part of a medical students' life. Yet, medical students are not provided with formal education on how to search for medical information on medical domains. Searching for medical information on medical domains is not a trivial task. It requires usage of appropriate terminology and the ability to comprehend returned results. The search behavior of medical students is rarely studied to understand how their information search goal is being satisfied or to identify specific search challenges faced by them. In this paper, we study interactive information searching behavior of medical students. Using simulated work task scenarios, we identify similarities and variability's in search patterns and analyzes search behavior traits demonstrated by medicals students. Based on our findings, we suggest intuitive methods medical search engines could adapt to improve medical students' search interaction to better support the search process.

22.1 Introduction

Medical students use medical search engines frequently through their course of study. They not only use these search engines during lecture sessions but also when they undergo practical training in hospitals. In both these scenarios, the search goal of the student is different. When the student is attending lectures the search goal is to obtain more information on a particular topic or to complete assignments. When the student is undergoing practical training, the search goal is to obtain information on a patient's disease, diagnosis and treatment options. Both these examples are typical searches performed by medical students. Although these types of searches seem simple, very little is known about search tactics employed by medical students. Similarly, we are also unaware of search challenges faced by medical students.

A. Inthiran (✉) · S.M. Alhashmi
School of Information Technology, Monash University, Sunway Campus, Bandar Sunway, Malaysia
e-mail: anushia.inthiran@monash.edu

P.K. Ahmed
School of Business, Monash University, Sunway Campus, Bandar Sunway, Malaysia

L. Cao, P.S. Yu (eds.), *Behavior Computing*,
DOI 10.1007/978-1-4471-2969-1_22, © Springer-Verlag London 2012

A junior medical student may not have sufficient medical knowledge to perform a search effectively. Similarly, they may also have trouble understanding search results presented to them. On the other hand, a senior medical student with significantly more knowledge than a junior medical student could perform searches more effectively. However, domain knowledge on its own will not guarantee a successful search session. The combination of domain knowledge and effective search expertise is essential for a successful search session. Medical students are rarely if ever taught information retrieval skills. Furthermore, the use of medical terminology is paramount when searching on medical search engines. Hence, medical students need to be assisted during a search session. While medical students prefer using medical search engines as opposed to other forms of resources, our survey results indicate they are dissatisfied with their search experience. Amongst issues reported by students are: search engine returned too many search results, having to perform too many search iterations, not satisfied with search results, difficulty in locating specific information and not returned with relevant results.

The search pattern of medical students is rarely studied to understand the interaction that takes place during a search session. The study of a user's interactive search behaviour provides deeper understanding of a search session. It also provides us with implicit information about a search session. For example, search behaviour can indicate if a user is having difficulties with the search or if the search engine is not providing users with relevant results. A user's behaviour captured during the search process is thought to be more representative of the user's true behaviour [5, 10]. Furthermore, since humans act and compensate in many ways while dealing with errors or failures during the search process [15] it is important that we study these interactions when developing effective retrieval strategies. Understanding a users search behaviour allows for the creation of various user models. The identification and classification of these user models allows for tailored assistance by search engines to improve a user's search session.

In this paper, we present our preliminary results of interactive search behaviour of medical students on *Medline Plus*.[1] To better understand the search behaviour of medical students, we emphasize on post-query search pattern. Through observation and logging of activities for 30 search sessions with 15 medical students, we gathered preliminary data for a focused study to analyse variability that exist in search behaviour of medical students. Understanding these differences has implications in areas such as the design of the search interface, identifying effective search strategies and automatic identification and classification of user models based on search behaviour. In particular, we focus on two research questions: (i) how variable are search interactions amongst medical students and (ii) is it possible to generally classifying the search behaviour of medical students based on search experience, medical knowledge, task complexity and topic familiarity. We believe by understanding a user's search behaviour, we can better support interactions at an appropriate level to improve a medical students' search experience. The remainder of this paper is

---

[1] http://www.nlm.nih.gove.medlineplus

structured as follows. Section 22.2 presents related work, Sect. 22.3 describes our research methodology, Sect. 22.4 provides findings of our experiment, Sect. 22.5 discusses the implications of this research and in Sect. 22.6 we conclude the paper.

22.2 Related Work

One of the earliest works in this area studied the search behaviour of medical students on MEDLINE.[2] This study reports how training and education using a system called *Grateful Med* over a period of 6 months improved medical students' search session. The study reports not only did students' search skills on *MEDLINE* improve but students managed to investigate overlapping topics and further researched other medical and genetic topics [4]. In a similar study, the *University of Michigan* provided an information retrieval course as part of their pharmacy course syllabus [9]. As a result of training, pharmacy students report significant improvement in search experience and continued usage of computerized literature searching even after completing their studies. While we welcome this initiative, we feel taking medical students off from their busy schedule to be trained to use a system may not be a feasible solution. We also believe a one-time training session may not be sufficient and repeated training is necessary. Instead, readily available automated methods would be a better solution.

Results of a study that investigated questioning techniques of first year medical students' based on clinical scenarios indicate students' questions broadly fell into 5 categories [20]. Most questions focused on the clinical aspect on the case and not on other related topics. In some cases, students' seemed to form a standard template of questions and applied it across scenarios. Limited information retrieval skills limit a students' ability to locate necessary information. In a related study, third year medical students medical retrieval skills were studied to inform the design of better medical retrieval systems [16]. This study specifically investigates the usage of *Medical Subject Heading Terms (MeSH[3])* terms on *Medical Literature Analysis and Retrieval System* (MEDLARS). Results show, students failed to use appropriate *MeSH* terms, did not retrieve relevant search results, even when students perceived they were retrieving relevant information they missed up to 100% of available citations, some retrieved far too many results. This led to frustration, incomplete searches and wasted time [16]. Although, we assumed third year medical students equipped with more medical knowledge would be better at medical retrieval, results of this study indicate otherwise. This indicates medical students at any level of study require assistance while searching on medical search engines.

As *MEDLINE* is a foremost medical domain used by most medical professionals, many researchers studied general information retrieval behaviours on this domain.

---

[2]http://www.nlm.nih.gov/bsd/pmresources.html

[3]http://www.nlm.nih.gov/mesh/

Outcome of these studies indicate users having problems using the *MeSH* index-ing feature [1]. In an effort to improve a users' search experience, end user search software was used to improve search strategies and query formulations. Results of a related study on the same domain reveal users failed to begin the search with a well-built query, failed to use *MeSH* headings and failed to apply proper limits to the search. As a result, users were either returned with too many or too few search results [9]. To improve the search session, this research points to the direction of 'thought and practice'. Whilst solutions suggest the use of intermediary software or require users perform searches more often, the schedule of a medical professionals does not permit them to search as often as they would like to. Moreover, using in-termediary software or using a medical librarian as an 'intermediary' does not teach the user to become a better searcher. Users will become dependent on these inter-mediaries and will not perform searches effectively without these intermediaries.

From the perspective of an information retrieval architect, better information re-trieval techniques should be developed to provide medical searchers with improved results. Popular methods include semantic, contextual and complex algorithms to expand a query or make a query more medically focused. The notion behind focus-ing on the query is based on the assumption that a medically focused query provides searchers with better search results. Jalali and Borujerdi [14] developed a method to match the query and documents using a combination of keywords and concept-based approach [17]. Although this technique improved retrieved results in comparison to using pure keyword and statistical based approaches, it does not necessarily make the search session a successful one. Medical students report being frustrated when too much information is provided [14]. Moreover, medical students may encounter difficulties in understanding returned search results that requires expert level knowl-edge to comprehend [17]. In another approach, algorithms were developed to adapt to a users browsing behavior [2]. This algorithm is based on the assumption that a user's previous searching behavior is an indication of future behavior. Although this behavior pattern is generally true, it is not valid when searching for medical informa-tion [18]. In *COMTN* [2], the strategy used attempts to bind the query and retrieved documents. *COMTN* automatically generates context-indicate query for a term des-ignated by an information consumer when reading a document. This technique of matching the query term to context in a medical domain is processing intensive and may also suffer medical related query issues such as: difficult general language words having the same meanings as technical terms, technical terms requiring do-main knowledge to understand and general language words with different technical meanings [8]. In this situation, although the search engine may have returned rele-vant results, a medical student is still required to filter irrelevant information.

Moving away from focusing on queries and documents, Luo and Tang [12] de-veloped a technique to assists users in formulating medical queries through inter-active questionnaires. This technique is demonstrated in *iMed*. After selecting one or more symptoms from a list of known symptoms and signs, *iMed* guides the user through an iterative question and answer session. Based on the searcher's answer to questions, *iMed* navigates the corresponding decision tree and automatically forms multiple queries. Unlike earlier techniques, *iMed* does not utilize any underlying knowledge base (*MeSH*). The users' involvement throughout the search session enables search goals to be incorporate into the search session. A medical students' information need requires non-restrictive search ability where multiple symptoms, patient history and medication can be considered in a single search session. In our opinion, the iterative technique used in *iMED* is suited for lay-users and not for medical students.

Unlike previous research efforts, our research study focuses on the search pattern of a medical student to infer implicit details from a search session. As medical students may fall into various categories depending on search skills, prior search knowledge and medical knowledge the study of their search behaviour allows information retrieval strategies to better support interactions at an appropriate level to improve their search experience. We focus on medical students as we believe this group of people require additional assistance when performing medical searches to progress in their study as well as to encourage the use of medical search engines for lifelong learning towards becoming better doctors. *Medline Plus* was selected as the domain of search as this search engine is suited for clinical based searching and it is developed and managed by the *American National Library of Medicine* and the *National Institute of Health*. Moreover, *Medline Plus* does not employ any type of information retrieval strategies hence it allows us to understand 'true' information search behaviour.

Our review of information retrieval strategies reveals interactive information searching has influenced the information retrieval domain. An early study looking at search behaviours of users the science department in *Oxford University* [19] provides us with evidence that based on the search goal and profession three distinct search behaviours were observed: medical practitioner, medical researcher and practitioner-researcher. Clearly, different users have different search goals thus requiring different types of results. Researchers have found that when searching on a medical search engine, medical experts issue more queries, longer queries, spend more time searching, issue queries with technical terminology and visit technical sites [3]. Identification and classification of such behaviours will enable us to tailor relevant search results to this group of people. In relation to domain knowledge and search experience, user studies report many interesting findings. Information retrieval experts who search in their domain of expertise use domain-specific search knowledge that enabled them to perform effective searching [19]. However, when a task is performed outside their domain expertise, only general purpose search strategies were utilized. A similar finding is also reported in a long-term study of medical students search tactics on a microbiology database. Results indicate users narrowing down search results by iteratively adding new concepts to an original query [6, 7] to obtain relevant results as they progressed to higher levels of study. This leads us to believe while medical students may have many years of search experience, domain knowledge is more important for a successful search. It is possible that a less experienced medical searcher could learn from the search behaviour of a more experienced medical searcher to achieve a successful search session. Additionally, a medical student with more knowledge and familiarity of a certain topic may demonstrate better search skills in comparison to students without this advantage.

In retrospect, since medical domains contain enormous amount of information [13] and a medical term could occur in hundreds of documents from ten diverse health topics [13] we feel newer retrieval methods must consider a user's context in the information retrieval process [11]. Previous research limits the inclusion of context to the query or to documents within a corpus. One possible method of including the user context into the search session is by investigating the search behaviour of a user. Observations of search behaviour not only allow us to identify the user context but also incorporate a user's search goal. In the next section, we describe our experiment methodology.

22.3 Experiment Methodology

We conducted an exploratory survey on a convenience sample of 15 participants with a total of 30 search sessions. We utilized the following data gathering methods for our experiment: pre-experiment interview, simulated work task scenario and a post-experiment interview. In addition, participants' activities were logged and observed.

22.3.1 Pre-experiment Interview

The pre-experiment interview is used to obtain demographic details, information on general search experience, medical knowledge, medical search experience and attitude towards computers and technology.

22.3.2 Simulated Work Task Scenario

Participants were handed a participation letter explaining their involvement in the experiment. Participants who had never used *Medline Plus* were given time to familiarize themselves with the search engine. Each participant was provided with three simulated work task scenarios: Task *A*, *B* and *C*. We tailored these scenarios to fit the description of simulated work task scenarios [10]. This was to ensure criteria's like realism and the ability for participants to engage in the task were incorporated. Simulated work task scenarios were also used to invoke a common information need amongst searchers. An example of a simulated work task scenario is provided in Table 22.1

In addition, participants were asked to prepare a personal task they wished to search on *Medline Plus*. The personal task was used to validate participants search behaviour against the simulated work task scenarios. We also wanted to observe any differences or similarities with the search behaviour of a personal task against the simulated work task scenarios. Tasks were rotated and participants were told they may stop searching once they have found satisfactory results.

Table 22.1 Simulated word task scenario

Simulated Work Situation: As part of your study of medicine it has been discussed why arterial blood gases (ABG's) are important. Now you want to learn more about this ABG's , why it is important, how test are these tests conducted, etc.

Indicative Request: Find for an instance, information about why do we need to perform ABG's, How would you conduct an ABG and what is included in ABG's.

22.3.3 Logging and Observation

Participant's search activity was observed by the researcher. We also utilised logging software to log activities performed by the user.

22.3.4 Post-experiment Interview

The post experiment interview took place after participants had completed each search task. At this stage, we obtained information about post-search task complexity, topic familiarity and comments about their search session.

22.4 Experiment Results

In this section, we present results of our experiment in the following sub-sections: demographic information, participants' perception of task complexity and general classifications of search pattern. In this paper, we present and discuss the search behaviour for the scenario in Table 22.1. In addition, we also provide the search behaviour based of participants personal task. Search patterns provided were generalised based on various classifications. In some cases, we provide individual search pattern to demonstrate variability in the search pattern.

22.4.1 Demographic Information

There were 11 females and 4 males in our sample. 5 participants were in the 1st year, 5 were in their 5th year, 3 participants were in their 3rd year and 2 participants were in their 3rd year of study. The average age for this sample was 21 years. The average general search experience for this sample was 7 years. All participants in this sample have searched for medical information online either using medical search engines or general search engines. They have spent an average of 4 years searching for medical information on medical search engines and 3 years searching for medical information on general search engines. All participants rated their medical knowledge as average except for 2 who rated their medical knowledge as poor. Participants indicate they prefer search results presented to them based on relevance and recently updated results. All participants had positive attitudes towards computer and technology. 7 participants were familiar with the topic described in Table 22.1 while 8 participants were not familiar with this topic.

22.4.2 Perception of Task Complexity

We asked our participants to rate task complexity twice for the scenario in Table 22.1. The first time we asked participants to rate task complexity was prior to commencing the search and the second time after completing the search. 5 participants rated pre-task and post-task complexity as easy. 5 participants changed their perception of task complexity from average to easy. 2 participants rated their pre-task and post-task complexity as average while 3 participants changed their perception of task complexity from difficult to average. Further investigation reveals participants in their 5th year of study rated the scenario is Table 22.1 as easy and did not change their perception of task complexity after completing the search. First year students either changed their perception of task complexity from average to easy or from difficult to easy. This coincides with our initial hypothesis that senior medical students equipped with more knowledge are more familiar with clinical topics as opposed to junior medical students. However, we also note after completing the search, junior medical students (1st–3rd year) did not perceive the task to be as complex as prior to commencing the search. We believe the search activity allowed junior medical students to learn a little more about the topic and this managed to change their perception of task complexity.

22.4.3 General Search Pattern

In this section, we analyse general search patterns based on task complexity and topic familiarity. We also compare the search pattern of the simulated work task against the participant's personal task. To better understand the search behaviour, we developed search pattern taxonomy in Table 22.2.

22.4.4 Task Complexity and Familiarity

We provide and compare the search behaviour of medical students based on task complexity and topic familiarity. We provide the search behaviour of participants who found the task complexity in Table 22.1 as easy and were familiar with the topic. We provide two search patterns for discussion. Both these patterns belong to medical students in their 5th year of study. The search behaviour provided was

Table 22.2 Taxonomy of search pattern

Code	Description
Q1...n	Query Number
L1...n	Result Link Number
LWL	Link Within Link (From Main Page)
SL	Sub-links from Link Within Links (LWL)
NR	No results
QR	Repeated Query
F	Find keyword within document
PB	Parallel Browsing
CL	Comparing Result Links
RP1...n	Results Page

Fig. 22.1 Search behaviour: task complexity easy and familiar with topic

based on participants' post task complexity. Figure 22.1 provides the search behaviour of participants who found this task easy and were familiar with the topic. Figure 22.2 also provides the search pattern of a participant with similar task complexity and familiarity, however the search behaviour in both figures are different. In the simpler search behaviour (Fig. 22.1), 1 query was issued and 2 links were clicked. In Fig. 22.2, double the number of queries and links were clicked on. Both these search patterns were completed in 5 minutes, yet the search behaviour was different. Generally, most participants search behaviour was similar to Fig. 22.1. The search behaviour in Fig. 22.2 is an exception. A second interview with the participant who demonstrated the search behaviour in Fig. 22.2 revealed this participant was not having difficulty in the search session, but was looking to see if better search results could be found for this task by issuing varying queries.

Participants' who found this task easy and were not familiar with the topic displayed search behaviour as shown in Fig. 22.3. When we compare the search pattern in Figs. 22.1 and 22.3, there is no difference in the search behaviour. This suggests topic familiarity does not affect search behaviour. However we note when junior medical students issued queries that returned no results, they tend to issue a previously issued query. On the other hand, when senior medical students are returned with no search results, they continued the search session with a new query. We suggest search engines provide alternatives when users are returned with no search results. Based on our observation this situation does not take place frequently, however we feel alternative measures should be in place to assist users to continue with the search session. An alternative measure is to provide query suggestions.

Figure 22.4 provides the search behaviour for participants' who found the task complexity as average and were familiar with the task. Figure 22.5 provides the search behaviour for participants who found the task complexity average and were

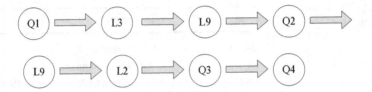

Fig. 22.2 Search behaviour: task complexity easy and familiar with topic (exception)

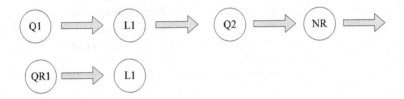

Fig. 22.3 Search behaviour: task complexity easy and not familiar with topic

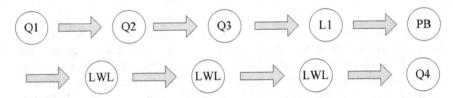

Fig. 22.4 Search behaviour: task complexity average and familiar with topic

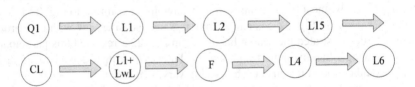

Fig. 22.5 Search behaviour: task complexity average and not familiar with topic

not familiar with the task. Both these search behaviours belong to students in their first year of study. We note both these search behaviours are similar. Many queries and result links were clicked. In Fig. 22.5, the search pattern demonstrates parallel browsing, frequently comparing search results and utilization of the 'find' command to locate the query keyword used in the document. Search engines could use this type of behaviour as indications that users require assistance during a search session. We also note when participants are provided with a very limited set of returned result (i.e. 1 or 2 links), participants immediately issued a new query without looking at the returned results. We note this behaviour is evident in medical students in their 1st and 2nd year of study. We suggest search engines provide alternate links or

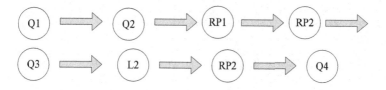

Fig. 22.6 Search behavior: personal task more difficult than simulated work task

perhaps consider providing additional information on returned results such as the number of times a link was clicked on for the same query and details of the page for example, the date when the page was last updated or relevance percentage of the page to the query issued. We feel additional information will encourage users to view these returned results. Our collection of search behaviour did not provide evidence of senior medical students encountering limited returned search results, hence we were not able to determine if senior medical students would demonstrate a different search behaviour when compared to junior medical students. Perhaps one method to improve a junior medical students' search experience is to provide them with queries used by senior medical students.

22.4.5 Personal Task vs. Simulated Work Task

In this section we compare the search behaviour of participants' personal task against our simulated work task scenario. In some cases, participants had previously searched for their personal task and in some cases they have not searched for their task. However, none of our participants had previously used *Medline Plus* to search for their personal task. Figure 22.6 provides the search behaviour for participants who found their personal task more difficult than the simulated work task scenario. We note there was only one participant that found their personal task more difficult than the simulated work task scenario. The participant had been searching on their personal task for about 2 months. A total of four queries were issued but only 1 link was clicked. Since the participant had prior search experience on the personal task the participant was interested in obtaining new information. Hence, more queries were issued to obtain different types of results and only one link looked interesting enough to be clicked on.

Figure 22.7 provides the general search pattern for participants who found the personal task easier in comparison to the simulated work task scenario. The participant had searched for this task before. The search behaviour reveals participants only issued one query but reviewed many result links. Figure 22.8 provides the search behaviour for participants found their personal task easier but had not searched for this task before. In both figures (22.7 and 22.8) the search pattern shows a very simple and straight-forward interaction. Although the participant had not searched for this task before, the search behaviour in Fig. 22.7 does not differ

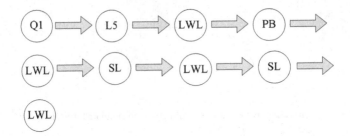

Fig. 22.7 Search behaviour: personal task easier than simulated work task (searched before)

Fig. 22.8 Search behavior: personal task: personal task easier than simulated work task (not searched before)

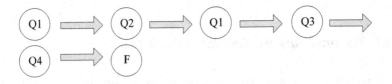

Fig. 22.9 Search behaviour: personal task similar with similar task complexity with simulated work task scenario (searched before)

from Fig. 22.8 except with deeper investigation of links. This suggests prior search experience had no effect on the search behaviour. However, we note this search behaviour is only valid when participants have not searched on *Medline Plus*. We believe, there will be difference in search behaviour due to prior search experience when the search is performed on a previously used domain.

We also managed to obtain the search pattern of participants who found their personal task and our simulated task of the same task complexity. Figure 22.9 provides the search pattern of a participant who had searched for the personal task before, while Fig. 22.10 provides the search pattern of a participant who had not searched for this task. Both search behaviours appear identical. We notice here again, prior search experience had no influence on the search behaviour. We also wanted to investigate if medical knowledge had in any way impacted the search pattern. We note in relation to personal task, junior and senior medical students demonstrated similar search behaviours. However, medical knowledge and search experience influenced the search behaviour of the simulated work task scenario. This provides us with evidence that unique search behaviours are demonstrated based on various external

Fig. 22.10 Search behaviour: personal task similar with similar task complexity with simulated work task scenario (not searched before)

settings. We summarise that the study of interactive search behaviour provides accurate results if the study is done by contextualising the user. Although there were distinct search behaviours demonstrated when the search was done on a simulated work task scenario, the search behaviour of personal tasks did not indicate any difference amongst medical students.

22.5 Discussion

Our study attempted to characterize differences in medical students' search behaviour on *Medline Plus*. Our findings indicate some consistency and variability in the search behaviour of medical students based on task complexity, medical knowledge, search experience and topic familiarity. Furthermore, we also analysed the search behaviour of medical students' based on their personal task. Medical students, who were familiar with the topic and found the task easy, demonstrated simple and straight forward interactions. Students who were not familiar with the topic but found the task complexity easy issued slightly more queries and viewed more results. However, when students were not familiar with the topic and found the task of average complexity, search behaviour revealed students clicking on many result links and comparing results. The search behaviour of personal tasks shows students tend to click on more links and issued more queries when searching for their personal task.

Generally, more interactions (query and clicking on results) indicate students experiencing difficulty with the search session. Junior medical students tend to repeat previously issued queries when not returned with search results. Junior medical students also searched for keywords within the returned results and continued to issue queries when limited search results were returned. Senior medical students did not demonstrate any of these search traits.

Students only clicked on result links after careful evaluation of search results. Hence, the presentation of results summary is important. Students only explored links within documents when information presented in the main result page was relevant. Students also tend to review/read the entire page before clicking on the back button when they find the search result interesting. This search behaviour has some design implications for medical search engines. The need to highlight keywords within search results, providing alternative query keywords when no search results are returned and methods to encourage students to click on search results when limited results were presented by providing information about the search results. We also note students tend to issue queries without clicking on search results when they are not satisfied with returned search results. Search engines should use this as an indication to provide a different set of results to users.

Although *Medline Plus* has many built-in features such as *medical dictionary, specialised tabs on health topics, drugs and supplements* and *images and videos.* None of our participants used these features. We also note participants' only viewing text based results. We hypothesis perhaps our simulated scenarios did not require the use of these features but participants did not use these features even when searching for their personal task. A new feature that could be incorporated is to provide various options to view search results based on relevance and recently updated results. As our survey results indicate medical students like this feature; *Medline Plus* could incorporate this to appear more user-friendly.

22.6 Conclusion

In this paper we described preliminary results of medical students search behaviour on *Medline Plus*. We studied 30 search sessions with 15 users. The findings suggest the possibility of developing user models based on search behaviour. Instead of using crude classifications we utilised various user implicit and explicit features such as task complexity, topic familiarity, previous search experience and medical knowledge. Although the number of search sessions studied were small, it was sufficient to provide us with initial understanding medical students search behaviour. These findings recommend several implications for design strategies for *Medline Plus*: tools to support junior medical students, features to assist users to continue with a search session when no or limited results are returned, provide additional information on search results and the ability to present search results using different views. We also provide generic search behaviour patterns to inform the development of user models based on interactive searches. Future work will include studying the search behaviour of medical practitioners. This will allow us to leverage on the search behaviour of medical practitioners to assist medical students develop effective search strategies for an improved search experience.

Acknowledgement   We thank medical students from Monash University Sunway Campus, University of Malaya and MAHSA University College.

References

1. Allison, J.J., Kiefe, C.I., Weissman, N.W., Carter, J., Centor, R.M.: The art and science of searching MEDLINE to answer clinical questions. Finding the right number of articles. Int. J. Technol. Assess Health Care **15**(2), 281–296 (1999)
2. Anagnostopoulous, I., Maglogiannis, I.: 'Adapting user's browsing behaviour and web evolution for effective search in medical portals. In: First International Workshop on Semantic Media Adaptation and Personalization, SMAP06, pp. 195–197 (2006)
3. Bhavnani, S.K.: Important cognitive components of domain specific search knowledge. In: Proceedings TREC, pp. 571–578 (2002)

4. Brember, V.L., Leggate, P.: Linking a medical user survey to management for library effectiveness: I, the user survey. J. Doc. **41**(1), 1–14 (1985)
5. Cao, L.: In-depth behavior understanding and use: the behavior informatics approach. Inf. Sci. **180**, 3067–3085 (2010)
6. Can, A.B., Baykal, N.: MedicoPort: A medical search engine for all. Comput. Methods Programs Biomed. **86**, 73–86 (2007)
7. Delozier, E.P., Lingel, V.A.: MEDLINE and MeSH: challenges for end users. Med. Ref. Serv. Q. **11**(3), 29–46 (1992)
8. Eysenbach, G., Kohler, C.: How do consumers search for and appraise health information on the World Wide Web? Qualitative study using focus groups, usability tests, and in depth interviews. Br. Med. J. **24**, 573–577 (2002)
9. Kelly, D.: Methods for evaluating interaction information retrieval with users. Found. Trends Inf. Retr. **3**, 1–2 (2009)
10. Lechani-Tamine, L., Boughanme, M., Daoud, M.: Evaluation of contextual information retrieval effectiveness: overview of issues and research. Knowl. Inf. Syst. **24**, 1–34 (2010)
11. Liu, R.L., Lu, Y.L.: Context based online medical terminology navigation. Expert Syst. Appl., (2009). doi:10.1016/j.eswa.2009.06.038
12. Mitchelle, J.A., Johnson, E.D., Hewitt, J.E., Proud, V.K.: Medical students using grateful med: analysis of failed searches and a six month follow-up study. Comput. Biomed. Res. **25**, 43–55 (1992)
13. Pluye, P., Grad, R.M., Dunikowsko, L.G., Stephenson, R.: Impact of clinical information retrieval technology on physicians: A literature review of quantitative, qualitative and mixed methods studies. Int. J. Med. Inform. **74**, 745–768 (2005)
14. Saracevic, T.: The notion of Context in Information Interaction in Context, Keynote Address, IIiX, New Jersey, USA, pp. 1–2 (2010)
15. Scott, N., Weiner, M.F.: Patientspeak: an exercise in communication. J. Med. Educ. **59**, 890–893 (1994)
16. Spink, A., Yang, Y., Jansen, J., Nykanen, P., Lorence, D.P., Ozmutlu, S., Ozmutlu, H.C.: A study of medical and health queries to web search engines. Health Libr. Rev. J. **21**, 44–51 (2004)
17. Tang, C., Luo, G.: On iterative intelligent medical search. In: Proceedings of the 31st Annual International ACM Special Interest Group on Information Retrieval, pp. 3–10 (2008)
18. White, R.W., Dumais, S.T., Teevan, J.: How medical expertise influences web search interaction. In: Proceedings of the 26th Annual SIGCHI Conference on Human Factors in Computing, pp. 179–181 (2008)
19. Wildermuth, B.M., Bliek, R., Miya, T.S.: Information-seeking behaviours of medical students: a classification of questions asked of librarians and physicians. Bull. Med. Libr. Assoc. **82**(3), 295–304 (1994)
20. Wildermuth, B.M.: The effects of domain knowledge on search tactic formulation. J. Am. Soc. Inf. Sci. Technol. **55**(3), 246–258 (2003)

Chapter 23
An Evaluation Scheme of Software Testing Strategy

K. Ajay Babu, K. Madhuri, and M. Suman

Abstract This paper briefly surveys the software testing techniques based on the works of classification and evaluation. In addressing the two major software testing issues, that is when should testing stop and how good the technique is after testing, I present a scheme by a data flow diagram for evaluating software testing techniques. Following this diagram step by step, all the activities involved and the relative techniques were described. Software testing has progressed through five major paradigms they are the debugging, demonstration, destruction, evaluation and prevention periods, as outlined by a number of authors. During its development, software testing has focused on two separate issues, verification and validation. A strategy proposal for software testing in the development of applications is advocated later.

23.1 Introduction

The history of software testing is as long as the history of software development itself. It is an Integral part of the software life–cycle and must be structured according to the type of product, Environment and language used. Software testing is a critical element of software quality assurance and represents the ultimate review of specification, design, and code generation. In the absence of feasible and cost-effective theoretical methods for verifying the correctness of software designs and implemen-

K. Ajay Babu (✉)
Informations & Communications Technology, Melbourne Institute of Technology, Melbourne, Australia 3011
e-mail: konatham.ajay@gmail.com

K. Madhuri · M. Suman
Department of Electronics and Computer Engineering, K.L.E.F University Vaddeswaram, Guntur (dist) 522502, A.P., India

K. Madhuri
e-mail: madhu.k19@gmail.com

M. Suman
e-mail: suman.maloji@gmail.com

L. Cao, P.S. Yu (eds.), *Behavior Computing*,
DOI 10.1007/978-1-4471-2969-1_23, © Springer-Verlag London 2012

tations, software testing plays a vital role in validating both. The goal is [1, 2]: to design a series of test cases that have a high likelihood of finding errors but how? That's where software testing techniques enter the picture. These techniques provide systematic guidance for designing tests that

1. Exercise the internal logic of software components.
2. Exercise the input and output domains of the program to uncover errors in program function, behavior and performance.

As an important software development behavior [11], software testing [5] has focused on two separate issues [3]: verification (static testing) and validation (dynamic testing). Verification [4]: refers to the set of activities that ensure that software correctly implements a specific function. Validation refers to a different set of activities that ensure that the software that has been built is traceable to customer requirements. In practice, the software development methodologies typically employ a combination of several software testing techniques.

23.2 A Framework for Software Testing Techniques

A standard for testing technique classification would allow testers to compare and evaluate testing techniques more easily when attempting to choose the testing strategy for the software not lose information of value during an attack, it loses time and money repairing the computer system as well as potential customers who are temporarily unable to access the system development. The software testing techniques can be classified according to the following viewpoints:

1. Does the technique require us to execute the software? If so, the technique is dynamic testing; if not, the technique is static testing.
2. Does the technique require examining the source code in dynamic testing? If so, the technique is white-box testing; if not, the technique is black-box testing.
3. Does the technique require examining the syntax of the source in static testing? If so, the technique is syntactic testing; if not, the technique is semantic testing.
4. How does the technique select the test data? Test data is selected depending on whether the technique refers to the function or the structure of the software, leading respectively to functional testing and structural testing, where as test data is selected according to the way in which software is operated with respect to random testing.
5. What type of test data does the technique generate? In deterministic testing, test data are predetermined by a selective choice according to the adopted criteria. In random testing, test data are generated according to a defined probability distributed on the input domain.

23.3 An Evaluation of Software Testing Techniques

With reference to this classification, the work on the evaluation of software testing techniques can be done in correspondence with the two major testing issues as shown:

1. When should testing stop? The exit criterion can be based on a reliability measure when the test data have been selected by random testing, whereas a test data adequacy criterion for determining whether or not a test set is sufficient for deterministic testing.
2. How good is the technique after testing? The definition of software reliability measure with failure rate can be applicable to test software with discrete or continuous test data [6]. Test data adequacy criteria are measures of the quality of testing. From this viewpoint, the classification of test adequate criteria can be divided into fault-based testing and error-based testing [7].

A framework for surveying software testing techniques based on the works of classification an evaluation is shown in Fig. 23.1.

23.4 Classification

According to the framework for surveying software testing techniques, we present an evaluation scheme of software testing techniques. The purpose of this evaluation scheme is to allow us to identify the strengths and weakness of current software testing techniques. This will provide the information for selecting the testing strategy in the development of applications. In addressing the two major testing issues, that is when should testing stop and how good the technique (or the software) is after testing; a Data Flow Diagram (DFD) depicting the evaluation scheme is shown in Fig. 23.2; the circles in the diagram correspond to the tasks that will be identified in the following sub sections.

Software testing techniques are divided into two categories static and dynamic testing [8, 9]: Static testing techniques were those that examined the software without executing it and encompassed activities such as inspection, symbolic execution and verification. Dynamic testing techniques are those that examined the software with a view to generating test data for execution by the software.

23.4.1 Static Testing

Static testing techniques are concerned with the analysis and checking of system representations such as the requirements documents, design diagrams and the program source code, either manually or automatically, without actually executing the

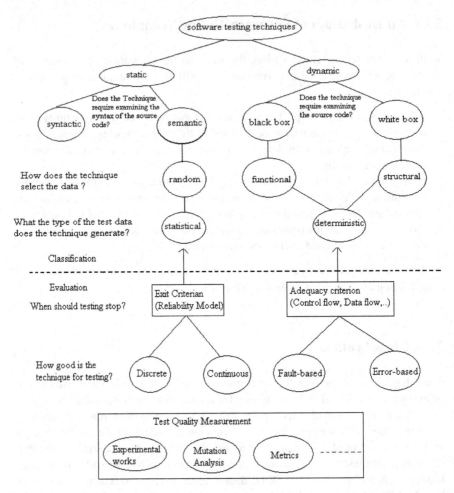

Fig. 23.1 A framework for surveying software testing technique

code [10]. During static testing, specifications are compared with each other to verify that errors have not been introduced during the process. In comparison to dynamic testing, static testing does not require input distributions, since they do not
require that the software be executed.

23.4.2 Dynamic Testing

Dynamic testing techniques are generally divided into the two categories black-
box and white-box testing [10] which correspond with two different starting
points for software testing the internal structure of the software and the requirements specification. They involve the execution of a piece of software with

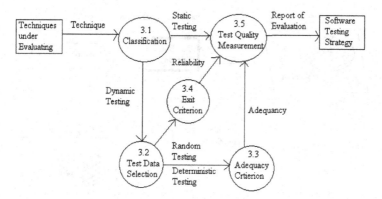

Fig. 23.2   An evaluation scheme of software testing techniques

Fig. 23.3   The process of
dynamic testing

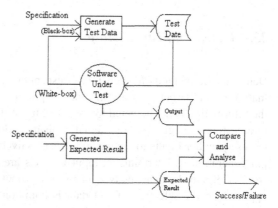

test data and a comparison of the results with the expected output which must
satisfy the users' requirements. The process of dynamic testing is shown in
Fig. 23.3.

23.4.3   Test Data Selection

Dynamic testing involves selecting input test data, executing the software on that
data and comparing the results to some test oracle, which determines whether the
results are correct or erroneous. To be sure of the certainty of the validity of software through dynamic testing, ideally we should try the software on the entire input
domain. In fact, due to the intrinsically discontinuous nature of software, given an
observation on any point of input domain, we can not infer any property for other
points in the input domain, see Figs. 23.4, 23.5.

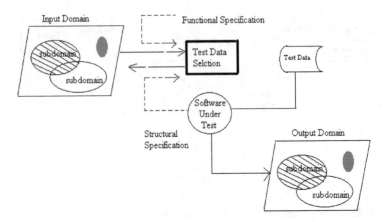

Fig. 23.4 The basic model of test data selection

23.4.4 Deterministic Testing

Deterministic methods for generating test inputs usually take advantage of information on the target software in order to provide guides for selecting test cases, the information being depicted by means of test criteria. Both functional and structural testing strategies use a systematic means to determine sub domains. They often use peculiar inputs to test peculiar cases. Given a criterion, the type of test input generation is deterministic: input test sets are built by selecting one element from each sub domain involved in the set proper to the adopted criterion. This approach to the selection of test data is commonly referred to as partition testing.

23.4.5 Random Testing

Random testing strategy is the conventional probabilistic method for generating test inputs. It is assumed that the tests are randomly selected according to the operational distributed for the software. This is the probability distribution describing the frequency with which different elements of the input domain are selected when the software is in actual use. Without this assumption the results are not useful. It consists in generating random test data based on a uniform distribution over the input domain this is an extreme case of the black-box testing method, since no information related to the target piece of software is considered except for the range of its input domain. An advantage of using random testing is that quantitative estimates of software's operational reliability may be inferred.

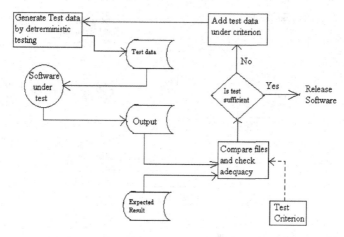

Fig. 23.5 A model of the test data adequacy

23.4.6 Adequacy Criterion

A software test data adequacy criterion is a means for determining whether a test set is sufficient or "adequate," for testing a given software or not by means of test criteria; if the test set is inadequate then more tests are added to the test set and the entire process is repeated. Otherwise, the software is released.

In the testing process, a test data adequacy criterion is only invoked when the tests no longer expose faults. Even though no faults are exposed by the test set, the software may not be correct.

An Example:

```
Display ``Did you pass your English test?''
Input score
If score > 70
Display ``Pass''
Else
Display ``Failure''
```

23.4.7 Exit Criterion

The exit criterion can be based on a reliability measure when the test set has been selected randomly from an appropriate probability distribution over the input domain.

The basic procedure shown in Fig. 23.6 is to determine the size n of test set and select test cases from an input distribution, to execute the software under test, to record the amount of execution time between failure or estimate the defective rate of the output population.

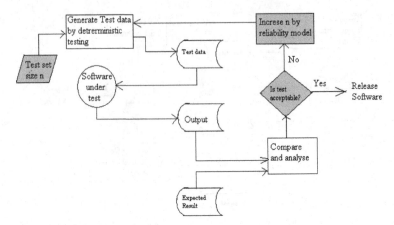

Fig. 23.6 A model of test data adequacy

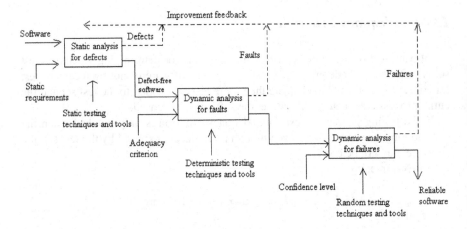

Fig. 23.7 The strategy proposal for software testing

23.5 Proposal of Strategy

There are two points of view about the relative merits restatic versus dynamic testing. Some researchers have suggested that static testing techniques should completely replace dynamic testing techniques in the verification and validation process
and that dynamic testing is unnecessary. However, static testing can only check the
correspondence between a program and its specification but it cannot demonstrate
that the software is operationally useful. Although static testing techniques are becoming more widely used, dynamic testing is necessary for reliability assessment,
performance analysis, user interface validation and to check that the software requirements are what the user really wants. The strategy proposal for software testing
is shown in Fig. 23.7.

23.6 Conclusions

To achieve software quality, software testing is an essential component in all software development. Software testing is characterized by the existence of many methods, techniques and tools that must fit the test situation, including technical properties, goals and restrictions. There is no single ideal software testing techniques for assessing software quality.

Therefore, we must ensure that the testing strategy is chosen by the combination of testing techniques at the right time on the right work products. From this viewpoint, a scheme for evaluating software testing techniques is presented to the classification and evaluation of software testing techniques. A strategy proposal for software testing is also discussed in this paper. We expect that the proposal will provide a guide-line to testers in the development of applications.

References

1. Myers, G.: The Art of Software Testing. Wiley, New York (1978)
2. Vliet, H.: Software Engineering: Principles and Practice. Wiley, New York (1994)
3. Gelperin, D., Hetzel, B.: The growth of software testing. Commun. ACM **31**(6), 687–695 (1988)
4. IEEE: IEEE Standard for Software Test Documentation: IEEE/ANSI Standard 829–1983. IEEE, New York (1983)
5. Kit, E.: Software Testing in the Real World: Improving the Process. Addison-Wesley, Reading (1995)
6. Demillo, R., McCracken, W., Martin, R., Passafiume, J.: Software Testing and Evaluation. The Benjamin/Cummings, Redwood City (1987)
7. Zhu, H., Hall, P., May, J.: Software test coverage and adequacy. Technical report, The Open University (1994)
8. Ould, M., Unwin, C.: Testing in Software Development. Cambridge University Press, Cambridge (1986)
9. Roper, M.: Software Testing. McGraw-Hill, New York (1994)
10. Sommerville, I.: Software Engineering. Addison-Wesley, Reading (1996)
11. Cao, L.: In-depth behavior understanding and use: the behavior informatics approach. Inf. Sci. **180**, 3067–3085 (2010)

Author Index

L. Cao, P.S. Yu (eds.), *Behavior Computing*, DOI 10.1007/978-1-4471-2969-1, © Springer-Verlag London 2012

Subject Index